AF389512

Selected Papers from 2018 IEEE International Conference on High Voltage Engineering (ICHVE 2018)

Selected Papers from 2018 IEEE International Conference on High Voltage Engineering (ICHVE 2018)

Editors

Issouf Fofana
Ioannis F. Gonos

MDPI • Basel • Beijing • Wuhan • Barcelona • Belgrade • Manchester • Tokyo • Cluj • Tianjin

Editors
Issouf Fofana
Research Chair on the Aging of
Power Network Infrastructure
(ViAHT), Université du Québec
à Chicoutimi (UQAC),
Department of Applied Sciences
Canada

Ioannis F. Gonos
School of Electrical and
Computer Engineering,
National Technical University of Athens
Greece

Editorial Office
MDPI
St. Alban-Anlage 66
4052 Basel, Switzerland

This is a reprint of articles from the Special Issue published online in the open access journal *Energies* (ISSN 1996-1073) (available at: https://www.mdpi.com/journal/energies/special_issues/ichve2018).

For citation purposes, cite each article independently as indicated on the article page online and as indicated below:

LastName, A.A.; LastName, B.B.; LastName, C.C. Article Title. *Journal Name* **Year**, *Article Number*, Page Range.

ISBN 978-3-03943-625-5 (Hbk)
ISBN 978-3-03943-626-2 (PDF)

Cover image courtesy of Ioannis F. Gonos.

Contents

About the Editors

Issouf Fofana Professor and IET Fellow, held the Canada Research Chair on insulating liquids and mixed dielectrics for electrotechnology (ISOLIME) from 2005 to 2015. At his university, he serves as Director of the International Research Centre on Atmospheric Icing and Power Network Engineering (CENGIVRE) and is the research chair on the Ageing of Power Network Infrastructure (ViAHT). Dr. Fofana is a member of a number of technical/scientific committees of international conferences (including IEEE ICDL, IEEE CEIDP, IEEE ICHVE, CATCON, and ISH), as well as a member of IEEE DEIS and CEIDP AdComs. He is also serving the scientific community as an Associate Editor for *IET GTD*, *Electrical Engineering* (Springer), and *Energies* and chair of the IEEE DEIS Technical Committee on Dielectric Liquids. He is also a member of a few working groups (CIGRE and ASTM). Dr. Fofana's research in the area of HV engineering has focused on insulation systems relevant to power equipment. His lifetime publication record includes more than 300 scientific publications and three patents.

Ioannis F. Gonos is Associate Professor at the School of Electrical and Computer Engineering of the National Technical University of Athens, where he has served as the Director of the High Voltage and Electric Measurements Laboratory since 2019. He is a member of the Technical Committees ELOT TE/80 "Electromagnetic Compatibility" and ELOT TE/82 "Electrical Installations of Buildings" of the Hellenic Organization for Standardization (ELOT). He is Lead Assessor of the Hellenic Accreditation Council (ESYD). Dr. Gonos is a member of IEEE and CIGRE. He is a member of the reviewer committee of the International Symposium on High Voltage Engineering (ISH) and the Steering Committee of the IEEE International Conference on High Voltage Engineering and Application (ICHVE). In 2018, he was the Chairman of the ICHVE 2018, which was held in Athens on 10–13 September. He is a reviewer for many scientific journals, Associate Editor in IET's *High Voltage*, and Guest Editor in MDPI's *Energies*. His research interests are high voltages, grounding, power systems protection, lightning, and electromagnetic compatibility. His lifetime publication record includes more than 250 scientific publications, 75 of which are in international scientific journals.

Editorial

Special Issue "Selected Papers from the 2018 IEEE International Conference on High Voltage Engineering (ICHVE 2018)"

Ioannis F. Gonos [1] and Issouf Fofana [2,*]

[1] School of Electrical and Computer Engineering, National Technical University of Athens, 9, Iroon Politechniou Street, Zografou Campus, GR 15780 Athens, Greece; igonos@cs.ntua.gr
[2] Research Chair on the Aging of Power Network Infrastructure (ViAHT), Department of Applied Sciences, Université du Québec à Chicoutimi (UQAC), 555, Boulevard de l'Université, Chicoutimi, QC G7H 2B1, Canada
* Correspondence: ifofana@uqac.ca

Received: 2 July 2020; Accepted: 8 September 2020; Published: 22 September 2020

---

1. An Outlook of the Special Issue

The 2018 IEEE International Conference on High Voltage Engineering and Application (ICHVE 2018) was organized by the National Technical University of Athens, Greece and endorsed by the IEEE Dielectrics and Electrical Insulation Society. The conference, chaired by Professor Ioannis Gonos and co-chaired by Professor Ioannis Stathopulos, Professor Jian Li and Professor Nicolas Younan, was held in Athens, Greece from 10 to 13 September 2018 and was the fifth of a series of successful conferences (Chongqing, China: 2008; New Orleans, USA: 2010; Shanghai, China: 2012; Poznan, Poland: 2014; and Chengdu, China: 2016). The 406 participants from 31 different countries had the opportunity to attend 32 oral sessions and eight poster sessions and get acquainted with the trending high voltage (HV) research in over 190 presentations. The main research areas covered by the conference papers were:

- Electromagnetic fields;
- Transients and electromagnetic compatibility (EMC);
- Aging, space charge and maintenance;
- Grounding systems;
- Monitoring and diagnostics;
- Power and industrial applications;
- HV insulation systems;
- HV testing and measurement.

Seventeen high quality papers from the 2018 IEEE International Conference on High Voltage Engineering and Application (ICHVE 2018) from all of the above research areas were presented in a Special Issue. The works included in the Special Issue are not the exact papers presented in ICHVE 2018 but enlarged versions of them, enriched with more research outcomes. The Special Issue was welcomed with great interest, as ICHVE attracts the recent advances in all fields of high voltage engineering and applications. We would like to thank all the researchers who made their contribution to this Special Issue and all the reviewers who enabled its publication through their efforts for a prompt and instructive revision procedure.

2. A Review of the Special Issue

The topics of HV testing and monitoring/diagnostics are covered by papers [1,2]. In [1], the authors discuss the use of damped AC voltages (DAC) for after-laying testing and diagnosis of submarine power cables—both the export and inter-array cables—according to the recommendations of standards IEEE 400 and IEEE 400.4 for partial discharge monitored testing. Within the absence of adequate relevant specifications in the International Electrotechnical Commission (IEC) standards, and the increased need for quality control, the advantages of this testing technique, in combination with actual testing examples, are highlighted.

A novel diagnostic method of measuring water content in transformer oil using multi-frequency ultrasonic, with a back propagation neural network that was optimized by the principal component analysis and genetic algorithm (PCA-GA-BPNN), is reported in [2]. Accurate prediction of water content in oil is important for the stability and security level of power systems as moisture is considered enemy "number one" of insulation. Investigations with 160 oil samples of different water content with multi-frequency ultrasonic detection technology enabled the development and training of a PCA-GA-BPNN model that was proven to be particularly robust. Indeed, the resulting mean squared error of the test sets was 8.65×10^{-5}, with a correlation coefficient of 0.98.

Further discussion on insulating oils is provided in [3,4]. The accumulation of various types of particles and their effect on the DC breakdown voltage of mineral oil is the topic of paper [3]. Simulation results aiming to explore the motion mechanism and accumulation characteristics of different particles are in agreement with experimental measurements of DC breakdown voltages. As a major conclusion, the properties of the impurities determine the bridge shape, conductivity characteristics and variation law of DC breakdown voltages. The increase of current and the electric field distortion which modify the DC breakdown voltage are mainly affected by metal and mixed particles.

The dielectric characteristics of two types of nanofluids modified natural esters were studied in [4]. The addition of colloidal Fe_2O_3 nanoparticles in natural ester oil revealed an increase of the AC breakdown voltage whereas the opposite behavior was observed after the addition of SiO_2 nanoparticles. The conductivity, along with the permittivity, of nanoparticles arises as a key parameter in the performance of the nanofluid. The experimental results point out that specific concentrations of nanoparticles (with electrical conductivity and permittivity different from those of the matrix oil) are required for the enhancement of the breakdown voltage strength.

Papers [5,6] are focused on outdoor HV insulation. The authors of paper [5] investigated the formation of dry bands on insulators as a function of the insulator design and pollution level. Artificial pollution tests were performed on a 4-shed 11 kV insulator with conventional and textured surface designs in a clean-fog test chamber with the application of a voltage ramp-shape source over three pollution levels (extremely high, high and moderate), and the appearance and extension of dry bands were automatically recognized via a newly developed MATLAB based program procedure. The resulting clear distinction between designs and pollution levels regarding the statistical location and extension development over time of the dry bands may serve as a basis of design guidelines to minimize dry band zones.

Trip accidents of transmission lines during forest fires and, specifically, the contribution of vegetation combustion particles to the triggering of discharges within the gap below the transmission line are investigated in [6]. A two-dimensional (2D) axisymmetric simulation model was established by simplifying the flame region and magnitudes, such as fluid temperature and velocity, fluid field and electric field. Moreover, the forces on particles and movement were calculated. The ultimate goal is a future analysis of the electric field distortion and a further study of the discharge mechanism of the gap under the condition of vegetation flames.

Effects regarding HV transmission lines are also presented in [7,8]. In [7], the authors elaborate on the non-uniformities of transmission line parameters that may affect the magnitude of the transferred transients. A frequency domain method was utilized to compute transient voltage and current profiles along nonuniform multiconductor transmission lines, including the effect of time-varying and nonlinear elements. Through the cascade connection of chain matrices, the line was subdivided into segments with different spatial and time-varying properties. The transition to the time domain was performed with the Laplace transform. The obtained results were compared to respective alternative transients program (ATP) simulations displaying a high accuracy.

The formation of a transmission line simulation model was also presented in [8] for the calculation of radio interference (RI) lateral profiles generated by corona discharge in high voltage direct current (HVDC) transmission lines. Both the RI and the maximum electric field were calculated as a function of sub-conductor radius, bundle spacing, number of sub-conductors in the bundle, soil resistivity and the radio interference voltage (RIV) frequency for the cases of a 500 kV and a 600 kV bipolar transmission line. Finally, optimal design values to minimize the RI levels were proposed after vector optimization.

In papers [9–11], the authors addressed issues regarding solid insulating materials: creeping discharges on insulating surfaces surrounded by gaseous insulation, electrical treeing in cross-linked polyethylene (XLPE) and partial discharge surface degradation on silicon rubber, respectively.

More specifically, paper [9] dealt with the electrical detection of creeping discharges over disc-shaped insulator samples of different dielectric materials (polytetrafluoroethylene (PTFE), epoxy resin and silicone rubber), using atmospheric gases (dry air, N_2 and CO_2) as insulation mediums in a point-plane electrode arrangement under AC voltage applications. According to the experimental and numerical results, the discharge activity depended highly on the geometrical and material properties of the dielectric and the solid/atmospheric gas interface.

In [10], the authors compared the effectiveness of three specific types of polycyclic compound fillers (compound A: 2-hydroxy-2-phenylacetophenone; compound B: 4-phenylbenzophenone; compound C: 4,40-difluorobenzophenone) in suppressing the electrical treeing growth in XLPE insulation, which can lead to the electrical failure of cables. Experimental results with DC impulse voltage at 30 °C, 60 °C and 90 °C revealed the great application prospects of compound A in HVDC cables as it presented the largest energy level and deep trap density, thus decreasing charge transportation and minimizing electrical tree growth.

An experimental procedure was proposed in [11] to evaluate long-term surface erosion of silicone rubber sheets caused only by partial discharge. The silicone rubber was subjected 50 or 100 times to a 24 h partial discharge cycle. This cycle consisted of an 8 h application of partial discharges using an electrode system with an air gap and a subsequent seizure of the partial discharges for 16 h for the recovery of hydrophobicity. The proposed method presented a satisfying performance in terms of no arc discharge occurrence, good repeatability of results and possible acceleration of erosion. Additionally, the authors stressed out the limited effectiveness of alumina trihydrate (ATH), an additive acting to avoid tracking and erosion by discharge, in the case of partial discharge erosion and the prevalence of the silicon rubber material characteristics.

Space charge accumulation in HV insulating systems was analyzed in papers [12,13]. An accurate description of the time dependent charge distribution within polymeric HVDC cables incorporating complex effects was performed in [12]. An empirical conductivity equation was developed for the accurate description of the stationary space charge and electric field distribution, where the bulk conductivity, found in literature, was extended with two sigmoid functions to represent a conductivity gradient near the electrodes, thus simulating the accumulated bulk space charges and hetero charges. Comparison to space charge measurements taken from the literature showed a good approximation of the space charges prediction with the proposed equation.

A statistical study on the space charge effects and the stages of the needle-plate corona discharge under DC voltage was reported in paper [13]. The statistical rules of repetition rate (n), amplitude (V) and interval time (D_t) were extracted, the corresponding space charge effects and electric field distributions on the PD process were analyzed and the discharge stages of corona discharge under DC voltage were separated.

The fields of HV testing, HV measurements and HV equipment were covered by papers [14–16]. In [14], the authors developed a cable termination model to reduce the electric field at cable ends and prevent surface and external discharges during standardized overvoltage tests. Simulations using the finite element method (FEM) on a 35 kV sample cable model with 300 kV RMS internal insulation breakdown voltage allowed verifying the sufficient reduction of the electric field at the cable ends.

A novel 363 kV/5000 A/63 kA sextuple-break vacuum circuit breaker (VCB) with a series-parallel structure was presented in [15]. The calculations of the voltage distribution and the electric field of each break at the fully open state via a 3D FEM model highlighted an uneven distribution of the applied voltage, with above 86% stressing the first break. An analysis of the distributed and stray capacitance parameters, along with the equivalent circuit simulation model of the VCB, resulted in an optimal grading capacitor value of 10 nF. The validity of the proposed voltage sharing design was proven through breaking tests of a single-phase unit, where the 363 kV VCB prototype broke both the 63 kA and the 80 kA short circuit currents successfully.

The accuracy and reliability of open-air capacitive sensors in measuring switching transients were demonstrated in [16]. The paper describes a method to calibrate a sensor to line coupling matrix based on assumed 50 Hz symmetric phase voltages. The network simulations indicated good agreement within seven percent with values reconstructed from measurements. Moreover, the comparison of the sensor with measurements of a transient overvoltage by a capacitive divider confirmed the capability of this differentiating/integrating method in interference rejection.

Finally, the scientific subfield of grounding is represented by paper [17], where the authors introduced a new approach to lightning transient studies by incorporating complex grounding grids modelled in MATLAB/Simulink based on the transmission line theory. An interface with electromagnetic transients program-restructured Version (EMTP-RV) allows co-simulation and the facilitation of the component libraries and network design provided by EMTP-RV for more accurate representation of the transmission network.

Author Contributions: The authors contributed equally to this work. All authors have read and agreed to the published version of the manuscript.

Funding: This research received no external funding.

Acknowledgments: Ioannis F. Gonos is thankful to Issouf Fofana for providing an opportunity to co-author this editorial.

Conflicts of Interest: The authors declare no conflict of interest.

References

1. Gulski, E.; Jongen, R.; Rakowska, A.; Siodla, K. Offshore Wind Farms On-Site Submarine Cable Testing and Diagnosis with Damped AC. *Energies* **2019**, *12*, 3703. [CrossRef]
2. Yang, Z.; Zhou, Q.; Wu, X.; Zhao, Z.; Tang, C.; Chen, W. Detection of Water Content in Transformer Oil Using Multi Frequency Ultrasonic with PCA-GA-BPNN. *Energies* **2019**, *12*, 1379. [CrossRef]
3. Dan, M.; Hao, J.; Liao, R.; Cheng, L.; Zhang, J.; Li, F. Accumulation Behaviors of Different Particles and Effects on the Breakdown Properties of Mineral Oil under DC Voltage. *Energies* **2019**, *12*, 2301. [CrossRef]
4. Charalampakos, V.; Peppas, G.; Pyrgioti, E.; Bakandritsos, A.; Polykrati, A.; Gonos, I. Dielectric Insulation Characteristics of Natural Ester Fluid Modified by Colloidal Iron Oxide Ions and Silica Nanoparticles. *Energies* **2019**, *12*, 3259. [CrossRef]
5. Albano, M.; Haddad, A.; Bungay, N. Is the Dry-Band Characteristic a Function of Pollution and Insulator Design? *Energies* **2019**, *12*, 3607. [CrossRef]
6. Pu, Z.; Zhou, C.; Xiong, Y.; Wu, T.; Zhao, G.; Yang, B.; Li, P. Two Dimensional Axisymmetric Simulation Analysis of Vegetation Combustion Particles Movement in Flame Gap under DC Voltage. *Energies* **2019**, *12*, 3596. [CrossRef]
7. Nuricumbo-Guillén, R.; Espino Cortés, F.; Gómez, P.; Tejada Martínez, C. Computation of Transient Profiles along Nonuniform Transmission Lines Including Time-Varying and Nonlinear Elements Using the Numerical Laplace Transform. *Energies* **2019**, *12*, 3227. [CrossRef]
8. Tejada-Martinez, C.; Espino-Cortes, F.; Ilhan, S.; Ozdemir, A. Optimization of Radio Interference Levels for 500 and 600 kV Bipolar HVDC Transmission Lines. *Energies* **2019**, *12*, 3187. [CrossRef]

9. Michelarakis, M.; Widger, P.; Beroual, A.; Haddad, A. Electrical Detection of Creeping Discharges over Insulator Surfaces in Atmospheric Gases under AC Voltage Application. *Energies* **2019**, *12*, 2970. [CrossRef]

10. Zhu, L.; Du, B.; Li, H.; Hou, K. Effect of Polycyclic Compounds Fillers on Electrical Treeing Characteristics in XLPE with DC-Impulse Voltage. *Energies* **2019**, *12*, 2767. [CrossRef]

11. Komatsu, K.; Liu, H.; Shimada, M.; Mizuno, Y. Assessment of Surface Degradation of Silicone Rubber Caused by Partial Discharge. *Energies* **2019**, *12*, 2756. [CrossRef]

12. Jörgens, C.; Clemens, M. Empirical Conductivity Equation for the Simulation of the Stationary Space Charge Distribution in Polymeric HVDC Cable Insulations. *Energies* **2019**, *12*, 3018. [CrossRef]

13. Wang, D.; Du, L.; Yao, C. Statistical Study on Space Charge effects and Stage Characteristics of Needle-Plate Corona Discharge under DC Voltage. *Energies* **2019**, *12*, 2732. [CrossRef]

14. Andrade, A.; Costa, E.; Andrade, F.; Soares, C.; Lira, G. Design of Cable Termination for AC Breakdown Voltage Tests. *Energies* **2019**, *12*, 3075. [CrossRef]

15. Yu, X.; Yang, F.; Li, X.; Ai, S.; Huang, Y.; Fan, Y.; Du, W. Static Voltage Sharing Design of a Sextuple-Break 363 kV Vacuum Circuit Breaker. *Energies* **2019**, *12*, 2512. [CrossRef]

16. Barakou, F.; Steennis, F.; Wouters, P. Accuracy and Reliability of Switching Transients Measurement with Open-Air Capacitive Sensors. *Energies* **2019**, *12*, 1405. [CrossRef]

17. Steinsland, V.; Sivertsen, L.; Cimpan, E.; Zhang, S. A New Approach to Include Complex Grounding System in Lightning Transient Studies and EMI Evaluations. *Energies* **2019**, *12*, 3142. [CrossRef]

Article

Offshore Wind Farms On-Site Submarine Cable Testing and Diagnosis with Damped AC [†]

Edward Gulski [1], Rogier Jongen [1,*], Aleksandra Rakowska [2] and Krzysztof Siodla [2]

[1] Onsite hv solutions AG, Lucerne, Toepferstrasse 5, 6004 Lucerne, Switzerland; e.gulski@onsitehv.com
[2] Institute of Electric Power Engineering, Poznan University of Technology, Piotrowo 3A, 60-965 Poznan, Poland; aleksandra.rakowska@put.poznan.pl (A.R.); krzysztof.siodla@put.poznan.pl (K.S.)
* Correspondence: r.jongen@onsitehv.com
† This paper is an extended and updated version of our conference paper published in 2018 IEEE International Conference on High Voltage Engineering and Application (ICHVE), 10–13 September 2018, Athens, Greece.

Received: 8 August 2019; Accepted: 25 September 2019; Published: 27 September 2019

Abstract: The current power cables IEC standards do not provide adequate recommendations for after-laying testing and diagnosis of offshore export and inter-array power cables. However the standards IEEE 400 and IEEE 400.4 recommend partial discharge monitored testing, e.g., by continuous or damped AC voltages (DAC). Based on the international experiences, as collected in more than 20 years at different power grids, this contribution focuses on the use of DAC for after-laying testing and diagnosis of submarine power cables both the export and inter-array cables. Higher risk of failure, long unavailability, higher repair costs, and maintenance costs imply that advanced quality control is becoming more important. The current state of the existing and drafting international standards are based on onshore experiences and not related to the actual serious problems experienced with failures on export up to 230 kV and inter-array cables up to 66 kV. The application of damped AC as a testing solution in this concern is specially discussed. The advantages of this testing technique, in combination with actual testing examples, show the findings on export and inter-array cables at offshore wind farms.

Keywords: offshore; export cables; inter-array cables; damped AC voltage (DAC), after-laying cable testing; on-site diagnosis; condition assessment; partial discharges; and dissipation factor

---

1. Introduction

Considering that reliable energy transport is fundamental for on- and offshore infrastructures, the aspects of maintaining the quality control regulations for newly installed and service aged cable connections are of importance. As a result, important questions about maintaining/updating internal procedures for a reliable network operation are:

1. How to perform, in a sensitive and non-destructive way, the detection of poor workmanship defects of newly installed cable circuits?
2. How to perform non-destructive diagnostics of cable circuits in service to determine the actual condition?

Following present IEC standards for power cables [1–3], the after-installation testing protocols for power cables are limited to manufacturer's minimum recommendations and therefore do not cover the present needs to keep possible failure risk during operation as small as possible. As a result considering responsible operation and asset management of offshore power cables the following aspects have to be considered:

1. After installation, testing of newly installed cable systems to find:

a. Manufacturing related defects → due to the high level of quality control less probable;
b. Accessories parts delivery problems → due to the diversification in the supply chains more probable;
c. Installation related defects → due to the diversification in the installation supply chains highly probable.

2. Maintenance and diagnostic testing of cable systems in operation to estimate:

a. Operational damages and electrical and thermal over-stresses → cannot be neglected e.g., transients and over-voltages;
b. Aging processes → depends on many operational and local factors, e.g., presence of installation defects, load constriction or load increase works;
c. The remaining life → goal of most asset managers to keep capital expenditure (CAPEX) and operational expenditure (OPEX) on an optimal level.

Unfortunately, regarding testing of offshore power cables, parties involved are not aware of risk management of those cable circuits and in their testing, specifications are simply referring to IEC procedures to ensure their quality testing procedures. About 30 years ago these standards were introduced for onshore application by manufacturers only e.g., the IEC 60502, IEC 60840, and IEC 62067 [1–3]. These cable manufacturers' standards are extensively discussing the factory testing aspects. However, only basic tests for the after-installation test are mentioned. Furthermore, no guidelines are provided regarding the maintenance of cable circuits and maintenance/diagnostic testing. In contrast to the IEC standards, the standards IEEE 400 [4] and IEEE 400.4 [5] recommend partial discharge monitored testing, e.g., by continuous or damped AC voltages (DAC).

The number of onshore and offshore wind farms is growing worldwide due to the increase in demand for renewable energy. An important aspect of an offshore wind farm is the submarine or subsea cables. They play a vital role in bringing the generated power from the wind turbine to the offshore substation and eventually to shore. In case of damage to this critical infrastructure repairing any damages can be challenging and costly [6].

In the past years, an average of at least 10 subsea cable failures is declared to wind farm insurers each year by the wind park owners. The financial severity of such a cable failure continues to grow. It is stated that the cable failures are accounting for 77% of the total global cost of offshore wind farm losses [7]. Almost 70% of the cable faults recorded in the claims database can be attributed to contractor errors during installation. However, those errors do not always become evident until the wind farms start operations or are operational for a certain time.

As the development of (offshore) wind turbines results in the increase of the physical dimensions as well as the generating capacity, the operating voltage of export cables up to 230 kV and the inter-array cable up to 66 kV networks needs to be increased. Both voltage classes have their own technical challenges. Besides the difference in the network components, also the quality assessment of the installed cable systems is different. High voltage cable testing is among the challenges facing this relatively new part of the offshore wind farm industry.

Advanced risk management and quality control are becoming more and more important. The main challenge during the installation phase is to have specialized teams for cable installation and testing activities. To reduce possible risks or to exclude failures during operation, it is important that systematic testing and diagnosis is performed during manufacturing, transportation, installation, and operation. It has been noticed that the after-installation testing protocol for offshore wind farms is still under consideration. The current state of the international standards is not based on long term experiences as obtained in the offshore sector, but more or less a copy of existing onshore standards and recommendations. As a result, the provided solutions do not cover the actual problems experienced on inter-array cables [6,7].

Evaluating the risks and the obtained valuable experiences, it shows that the offshore industry needs to set up their own reliable specifications for submarine cable testing and diagnosis. The first steps for this are done, for instance, in the new published CIGRE Technical Brochure 722 and the upcoming IEC 63026. Unfortunately, a serious gap in those recommendations is present in the practical implementation regarding offshore after installation testing.

2. Integral Quality Fingerprint

As offshore cables are often produced in very long lengths compared to onshore cables and a lot of typical handling is taken place from manufacturing up to the offshore installation, there are the risks of cable damages. Those can be related to manufacturing, transportation, and the installation stages of the cable. Therefore, there is a need to use an integral approach where the condition of the cables will be verified. Having such fingerprints as obtained during these stages, a comparison could be made to verify the cable quality. This can also be used as a good basis for the cable maintenance activity later on, see Figure 1.

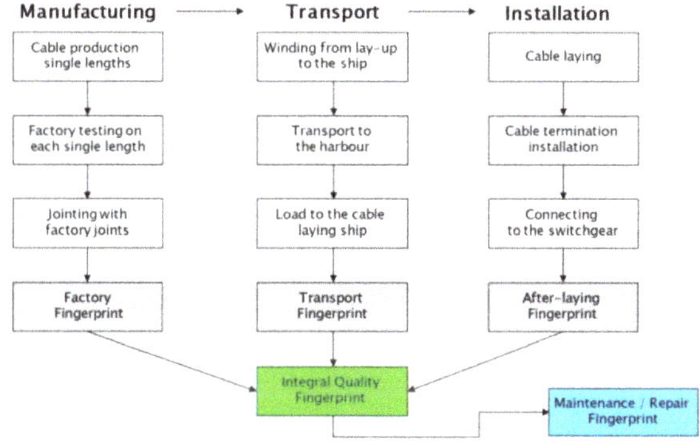

Figure 1. Integral fingerprint for quality control during the manufacturing, transportation, and installation of offshore power cables.

The applied technologies for such fingerprinting for quality control have to be suitable [8,9]:

- For offshore testing, taking into account restrictions regarding e.g., size, weight, weather protection;
- To provide adequate information: voltage testing and fingerprinting (e.g., partial discharges (PD) and dissipation factor (tan δ TD)) during the whole installation and operation process;
- To ensure reliable operation of the submarine power cables;
- To provide contractors the basis for lowering the risks during the warranty period;
- To enable service providers during operation a good basis for condition-based maintenance.

Moreover, the criteria for the risk management for the contractor (e.g., 5-year warranty), system operators and insurers have to be related to the quality control system applied during construction. This means that the various testing techniques provided in the international standards differ significantly in the effectiveness of cable quality assessment after the installation. Therefore the indicated testing options could be classified in different warranty levels:

a. Soak Test ((non-) monitored): Due to lack of information about operational reliability = No warranty.

b. Non-monitored voltage withstand test only: due to showing extreme defects only = Limited warranty.

c. Monitored voltage withstand test (damped AC voltage (DAC) resp. AC Voltage test) with sensitive (PD, TD) fingerprinting: Providing complete information = Full warranty.

Several testing technologies are available to perform on-site testing of power cables. In Table 1 an overview of several test techniques is given including weighting of important aspects. As can be seen from the table DAC monitored testing of offshore cables has several advantages.

Table 1. Overview of test technologies for monitored withstand test of 66 kV subsea inter-array cables.

Test Technology	Technical Acceptance (Meets Offshore Wind Recognized International Standards, e.g., IEC, IEEE)	Capability of Testing High Capacitive Load (Long and Multiple Cables)	Offshore Application (Size, Weight, Weather Protection)	Sensitive PD Detection (System PD level, PD Characteristics)	Costs (Investment /Rental Costs)
ACRT + PD	++	+/− [1]	−−	− [3]	−−
VLF + PD	+	+ [2]	+	− [3]	++
DAC + PD	+	++	+	++	+

ACRT—AC Resonance Test; VLF—Very Low Frequency; DAC—Damped AC; PD—Partial Discharges. [1] when connecting a second resonance reactor; [2] when lowering the VLF test frequency to e.g., 0.01 Hz; [3] Voltage source produces high interference during the PD measurement (>10 pC).

3. Damped AC

Damped AC (DAC) is a technology that is already for 20 years commercially available for testing all types of distribution and transmission cables [9–19]. With DAC it is possible to energize very long lengths of power cables with high capacitance, due to its low input power demand. Moreover this technology can be combined with diagnostic measurements like partial discharges (PD). Damped AC is applicable for factory PD monitored acceptance testing and it is already in use for on-site after-laying/commissioning, maintenance, and diagnostic testing. It is an approved testing methodology, in accordance with relevant testing parameters from international standards and recommendations (IEEE, IEC, and CIGRE).

3.1. DAC Principle

The application of damped AC (DAC) voltages including standardized conventional PD detection and analysis is accepted worldwide for on-site testing and diagnosis of (Extra) High Voltage power cables [5]. The DAC technology has been first introduced on the Jicable conference in 1999 and is now already 20 years worldwide in use to test and diagnose MV and HV power cables [19]. In addition to the equivalence of sinusoidal DAC voltages (in the frequency range of 20–300 Hz) compared to the 50/60 Hz network stresses, the characteristics of the applied technology meet the specification of modern on-site testing system:

- Lightweight modular system;
- Compactness in relation to the output voltage;
- Low effort for system assembling;
- Low power demand incl. long cable lengths;
- Low level of EM noises and the possibility of sensitive PD detection and localization as well as dissipation factor measurements.

DAC testing is used almost always in combination with partial discharge (PD) and dissipation factor (tan δ TD) measurements for new installed and service-aged cables. The use of DAC voltages for testing power cables is in compliance with relevant testing parameters derived from IEC, IEEE, and CIGRE international standards and guidelines. The DAC method is used to energize and to test on-site power cables with sinusoidal AC frequencies [5]. The system consists of a digitally controlled high voltage power supply to energize capacitive load of power cables with large capacitance (e.g., 10 µF), see Figure 2.

The energizing time depends on the maximum available load current of the high voltage power supply, the test voltage, and the capacitance of the test object and has to stay below 100 s [2]. During a number of AC voltage cycles (of several hundred milliseconds), the PD signals are initiated in a way similar to 50/60 Hz inception conditions [5,20–22]. In accordance with [5], no DC stresses are applied to the test object, and the DAC stress can be considered similar to factory partial discharge testing conditions, i.e., a 50 Hz AC test combined with a PD measurement. Due to the continuous voltage increase and immediate transition to the DAC voltage after the maximum test voltage is reached, no steady-state condition occurs, and the low electric field strength in the insulation (typically <20 kV/mm), and short durations (less than a second up to tens of seconds) of bipolar stresses ensures no space charge accumulation.

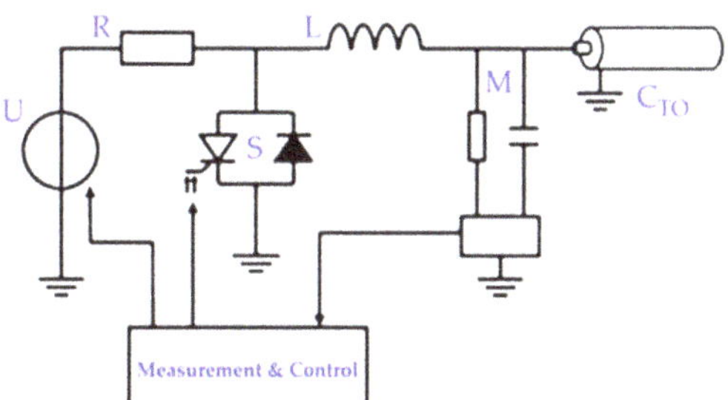

Figure 2. Schematic overview of a damped AC (DAC) system connected to a power cable, with U the high voltage source, R the protective resistor, S the semiconductor switch, L the air core inductor, M the voltage divider, and coupling capacitor and C_{TO} the cable under test.

During the AC resonance phase, the DAC voltage is characterized by a decaying sine wave with a frequency given by the inductance used and the capacitance of the cable under test. Inductor values of DAC systems are chosen such that the DAC voltage frequency is in the near power frequency range of 10-500 Hz, see Figure 3. The maximum cable length is only limited by the energizing time and the maximum current capabilities of the semiconductor switch, therefore DAC systems can be easily used to test cable lengths up to tens of kilometers.

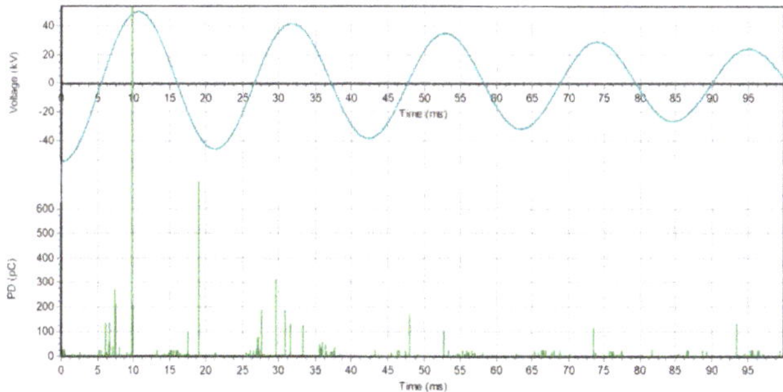

Figure 3. Example of damped AC voltage excitations monitored by partial discharge (PD) detection. The PD activity can be used to localize the breakdown site.

By applying DAC voltage-sensitive PD detection and PD localization of the fault in the power cables is possible. Using time-domain reflectometry (TDR), PD presences in cable terminations, joints, or cable parts can be localized, see Figure 4. According to IEEE 400.4 [5] to execute voltage withstand test this procedure should be repeated for 50 excitations followed after each other to perform a voltage withstand test on the maximum test voltage. Considering the time from the PD initiation until breakdown and the shorter duration of the excitation and decaying characteristics of the voltage, DAC test results obtained may differ from those obtained by continuous AC withstand voltage testing [5]. Assuming that testing should not necessarily be destructive and that the PD inception indicates presence of defects, this difference in the time until breakdown has to be considered as an advantage of DAC testing. In practice, a monitored damped AC hold test is performed to determine whether the cable passes or fails the damped AC test. Due to additional information as provided by PD detection, the monitoring insulation properties during a damped AC withstand test, and the effect of the test voltage during its application can improve the evaluation of the insulation condition.

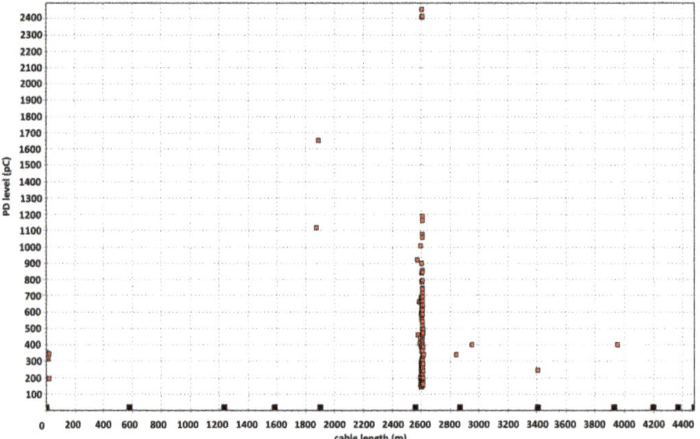

Figure 4. Monitored voltages withstand testing of a 150 kV XLPE cable underground circuit (4.5 km) up to 1.7 *Uo*. Example of PD mapping as obtained during DAC voltage testing up to 1.0 *Uo* shows that joint No. 5 has concentrated PD activity at operating voltage level.

3.2. DAC Diagnostic Testing

3.2.1. Partial Discharge Measurement (PD)

The DAC voltage source does not produce EM interference signals during the PD measurements, thus a sensitive PD detection and localization of the fault in power cables is possible. Therefore the PD monitored voltage withstand testing using damped AC voltage is a very effective method to detect most insulation weak-spots. The PD measurement can be used to pinpoint the exact location of insulation defects at an early stage [19–21], by means of time-domain reflectometry (TDR). With TDR, the PD presences in cable terminations, joints, or cable parts can be made, see Figure 4.

3.2.2. Dissipation Factor Estimation (tan δ TD)

DAC systems are able to estimate the dissipation factor from the damping of the decaying sine wave during the LC resonant phase [22,23]. The degradation of oil-impregnated insulation of HV power cable can be investigated with this parameter.

Applying DAC voltages the dissipation factor tan δ can be estimated using the decay characteristics of the damped AC voltage at different testing voltage levels and the change in the dissipation factor (delta tan delta) in relation to the increasing voltage can be especially valuable for finding insulation ageing development in power cables [5], see Figure 5.

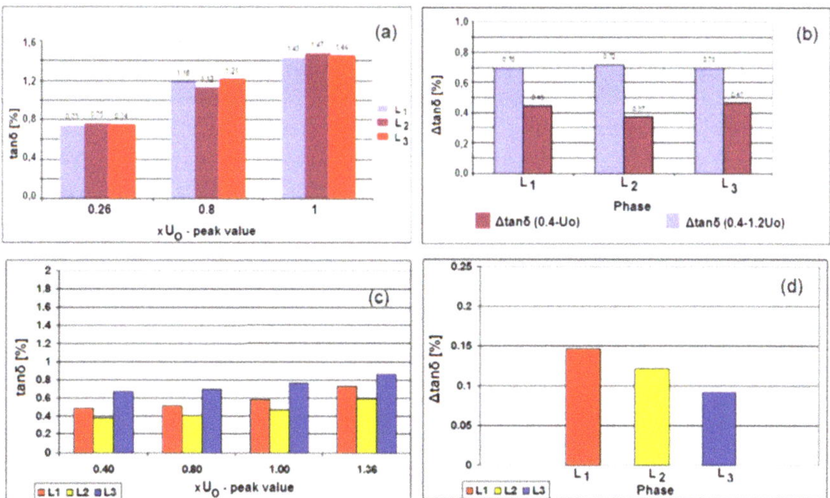

Figure 5. Example of dissipation factor diagnosis data as obtained for two different cable circuits: (**a**,**b**) A 150 kV power cable with self-contained fluid-filled (SCFF) insulation, length: 850 m, aged 49 years, (**c**,**d**) A 230 kV power cable with low-pressure fluid-filled (LPFF) insulation, length 13.3 km, aged 33 years.

4. On-Site Testing and PD Detection of Long Lengths Export Cables up to 230 kV

Detection and localization of PD in export cable system with long lengths e.g., 30 km can be improved by performing PD measurements at both sides of the cable circuit. This will, for the worst-case situation (PD at the near end), reduce the traveling distance for PD pulses by a factor of 2. In a single-sided measurement setup, a near end partial discharge has to travel through the whole cable length to the far end and the whole cable length back to the near end. The overall traveling distance is, therefore, two times the cable length. In a dual-sided measurement, the near end PD only has to travel to the far end to be detected there, so it only has to travel the cable length once. For this dual-sided measurement system, a damped AC system for energizing the cable system is used, see Figure 6. This system uses a coupling capacitor with PD detector on the near end side and a second PD detector at the far end side [11].

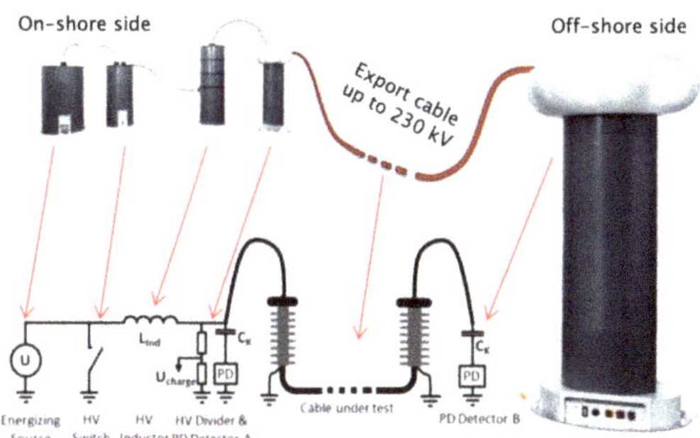

Figure 6. Setup for dual-sided PD monitored on-site testing techniques for offshore export cables of long lengths: DAC excitation circuit and synchronized PD detector units supported by PD measurement with localization feasibility with C_k the coupling capacitor (1–2 nF), L_{ind} is the air core inductance (5–9 H).

The main advantage of the dual-sided measurement is the better sensitivity compared to the single-sided measurement. The PD pulses have to propagate only to the ends of the cable. No reflection at the far end is necessary. As a result there is no additional attenuation due to imperfect reflection or attenuation for traveling through the whole cable once again. The attenuation decreases the amplitude of measurable partial discharge pulses.

Due to the attenuation, the decreasing PD amplitude has about 70% of the original amplitude. In a single-sided measurement setup, the reflected pulse over the far end decreases to 5%. A dual-sided measurement increases the remaining PD amplitude to about 26% [24].

5. On-Site Testing of HVDC Cables

Besides testing HVAC cables with DAC, the technology is also suitable for testing HVDC cables. In particular, the HVDC (submarine) power cable circuits have some typical characteristics:

- Installing a (submarine) HVDC cable is a costly and challenging activity and the technical interventions for its repairing in the case of faults are also costly and difficult;
- High capacitance that requires extremely high-power demand for conventional HVAC test systems;
- A large number of (factory) joints are installed.

Besides the enormous and unrealistic effort required to generate, on-site, the requested power (numbers of required AC resonance test (ACRT) sets) the testing with AC resonant systems only provides a go/no go based on the occurrence of a breakdown and does not provide the desired selection criteria to obtain an overall cable condition assessment where diagnostic parameters like PD and dissipation factor are included. When compared to HVAC cable systems and accessories, the HVDC cables are designed differently, which might result in the damage of the insulation of the HVDC cable and accessories in case of a defect breakdown under the regular AC voltage over-stresses. As the cable field design of HVDC cable insulation is based on the resistive field distribution, which is strongly temperature-dependent, the use of continuous AC stresses as produced by the AC resonant testing is related to a temperature increase (in particular by long lengths), which in its turn could be destructive for the HVDC insulation.

DAC testing makes it possible to energize very long lengths of AC as well as HVDC power cable with high capacitance, due to its low input power demand. Actually, possible defects can be detected in the installed HVDC cable length, as introduced in the factory or after the installation and transportation, and located by means of DAC voltage testing including single- or dual-side partial discharge detection at the cable terminations [19,22]. It is known that about 80% of the insulation defects in power cable circuits are visible through to partial discharges as detected at AC voltage stresses. As a result non-destructive DAC testing avoids the unnecessary temperature load and providing IEC conform sensitive PD detection to detect the insulation defects [11].

6. Field Example Diagnostic Test Offshore Export and Inter-Array Cables

During a maintenance outage of an offshore wind farm, diagnostic testing was performed using DAC testing. Ten complete strings of inter-array cables from the offshore substation (OSS) to the wind turbine generator (WTG) were DAC tested. For this purpose, the switchgear in each wind turbine was switched in such a way that all the inter-array cables between the wind-turbines were connected with each other, see Figure 7. The DAC test system was connected via a special test adapter in the switchgear on the offshore substation (OSS) to the cable string under test. The maximum total lengths of the strings were up to almost 10 km length.

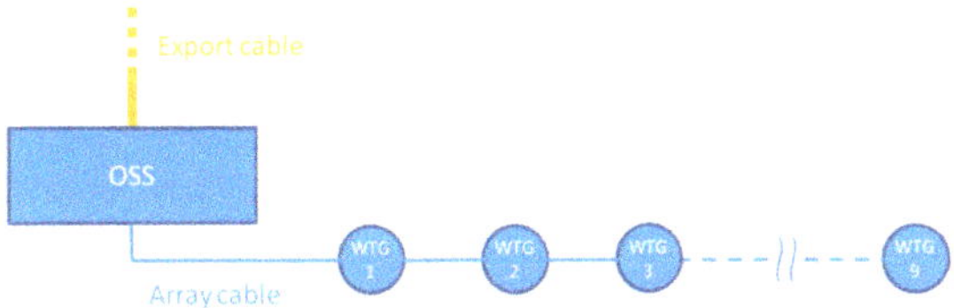

Figure 7. The layout of the array cable strings, each string consist of a total of nine wind turbine generators (WTGs). DAC test was performed from the offshore substation (OSS).

The benefit, in this case, was that the test system did not need to be transported between the tests, as all tests could be performed from the OSS, which required only a single mobilization of the test equipment.

In a string of 8.4 km length of inter-array cables between the OSS and nine wind turbines, partial discharges were detected and localized. Each phase was individually tested using a damped AC offshore system with a DAC frequency of 100 Hz (total cable capacitance 1.85 μF). The test voltage level was ramped up in steps up to the maximum applied test voltage level of 1.4 U_o. PD was monitored during the voltage ramp-up phase and during the withstand testing of 50 DAC excitations at the maximum test voltage. No breakdown was observed during the withstand test, however as the damped AC test voltage was increased up to 1.4 U_o, PD activity was observed in phases L2 and L3 of the string, see Figure 8. PD mapping revealed the PD concentration at WTG 5 in phase L3 and at WTG 6 km in phase L2.

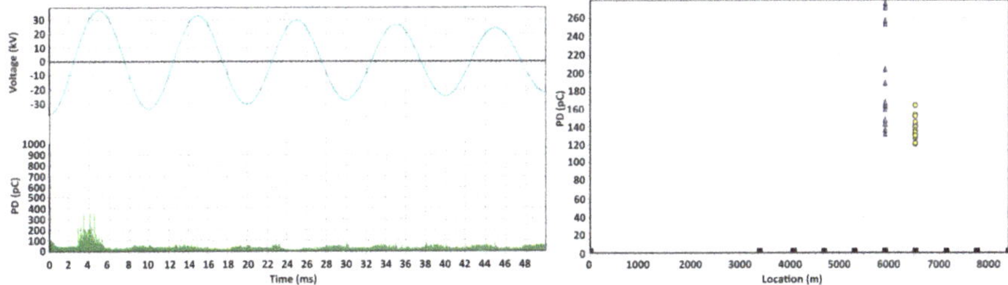

Figure 8. Phase-resolved PD pattern at 1.4 U_o (left), PD mapping showing the complete string length with the localized PD concentrations at the wind turbines (WTG) at WTG 5 and WTG 6. The squares are the locations of the WTGs in the string (right).

A second field example is from an after-laying test of a newly installed 13.3 km long, 220 kV XLPE insulated submarine cable circuit. This cable was tested using a damped ac system at 49 Hz, applying up to 1.3 U_o, (Figures 9–11). Monitored withstand testing was performed. As the damped ac test voltage was increased starting from 0.2 U_o, PD activity was observed in phase L1. An increase in the test voltage resulted in an increase of PD activity, At 0.4 U_o test voltage, a breakdown at the discharging site occurred. PD mapping revealed the PD concentration at 5.3 km indicated the breakdown position in the cable. The defect produced PD before an actual breakdown occurred, and with TDR analysis, the PD defect location could be determined. The other two phases fulfilled the after laying conditions and successfully passed the test. No internal PD activity in the cable insulation and accessories and no breakdown occurred during the tests of the other phases. The breakdown occurred during the first test, however the failure was high ohmic and the PD test could be continued on a lower voltage. Therefore the measurement could be repeated from the other end of the cable (phase L1). The PD activity occurring before the breakdown could be localized at 8 km, which is the same location seen from the original tests (13.3 − 8 = 5.3 km).

Figure 9. On-site testing of a 220 kV 13.3 km long XLPE cable circuit: The damped AC system HV300 is connected to one of the cable section phases (left) and the transition point between land and submarine cable (right).

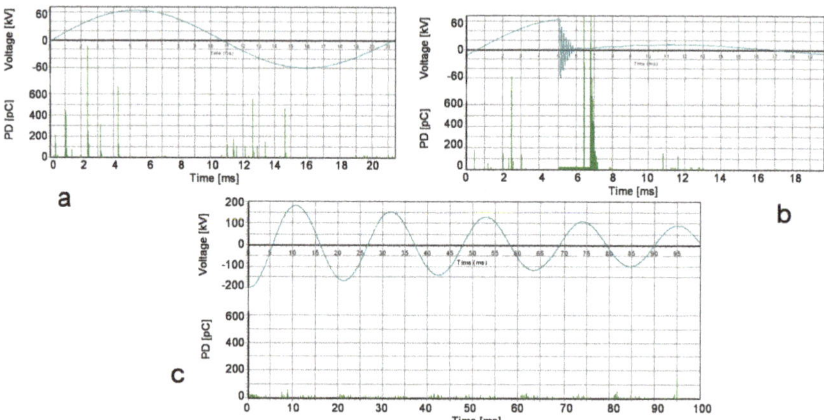

Figure 10. Damped AC voltages and PD patterns as observed during withstand testing of a 220 kV XLPE cable underground circuit (13.3 km): (**a**) example of PD pattern at 0.2 U_o of phase L1, (**b**) example of PD pattern at breakdown voltage of 0.4 U_o of phase L1, (**c**) PD pattern at 1.3 U_o of phases L2 and L3.

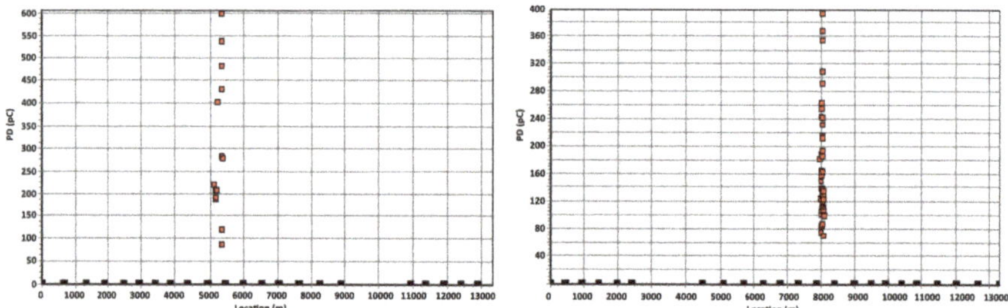

Figure 11. PD Mapping as made up to 1.3 U_o during DAC testing of a 220 kV 13.3 km long cable circuit. The PD concentration at 5.2 km distance indicates the breakdown site of phase L1 (left). Measurement from the other end confirmed this location at 8.1 km (right).

The cable was repaired at the indicated location. After repair, the cable was re-tested and no further PD activity was detected.

7. Failure Statistics and Financial Impact

On an offshore wind farm (OSWF), partial discharge was monitored using voltage withstand tests using DAC and was performed to obtain a full condition assessment. During the test sequence on both wind farms, PD activity was detected and localized in the inter-array cables. It could be verified that on multiple cable terminations, installation issues were the cause for the PD activity. If the number of affected cable terminations in relation to the total number of terminations installed is observed, it can be seen that in this case up to 3.7% of the terminations are affected, see Table 2.

Table 2. Results of DAC condition assessment on an offshore wind farm (OSWF).

No. WTGs	48
Strings	8
No. Terminations in WTGs	268
Terminations with PD	10
Percentage	3.7%

It can be concluded that a considerable number of wind turbines have increased risk of a failure due to the presence of PD at voltages higher than Uo. These PD sites are probably related to poor workmanship during the installation.

The costs for the testing and the costs for a possible failure of a cable termination during operation can be evaluated. The typical costs for testing are in the range of € 10,000 to € 15,000 per turbine dependent on the scope of works of testing; this cost reduces with volume.

Financial losses due to turbines out of operation can very quickly reach more than 100,000 € within three days, i.e., the mean time to repair (MTTR), which depends on fault location in a string (number of wind turbines which are affected), inter-array cable topology, and the weather conditions during the fault, see Table 3. With consideration of failure rates for terminations and other relevant factors, e.g., the MTTR for a 33 kV cable termination, an after-installation test can have a positive financial effect if it prevents the first cable termination fault in a wind farm.

Table 3. Losses of a single inter-array 33 kV or 66 kV termination failures. Approximate failure costs for a 100 WTG (800 MW) wind farm: (German calculation taking average 50% efficiency of the wind farm with a price of EUR 0.16 kWh).

Outage	1 WTG Affected (End of String)	8 WTG Affected (Start of String)
2 weeks stop	EUR 224,000	EUR 1,568,000
3 weeks stop	EUR 336,000	EUR 2,352,000
4 weeks stop	EUR 448,000	EUR 3,136,000
Repair of 1 termination	EUR 152,000–303,000 *	EUR 152,000–303,000 *

* Depending on the complexity of the repair.

However, the size (i.e., the power rating) of the turbine also plays an important role. In the future, larger turbines will be built were a termination fault will lead quickly to higher financial loss due to the greater loss of energy. As a result, a proper after-installation testing program is beneficial to verify the cable quality and can prevent costly failures during operation. The economic impact of an HVAC or HVDC export cable failure can be examined under the following categories:

1. The cost of repairing the cable;
2. The cost of lost electric power transmission over a period that can range from several weeks (for a pre-emptive repair) up to 3 to 9 months for an unexpected fault.

   The cost of cable repair includes:

(a) The precise location of the fault;
(b) Mobilization of a suitable repair vessel, equipment, and personnel;
(c) de-burial of the failed cable;

(d)   Removal of a length of cable (often several hundred meters long—that includes the fault location);

(e)   Jointing in a length of spare cable to replace the removed section;

(f)   Reburying or otherwise protecting the repaired cable.

HVAC and HVDC export cable links costs will depend on market conditions and weather conditions during the repair, but can be extremely high. Based on several export cable failures (which can be compared to a submarine HVDC cable) in the UK, it is calculated that the average repair cost from those failures is £12.5 million (ranging from £5.3 million to £15.5 million).

It has been estimated that the cost of a pre-emptive repair is substantially less than the cost of an unplanned repair. It is so much lower than the costs for the unplanned repairs that it is safe to state that the cost of a pre-emptive (estimated on £ 3.5 million) repair will be much lower than that of an unplanned repair.

As a result, based on the repair costs, the actual repair times, and the typical price of a loss of supplied energy (£/MWh), a rough estimate of the total cost to the industry due to cable failures can be calculated. This means that the above-stated costs for a failure repair can easily be doubled if the loss of supplied energy is also taken into consideration.

8. Conclusions

In the past 20 years field testing experiences have been obtained with damped AC testing. For several years now, this technology has also been applied to testing submarine cables at offshore wind farms. Besides this, the basic aspects of failure risks of export and inter-array cables for offshore wind farms have been discussed in this contribution. The following conclusions may be drawn after-installation testing:

- Based on the international failure statistics, it can be seen that the cable systems are the most critical parts of complete wind farm installations with a failure impact rate of 30%. However, up until now, the present international standards guidelines are not covering the needs of high-quality after-installation testing of export and inter-array cabling;

- Performing non-destructive PD monitored sensitive testing is a good basis for condition assessment and future life estimation of export and inter-array cables. With this after-installation testing, the faults that can occur as a result of stresses during manufacturing, transportation, and installation can be detected;

- For on-site cable testing, the PD monitored DAC voltage withstand testing is, in many countries, a common practice. PD measurement, including PD-pattern information and time domain reflectometry (PD localization), helps to detect and locate discharging defects in the insulation and accessories of power cables;

- DAC is a very suitable test technology to test long lengths of onshore and offshore export HVAC and HVDC power cable with a low demand of on-site power needed and with the possibility of sensitive PD detection and localization. This testing method makes it possible to obtain an integral cable fingerprinting by damped AC testing, which provides an assessment of the cable circuit integrity for the installation of both newly installed and under operation cables;

- Presented case studies have shown the value of applying the damped AC testing including PD detection and dissipation factor estimation to find upcoming failures prior to service operation. It is shown in the presented example that up to 3.7% of the installed cable terminations had partial discharges in this particular case. Those defects could not be found during traditional un-monitored (i.e., without PD measurements) voltage withstand testing, at which is only tested if the cable system withstands the over-voltage for a certain duration;

- Although testing of cables involves costs, it has shown that the testing with DAC is a cost-effective solution compared to the costs, for example, of a termination failure in an operational offshore wind farm.

Author Contributions: Conceptualization, R.J. and E.G.; validation, R.J. and E.G.; investigation, R.J.; data curation, R.J.; writing—original draft preparation, K.S., R.J., and E.G.; writing—review and editing, A.R. and K.S.; visualization and data submission, R.J.; supervision, E.G.

Funding: This research received no external funding.

Conflicts of Interest: The authors declare no conflict of interest.

References

1. IEC. *IEC 60840: Power Cables with Extruded Insulation and the Accessories for Rated Voltages Above 30 kV up to 150 kV Test Methods and Requirements*; IEC: Geneva, Switzerland, 2011.
2. IEC. *IEC 62067: Power Cables with Extruded Insulation and the Accessories for Rated Voltages Above 150 kV*; IEC: Geneva, Switzerland, 2011.
3. IEC. *IEC 60502: Power Cable with Extruded Insulation and the Accessories for Rated Voltages from 1 kV up to 30 kV*; IEC: Geneva, Switzerland, 2011.
4. IEEE. *IEEE 400-2012: Guide for Field Testing and Evaluation of the Insulation of Shielded Power Cable Systems Rated 5 kV and Above*; IEEE: Piscataway, NJ, USA, 2012.
5. IEEE. *IEEE 400.4-2015: Guide for Field-Testing of Shielded Power Cable Systems Rated 5 kV and Above with Damped Alternating Current Voltage (DAC)*; IEEE: Piscataway, NJ, USA, 2015.
6. Hodge, N.; Maurer, R. Power under the Sea, Allianz Global Risk Dialogue, Autumn 2014, pp. 26–29. Available online: https://www.agcs.allianz.com/assets/PDFs/GRD/GRD_02_2014_EN.pdf (accessed on 8 August 2018).
7. Tisheva, P. Cable Failures Account for Most of Offshore Wind Losses, June 2016. Available online: https://www.renewablesnow.com/news/cable-failures-account-for-most-of-offshore-wind-losses-528959 (accessed on 26 June 2018).
8. Gulski, E.; Jongen, R.; de Heus, M.; Rakowska, A.; Siodla, K.; Gaal, H. On-Site Acceptance and Diagnostic Testing of Submarine Inter-Array Cables at Offshore Wind Farms using Damped AC. In Proceedings of the 2018 IEEE International Conference on High Voltage Engineering and Application (ICHVE), Athens, Greece, 10–13 September 2018.
9. Jongen, R.; Gulski, E.; Rakowska, A.; Siodla, K.; Gaal, H. After Installation Testing of Inter-Array Cables at Offshore Wind Farms using Damped AC Voltages. In Proceedings of the 10th International Conference on Insulated Power Cables, Versailles, France, 23–27 June 2019.
10. Cejka, G.; Gulski, E.; Jongen, R.; Quak, B.; Parciak, J.; Rakowska, A. Integrated Testing and Diagnosis of Distribution Cables using Damped AC and Very Low Frequency Voltages. In Proceedings of the 10th International Conference on Insulated Power Cables, Jicable 2019, Versailles, France, 23–27 June 2019.
11. Gulski, E.; Jongen, R.; Quak, B.; Parciak, J.; Rakowska, A.; Siodla, K. Fifteen Years Damped AC On-site Testing and Diagnosis of Transmission Power Cables. In Proceedings of the 10th International Conference on Insulated Power Cables, Versailles, France, 23–27 June 2019.
12. Gulski, E.; Wester, F.J.; Smit, J.J.; Seitz, P.N.; Turner, M. Advanced PD diagnostic of MV power cable system using oscillating wave test system. *IEEE Electr. Insul. Mag.* **2000**, *16*, 17–25. [CrossRef]
13. Gulski, E.; Smit, J.J.; Petzold, F.; Seitz, P.P.; Quak, B.; de Vries, F. Advanced Solution for On-Site Diagnosis of Distribution Power Cables. In Proceedings of the Jicable 2007, Versailles, France, 24–28 June 2007.
14. Gulski, E.; Wester, F.J.; Schikarski, P.; Seitz, P.N. PD diagnoses and condition assessment of distribution power cables using damped AC voltages. In Proceedings of the XIII International Symposium on HV, Delft, The Netherlands, 25–29 August 2003; p. 776.
15. Gulski, E.; Jongen, R.; Patterson, R. *Modern Testing and Diagnosis of Power Cables using Damped AC Voltages*; NETA World: Portage, MI, USA, 2015.
16. Gulski, E.; Rakowska, A.; Siodla, K.; Jongen, R.; Minassian, R.; Cichecki, P.; Parciak, J.; Smit, J. On-Site Testing and Diagnosis of Transmission Power Cables up to 230 kV Using Damped AC Voltages. *IEEE Electr. Insul. Mag.* **2014**, *3*, 27–38. [CrossRef]
17. Gulski, E.; Chojnowski, P.; Rakowska, A.; Siodla, K. Importance of sensitive on-site testing and diagnosis of transmission power cables. *Przeglad Elektrotechniczny* **2009**, *2*, 171–176.
18. Gulski, E.; Rakowska, A.; Siodla, K.; Chojnowski, P. On-site testing and diagnosis of transmission power cables. *Przeglad Elektrotechniczny* **2009**, *4*, 195–200.

19. Wester, F.J.; Gulski, E.; Smit, J.J. Electrical and acoustical PD on-site diagnostics of service aged medium voltage power cables. In Proceedings of the 5th International Conference on Power Insulated Cables, Jicable 1999, Versailles, France, 20–24 June 1999.

20. Bodega, R.; Morshuis, P.H.; Lazzaroni, M.; Wester, F.J. PD recurrence in cavities at different energizing methods. *IEEE Trans. Instrum. Meas.* **2004**, *53*, 251–258. [CrossRef]

21. Wester, F.J.; Guilski, E.; Smit, J.J. Detection of partial discharges at different AC voltage stresses in power cables. *IEEE Electr. Insul. Mag.* **2007**, *23*, 28–43. [CrossRef]

22. Wester, F.J. *Condition Assessment of Power Cables using PD Diagnosis at Damped AC Voltages*; Optima Grafische Communicatie: Rotterdam, The Netherlands, 2004; ISBN 90-8559-019-1.

23. Houtepen, R.; Chmura, L.; Smit, J.J.; Quak, B.; Seitz, P.P.; Gulski, E. Estimation of dielectric loss using damped AC voltages. *IEEE Electr. Insul. Mag.* **2011**, *27*, 20–25. [CrossRef]

24. Wild, M.; Tenbohlen, S.; Gulski, E.; Jongen, R. Basic aspects of partial discharge on-site testing of long length transmission power cables. *IEEE Trans. Dielectr. Electr. Insul.* **2017**, *24*, 1077–1087. [CrossRef]

Article

Detection of Water Content in Transformer Oil Using Multi Frequency Ultrasonic with PCA-GA-BPNN [†]

Zhuang Yang [1], Qu Zhou [1,*], Xiaodong Wu [1], Zhongyong Zhao [1], Chao Tang [1] and Weigen Chen [2]

[1] College of Engineering and Technology, Southwest University, Chongqing 400715, China; 15520004027@163.com (Z.Y.); wuxiaodong5203@163.com (X.W.); zhaozy1988@swu.edu.cn (Z.Z.); tangchao_1981@163.com (C.T.)

[2] State Key Laboratory of Power Transmission Equipment & System Security and New Technology, Chongqing University, Chongqing 400030, China; weigench@cqu.edu.cn

* Correspondence: zhouqu@swu.edu.cn; Tel.: +86-23-68251265

† This paper is an extended version of our paper published in the 2018 IEEE International Conference on High Voltage Engineering and Application (ICHVE), Athens, Greece, 10–13 September 2018.

Received: 11 March 2019; Accepted: 4 April 2019; Published: 10 April 2019

Abstract: The water content in oil is closely related to the deterioration performance of an insulation system, and accurate prediction of water content in oil is important for the stability and security level of power systems. A novel method of measuring water content in transformer oil using multi frequency ultrasonic with a back propagation neural network that was optimized by principal component analysis and genetic algorithm (PCA-GA-BPNN), is reported in this paper. 160 oil samples of different water content were investigated using the multi frequency ultrasonic detection technology. Then the multi frequency ultrasonic data were preprocessed using principal component analysis (PCA), which was implemented to obtain main principal components containing 95% of original information. After that, a genetic algorithm (GA) was incorporated to optimize the parameters for a back propagation neural network (BPNN), including the weight and threshold. Finally, the BPNN model with the optimized parameters was trained with a random 150 sets of pretreatment data, and the generalization ability of the model was tested with the remaining 10 sets. The mean squared error of the test sets was 8.65×10^{-5}, with a correlation coefficient of 0.98. Results show that the developed PCA-GA-BPNN model is robust and enables accurate prediction of a water content in transformer oil using multi frequency ultrasonic technology.

Keywords: transformer oil; multi frequency ultrasonic; water content; back propagation neural network; genetic algorithm

1. Introduction

In different insulation systems, transformer oil is an important insulating medium in power transformers, and the water content in oil is an important factor in determining the insulation life of transformers [1–3]. Many problems in the insulating system, such as breakdown voltage reduction, dielectric loss increase and the acceleration of the chemical reaction of organic matter, are caused by a higher water content in oil [4–7]. Therefore, the detection of water content in transformer oil is of great significance to ensure the safe and stable operation of the transformer.

The technology and methods of water content detection in transformer oil have been studied by many scholars domestically and abroad [8–12]. Detection methods of water content in transformer oil, including off-line and on-line, have been widely reported. The measurement of water content based on humidity sensors and the reduction of temperature error have been studied by some scholars [8]. Martin et al. predicted water content by means of establishing a mathematical model of the water dynamic [9]. But this measurement should be corrected for by any difference in oil temperature between the water activity probe location and the insulation hot-spot location. The measurement of water content using direct optical techniques has been proposed in [10]. But the polar molecules in oil can affect optical measurement accuracy. The authors in [11] proposed to predict the water content in oil from the curve of water equilibrium. In order for this method to yield accurate data, the water content and temperature must have reached equilibrium. In conclusion, more or less errors were caused when water content was tested using the above method, because the inner part of the transformer in operation is an extremely complex movement process and the transformer oil is a very complex mixture [13–15]. Since the multi frequency ultrasonic detection technology is a non-destructive testing technology, and has many advantages, such as increasing cavitation events, reducing the dead angle caused by standing wave, and improving the sonochemical yield [16–18]. It has been implemented for various applications, e.g., distance measurement [19], medical examination [20], partial discharge localization in power transformers [21] and quality inspection [22]. The multi-frequency ultrasonic detection technology reflects the internal information of the measured object at the molecular level, and can avoid the interference of external environmental factors such as temperature to a great extent, thereby realizing high-precision detection.

In this article, an alternative method of measuring the water content in transformer oil, using multi-frequency ultrasonic detection technology and PCA-GA-BPNN was proposed. The multi-frequency ultrasound data was obtained by the experiment was reduced into dimensions by PCA and input into the BPNN, then the Carle Fischer method was used to measure the water content as BPNN output. The number of neurons in the hidden layer was determined by the test method, the weight and the threshold of the BPNN was optimized by GA which improved the prediction accuracy. Finally, a case study of test set was carried out to verify the validity of the proposed empirical model.

The remainder of this paper is organized as follows. The Multi Frequency Ultrasonic (MFU) testing system is described in Section 2. Experiments are discussed in Section 3. The PCA, GA and BPNN algorithms are represented in Section 4. The prediction model of water content in transformer oil and the prediction results are presented in Section 5. Finally, a conclusion is provided in Section 6.

2. MFU Testing System

As shown in Figure 1, the multi frequency ultrasonic detection system consists of three parts, namely, an Ultrasonic Measurement Device (Yucoya Energy Safety GmbH, GER), an Ultrasonic Sensor and Measurement Software (Yucoya Ultrasound Manager, Yucoya Energy Safety GmbH, GER). The internal state information of transformer oil was reflected by these ultrasonic parameters that were obtained by continuous scanning at the molecular level.

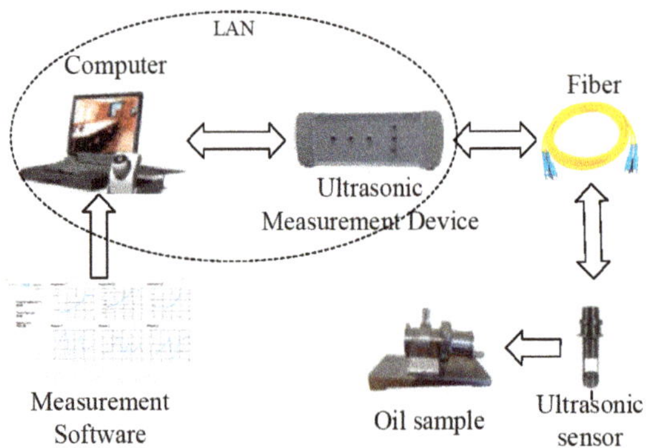

Figure 1. Structure chart of the Multi-frequency ultrasonic device.

The structure of the ultrasonic sensor is shown in Figure 2. At the time of detection, the ultrasonic sensor was dipped into the transformer oil, except for the mounting bracket and the external connectors, to fill the measurement chamber with the oil. Meanwhile, the ultrasonic transmitter Tx emits an ultrasound signal, including 20 frequencies within the range of 600 kHz–1000 kHz and the central resonance frequency was about 750 kHz. A part of this signal is first reflected at the interface between the reference medium and the measurement chamber. This reflected signal travels back to ultrasonic receiver Rx1 where it is measured. This signal is called L1. The other part of the signal emitted by ultrasonic transmitter Tx is transmitted through the interface of the reference medium and the measurement chamber. It travels to ultrasonic receiver Rx2, where it is measured. This part of the signal is called L3. Finally, at ultrasonic receiver Rx2, a part of the signal is reflected again and travels back to ultrasonic receiver Rx1. This signal is called L2.

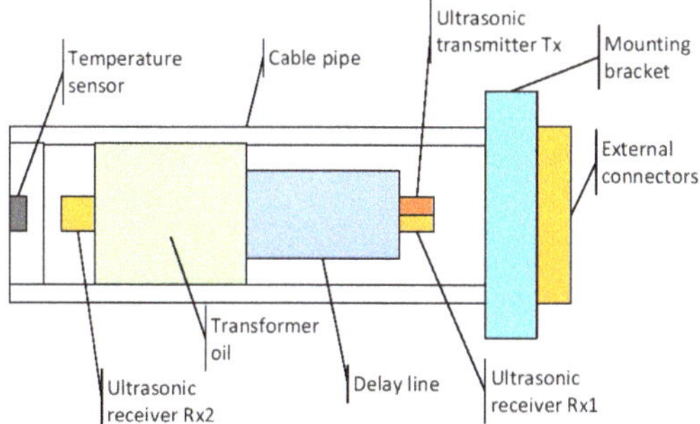

Figure 2. Principle structure diagram of ultrasonic sensor.

The received signals which are called L1, L2 and L3 can be written as:

$$x(t) = A \sin(\omega t + \varphi) \tag{1}$$

where A is signal amplitude, φ is signal phase, $\omega = 2\pi f$ and f is signal frequency.

This signal can be written as:

$$x(t) = A \cos(\varphi) \sin(\omega t) + A \sin(\varphi) \cos(\omega t) \tag{2}$$

$$x(t) = C_0 \sin(\omega t) + C_1 \cos(\omega t) \tag{3}$$

where:

$$C_0 = A \cos(\varphi) \tag{4}$$

$$C_1 = A \sin(\varphi) \tag{5}$$

The requested amplitude and phase can be determined as follows:

$$A = \sqrt{C_0^2 + C_1^2} \tag{6}$$

$$\varphi = \arctan\left[\frac{C_1}{C_0}\right] + 1 - \mathrm{sgn}(C_0)\frac{\pi}{2} \tag{7}$$

3. Experimental Results

A series of measurements were conducted on 160 transformer mineral oil samples, which were made up of 10 new oils, 10 drying oils from new oils and 140 service-aged oils. The Carle Fischer method was used to identify the water content of each sample. The same oil samples were tested using multi-frequency ultrasonic transformer detection device for the acoustic wave frequency spectrum.

The six samples oil response spectrum are shown in Figure 3, where the legend is the moisture content of the sample in mg/L. It can be seen that, as the moisture content fell, the amplitude response of each frequency fell, and the increase in amplitude response was maintained in L1 phase and L2 phase. In L3 phase, there was a "basin" inside the amplitude response within the scope of the frequency range 700 kHz to 850 kHz. An oil sample with a moisture content of 1.51 mg/L exhibited the lowest amplitude response in L1 phase, L2 phase and In L3 phase.

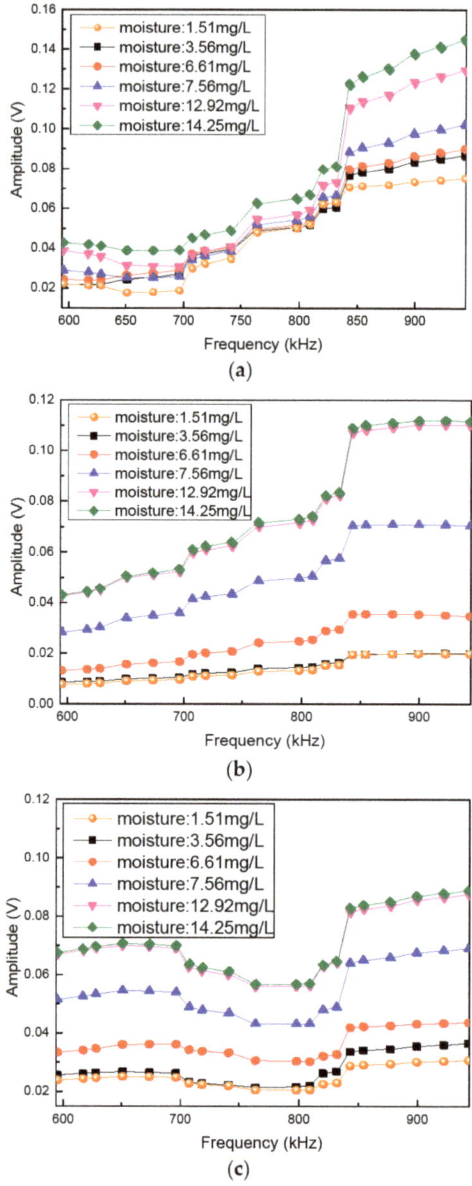

Figure 3. Spectrum curve of ultrasonic signal. (**a**) Amplitude response of L1 phase of multi-frequency ultrasonic detection; (**b**) Amplitude response of L2 phase of multi-frequency ultrasonic detection; (**c**) Amplitude response of L3 phase of multi-frequency ultrasonic detection.

4. Artificial Neural Network

4.1. Principal Component Analysis

According to [23,24], as a kind of multivariate statistical analysis, PCA is successfully used in various applications such as picture processing, face recognition and so on. As a model of dimension reduction, it operates mainly though creating a few new variables, which are uncorrelated, from the original variables, and these new variables retain the maximum information of original variables as much as possible.

The dimension reduction matrix was obtained by PCA, as described below.

Step 1: Standardize the original matrix:

$$x_{ij}^* = \frac{\left(x_{ij} - \overline{x}_j\right)}{S_j} \tag{8}$$

where x_{ij}^* is a variable of standard matrix, x_{ij} is a variable of original matrix, i (=1, 2, … , N) is the number of variable, j (=1, 2, … , m) is the dimensions of each sample, and $\overline{x}_j$, S_j are the mean and variance of the indicator variable x_j, respectively.

Step 2: Calculate the correlation matrix, the eigenvectors and the eigenvalues of the correlation matrix:

$$R = \frac{X^{*T} \times X^*}{(N-1)} \tag{9}$$

$$R \times \lambda_j = \lambda_j \times u_j \tag{10}$$

where R is the correlation matrix, X^* is the standard matrix, and λ_j, u_j are the eigenvalues and the eigenvectors of the correlation matrix, respectively.

Step 3: Calculate the contributing rate of cumulative variance and the contributing rate of variance:

$$\eta_j = \frac{\lambda_j}{\sum_j^m \lambda_j} \times 100\% \tag{11}$$

$$\eta_\Sigma(p) = \sum_j^p \eta_j \tag{12}$$

where η_j is the contributing rate of variance of the jth principal component, and $\eta_\Sigma(p)$ is the accumulative variance contribution of the first p principal components.

Step 4: Calculate the projection of original matrix:

$$Z_{N \times p} = X_{N \times m}^* U_{m \times p} \tag{13}$$

where $Z_{N \times p}$ is the dimension reduction matrix, and $U_{m \times p} = \left[u_1, u_2, \ldots, u_p\right]$.

In this study, the original data matrix is the multi frequency ultrasonic detection data which is a 242-dimensional space including the amplitude and phase of 20 frequencies, time of flight (TOF) and velocity, which could be reduced by PCA. The information retention ratio of the original data matrix after the dimension reduction is shown in Figure 4. It was observed as conspicuous from Figure 4 that the first principal component encompassed only about 35% of the total variation, and the information retention rate increased with the increase in the dimensions, and the information retention rate of the first seven principal components reached 90%. The number of neurons in the input layer of BPNN model were determined by the number of principal components that had a high information retention rate. In order to ensure performance of BPNN model, the first eight principal components, which the information retention rate hit 95%, were used as the inputs of the model.

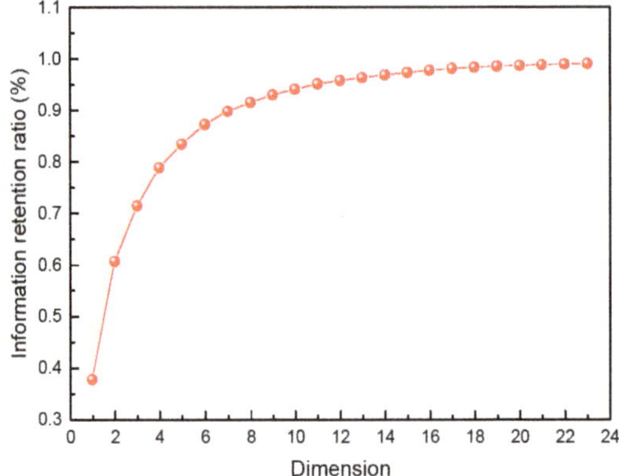

Figure 4. Information retention ratio.

4.2. BP Neural Network

An artificial neural network model has been developed to investigate the correlation between the water content in oil samples and their multi frequency ultrasonic spectrum data. According to the authors in [25], when there are enough neurons in the hidden layer, a three-layer BPNN can realize the mapping of an arbitrary I-dimension (input layer) to any k-dimension (output layer). Therefore, in this paper, a three-layer BPNN was chosen, and the input variables of the BPNN with PCA were the first eight principal components from the analysis of Section 4.1, and the output layer consisted of 1 neuron, corresponding to the water content.

The flow chart of model training and learning is shown in Figure 5.

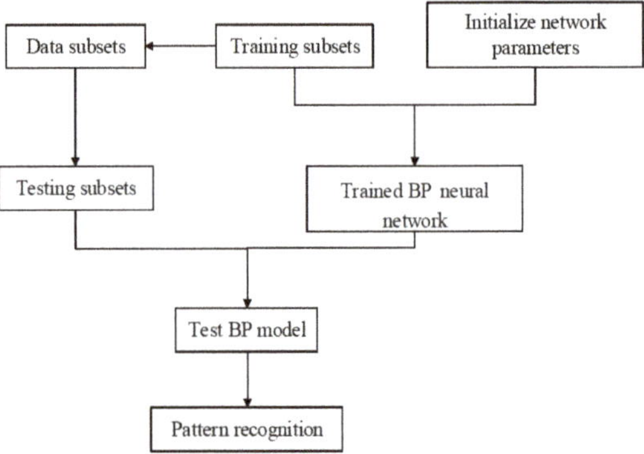

Figure 5. Flow chart of back propagation neural network training.

4.3. Genetic Algorithm

It is difficult for traditional BPNNs to find out the global optimum solution of the prediction application [26,27]. A genetic algorithm (GA) is a method to obtain global optimum solution of the proposed problems based on a natural selection process which mimics the biological evolution process [28–31]. In this article, GA was used to optimize the weight and the threshold of the BPNN. The BPNN flow chart of the GA-BPNN is shown in Figure 6.

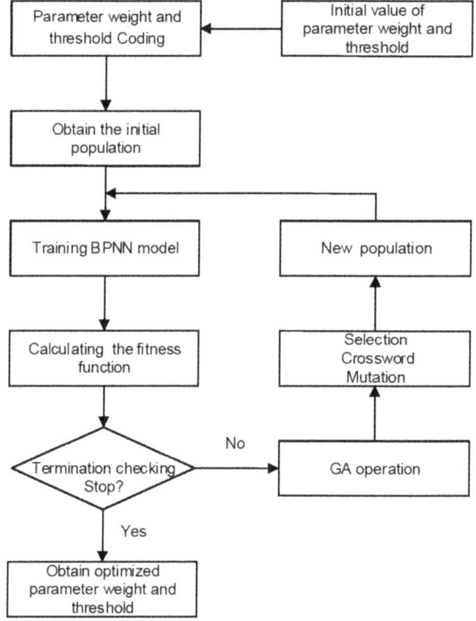

Figure 6. Flow chart of optimizing the BP neural network parameters with a GA.

4.4. PCA-GA-BPNN Prediction Model

The working process of the prediction model was divided into three stages.

The first stage: create a database module. The database module not only matched the multi-frequency ultrasonic parameters of oil with water content in oil, but also divided it into the training sample collection and test sample set in certain proportions.

The second stage: create the prediction model. The prediction model first read out training samples from the database module, and combined them with the PCA to get the input matrix, which was composed of the first eight principal components. Then, by means of the initial GA-BPNN parameters, the initial forecast model, and the initial forecast results were produced. The fitness function of the model was described as:

$$F(x) = \sum_{i=1}^{n} |y_i - \hat{y}_i| \tag{14}$$

where y_i is predicted value, $\hat{y}_i$ is observed value, i and is sample size.

Using the fitness function of the genetic algorithm to calculate the fitness value each individual in each generation, if it meets fitness convergence conditions, the initial forecast model is the final prediction module, or it will perform the operation of selection, crossover and mutation, and pass the new parameters to the GA-BPNN, then the second generation prediction model combines with the database module to get the second generation forecast results and so on, until it finally meets the prediction model of the fitness convergence conditions. The roulette method was used as the selection of GA, the probability of selection, P_x, for each individual, x, was described as:

$$P_x = \frac{f_x}{\sum_{j=1}^{N} f_j} \tag{15}$$

$$f_x = \frac{k}{F_x} \tag{16}$$

where F_x is the fitness values of individual x, N is the individual number and k is the coefficient.

The crossover operation method of the kth chromosome a_k and the lth chromosome a_l in the j position was described as:

$$\begin{cases} a_{kj} = a_{kj}(1-b) + a_{lj}b \\ a_{lj} = a_{lj}(1-b) + a_{kj}b \end{cases} \tag{17}$$

where b is a random number in the range of 0 to 1.

The mutation operation was determined by Formulas (18) and (19):

$$a_{ij} = \begin{cases} a_{ij} + \left(a_{ij} - a_{max}\right) \times f(g) \; r > 0.5 \\ a_{ij} + \left(a_{min} - a_{ij}\right) \times f(g) \; r \leq 0.5 \end{cases} \tag{18}$$

$$f(g) = r_2 \left(1 - \frac{g}{G_{max}}\right)^2 \tag{19}$$

where a_{max} is the upper bound of a_{ij}, a_{min} is the lower bound of a_{ij}, r_2 is a random number, g is the current iterations, G_{max} is the maximum number of evolution generations, and r is a random number in the range of 0-1.

The third stage: the water content in oil is forecast. According to the final prediction model, it will accurately predict the water content in oil.

5. Results and Discussion

In the simulation and compiled environment of Matlab (2014a), the prediction model of water content in transformer oil was established. The model was trained with the training data which included 150 random sets, and the prediction accuracy of the model was tested with the remaining 10 sets. Since the number of optimal hidden layer neurons gives uncertainty in the initial modeling, the range of the number of hidden layer neurons was determined by empirical formula [32–34]:

$$n < \sqrt{m + l} + a \tag{20}$$

$$n < \log_2 l \tag{21}$$

where l is the number of input layer neurons, n is the number of hidden layer neurons, and m is the number of output layer neurons. a is in the range of 1-10. Therefore, in this paper, the number of hidden layer neurons was between 1 and 16. The back-check diagnosis was performed using the network of 1-16 hidden neurons, and the mean square error (MSE) of recheck was calculated. The variation of MSE with the number of neurons in the hidden layer as shown in Figure 7.

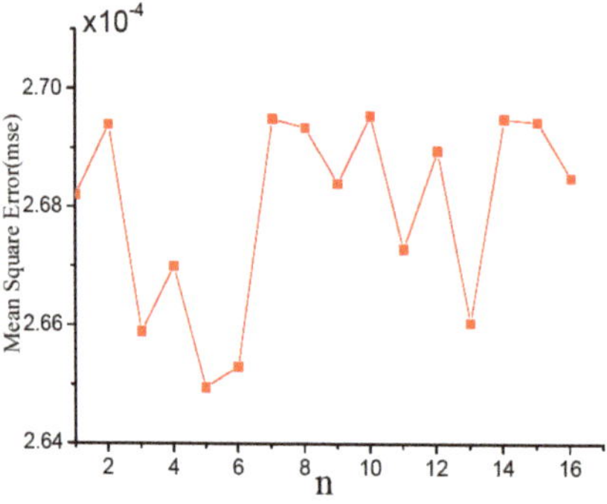

Figure 7. The MSE of BP neural network recheck.

As shown in Figure 6, when the number of hidden layer neurons is 5, the MSE of the model was at the minimum. Therefore, according to Section 4.2, the topological structure of the BPNN model was determined as "8-5-1" by many experiments. In order to improve the convergence speed and prediction accuracy of the BPNN, the weight and the threshold of the BPNN was optimized by GA.

The quality of the solution evaluated by the genetic algorithm depends on the fitness value of the solution. In this model, the sum of the absolute value of the error between the predicted output and the expected output was the individual fitness value, so for the individual fitness, a lower value is better. Figure 8 shows the model of optimal fitness is declining in the process of evolution, finally the optimal individual fitness value was 9.6.

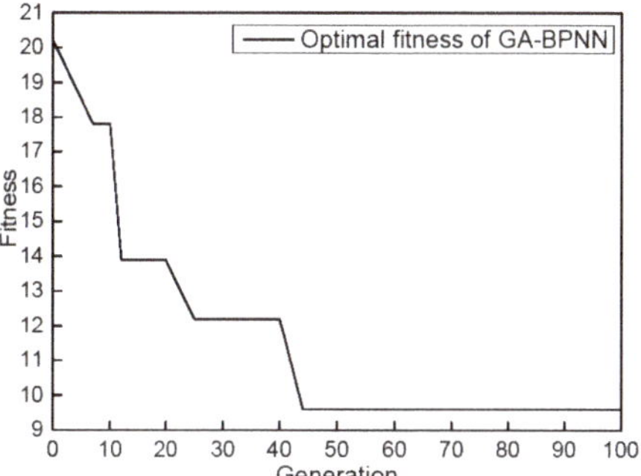

Figure 8. The fitness curve of Genetic algorithms.

The optimal weights and the optimal thresholds obtained from the optimization of GA were assigned to the BPNN. Then, the model was trained with the training set. The regression curve of the model is shown in Figure 9. Figure 9 shows that the correlation coefficient of the PCA-GA-BPNN model was about 0.98, indicating that the model has a good regression fit.

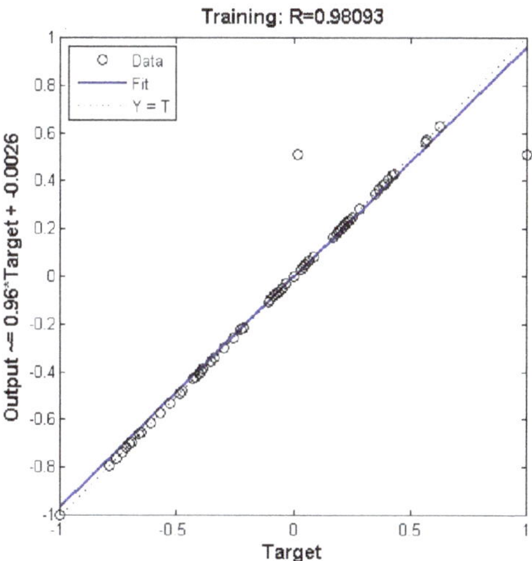

Figure 9. The regression fitting curve of GA-BPNN.

The MSE of the training process of the PCA-GA-BPNN is shown in Figure 10. The MSE of this model gradually decreased with the increase of training times. The minimum MSE (8.65×10^{-5}) was obtained by the PCA-GA-BPNN at the 18th iteration.

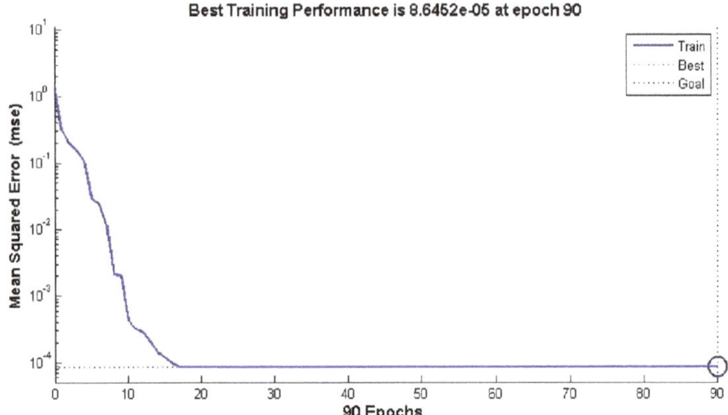

Figure 10. The training process of GA-BPNN.

In this paper, in order to demonstrate the superiority of the prediction model of water content in transformer oil based on the PCA-GA-BPNN, the experiment compared the prediction accuracy with the BPNN and GA-BPNN. The three models were used to identify the test set, the prediction results are shown in Table 1 and Figure 11.

Table 1. Prediction results of water content.

Sample Number	Test Value	BPNN		GA-BPNN		PCA-GA-BPNN	
		Predicted Value	Error	Predicted Value	Error	Predicted Value	Error
1	14.3	16.1	12.59%	13.1	8.39%	13.56	5.17%
2	11.32	13.22	16.78%	12.26	8.30%	10.52	7.07%
3	16.48	18.49	12.20%	14.95	9.28%	15.63	5.16%
4	6.44	5.11	20.65%	5.98	7.14%	5.98	7.14%
5	11	9.93	9.73%	12.14	10.36%	10.25	6.82%
6	2.29	3.11	35.81%	2.67	16.59%	2.55	11.35%
7	4.22	3.56	15.64%	3.79	10.19%	3.99	5.45%
8	7.49	9.01	20.29%	6.44	14.02%	6.77	9.61%
9	22.16	18.98	14.35%	20.01	9.70%	21	5.23%
10	18.43	20.69	12.26%	20.11	9.12%	17.02	7.65%

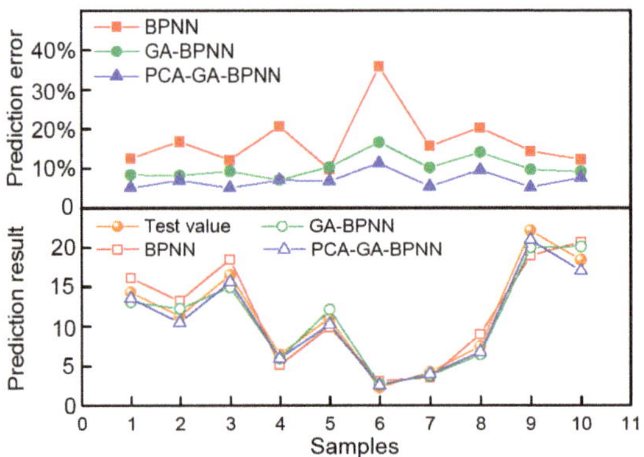

Figure 11. Prediction results.

In order to quantitatively analyze the predictive effect of the three models, the mean absolute percent error (MAPE) was used to compare the prediction errors of the BPNN model, the GA-BPNN model and the PCA-GA-BPNN model. According to Table 1, it was concluded that the MAPE of the BPNN was 17.03%, the MAPE of the GA-BPNN was 10.31%, and the MAPE of the PCA-GA-BPNN was 7.07%.

6. Conclusions

A prediction model of water content in transformer oil using multi frequency ultrasonic with a PCA-GA-BPNN was established. The topological structure of the model was 8-5-1, and the generalization ability of the model was tested with test sets. The experimental results show that the accuracy rate of this model is higher than 90%.

Different structured networks have a different prediction performance. Compared with the BPNN and GA-BPNN models, the PCA-GA-BPNN model can more accurately predict water content in transformer oil according to multi frequency ultrasonic data.

The predictive model of water content in transformer oil using multi frequency ultrasonic with PCA-GA-BPNN, which was proposed in this paper, provides a new online detection method for transformer oil for the power industry. In addition, the application of multi frequency ultrasonic testing technology to detect other parameters of transformer oil is the key point of future research.

Author Contributions: Conceptualization, Z.Y. and Q.Z.; methodology, Z.Y. and Q.Z.; validation, X.W., Z.Z. and C.T.; investigation, Z.Y. and X.W.; resources, C.T.; data curation, Q.Z.; writing—original draft preparation, Z.Y.; writing—review and editing, Z.Y., Q.Z. and W.C.; visualization, Z.Z.; supervision, Q.Z.; project administration, Q.Z.

Funding: This work has been supported in part by the National Natural Science Foundation of China (No. 51507144), Fundamental Research Funds for the Central Universities (No. XDJK2019B021), the China Postdoctoral Science Foundation funded project (Nos. 2015M580771, 2016T90832) and the Chongqing Science and Technology Commission (CSTC) (No. cstc2016jcyjA0400).

Conflicts of Interest: The authors declare no conflict of interest.

References

1. Lundgaard, L.; Hansen, W.; Ingebrigtsen, S. Ageing of mineral oil impregnated cellulose by acid catalysis. *IEEE Trans. Dielectr. Electr. Insul.* **2008**, *15*, 540–546. [CrossRef]

2. Pradhan, M. Assessment of the status of insulation during thermal stress accelerated experiments on transformer prototypes. *IEEE Trans. Dielectr. Electr. Insul.* **2006**, *13*, 227–237. [CrossRef]

3. Zhou, Q.; Chen, W.G.; Xu, L.N.; Kumar, R.; Gui, Y.G.; Zhao, Z.Y.; Tang, C.; Zhu, S.P. Highly sensitive carbon monoxide (CO) gas sensors based on Ni and Zn doped SnO_2 nanomaterials. *Ceram. Int.* **2018**, *44*, 4392–4399. [CrossRef]

4. Rubio-Serrano, J.; Rojas-Moreno, M.V.; Posada, J.; Martínez-Tarifa, J.M.; Robles, G.; Garcia-Souto, J.A. Electro-acoustic detection, identification and location of partial discharge sources in oil-paper insulation systems. *IEEE Trans. Dielectr. Electr. Insul.* **2012**, *19*, 1569–1578. [CrossRef]

5. Martin, D.; Perkasa, C.; Lelekakis, N. Measuring Paper Water Content of Transformers: A New Approach Using Cellulose Isotherms in Nonequilibrium Conditions. *IEEE Trans. Power Deliv.* **2013**, *28*, 1433–1439. [CrossRef]

6. Jadav, R.B.; Ekanayake, C.; Saha, T.K. Understanding the impact of moisture and ageing of transformer insulation on frequency domain spectroscopy. *IEEE Trans. Dielectr. Electr. Insul.* **2014**, *21*, 369–379. [CrossRef]

7. Zhou, Q.; Xu, L.N.; Ahmad, U.; Chen, W.G.; Rajesh, K. Pt nanoparticles decorated SnO_2 nanoneedles for efficient CO gas sensing applications. *Sens. Actuators B* **2018**, *256*, 656–664. [CrossRef]

8. Szepes, L.; Torkos, K.; Dobo, R.; Szekely, A. A New Analytical Method for the Determination of the Water Content of Transformer Oils. *IEEE Trans. Electr. Insul.* **1982**, *17*, 345–349. [CrossRef]

9. Martin, D.; Saha, T.; Perkasa, C.; Lelekakis, N.; Gradnik, T. Fundamental concepts of using water activity probes to assess transformer insulation water content. *IEEE Electr. Insul. Mag.* **2016**, *32*, 9–16. [CrossRef]

10. Rodríguezrodríguez, J.H.; Martínezpiñón, F.; ÁlvarezChávez; José, A.; Jaramillo-Vigueras, D.; Robles-Pimentel, E.G. Direct optical techniques for the measurement of water content in oil–paper insulation in power transformers. *Meas. Sci. Technol.* **2011**, *22*, 2572–2575.

11. Sarfi, V.; Mohajeryami, S.; Majzoobi, A. Estimation of water content in a power transformer using moisture dynamic measurement of its oil. *High Volt.* **2017**, *2*, 11–16. [CrossRef]

12. Martin, D.; Krause, O.; Saha, T. Measuring the Pressboard Water Content of Transformers Using Cellulose Isotherms and the Frequency Components of Water Migration. *IEEE Trans. Power Deliv.* **2017**, *32*, 1314–1320. [CrossRef]

13. Belanger, G.; Duval, M. Monitor for Hydrogen Dissolved in Transformer Oil. *IEEE Trans. Electr. Insul.* **2007**, *12*, 334–340. [CrossRef]

14. Yang, Y. Fuzzy Set Pair Analysis in Transformer Condition Evaluation. *Electr. Eng.* **2013**, *14*, 30–35.

15. Lesieutre, B.C.; Hagman, W.H.; Kirtley, J.L. An improved transformer top oil temperature model for use in an on-line monitoring and diagnostic system. *IEEE Trans. Power Deliv.* **1997**, *12*, 249–256. [CrossRef]

16. Ben, B.S.; Yang, S.H.; Ratnam, C.; Ben, B.A. Ultrasonic based structural damage detection using combined finite element and model Lamb wave propagation parameters in composite materials. *Int. J. Adv. Manuf. Technol.* **2013**, *67*, 1847–1856. [CrossRef]

17. Xie, Q.; Tao, J.; Wang, Y.; Geng, J.; Cheng, S.; Lü, F. Use of ultrasonic array method for positioning multiple partial discharge sources in transformer oil. *Rev. Sci. Instrum.* **2014**, *85*, 084705.

18. Kweon, D.J.; Chin, S.B.; Kwak, H.R.; Kim, J.C.; Song, K.B. The analysis of ultrasonic signals by partial discharge and noise from the transformer. *IEEE Trans. Power Deliv.* **2005**, *20*, 1976–1983. [CrossRef]

19. Khyam, M.O.; Ge, S.S.; Li, X.; Pickering, M.R. Highly Accurate Time-of-Flight Measurement Technique based on Phase-correlation for Ultrasonic Ranging. *IEEE Sens. J.* **2017**, *17*, 434–443. [CrossRef]

20. Jacquemart, P. A progressive method of medical examination: Ultrasonic echotomography. *Inf. Dent.* **1974**, *56*, 137–140.

21. Tang, L.; Luo, R.; Min, D.; Su, J. Study of Partial Discharge Localization Using Ultrasonics in Power Transformer Based on Particle Swarm Optimization. *IEEE Trans. Dielectr. Electr. Insul.* **2008**, *15*, 492–495.

22. Awad, T.S.; Moharram, H.A.; Shaltout, O.E.; Asker, D.; Youssef, M.M. Applications of ultrasound in analysis, processing and quality control of food: A review. *Food Res. Int.* **2012**, *48*, 410–427. [CrossRef]

23. Nie, P.; Chen, X. Prediction of tool VB value based on PCA and BP neural network. *J. Beijing Univ. Aeronaut. Astronaut.* **2011**, *37*, 364–367.

24. Fei, H.; Zhang, L. Prediction model of end-point phosphorus content in BOF steelmaking process based on PCA and BP neural network. *J. Process Control* **2018**, *66*, 51–58.

25. Ding, S.; Su, C.; Yu, J. An optimizing BP neural network algorithm based on genetic algorithm. *Artif. Intell. Rev.* **2011**, *36*, 153–162. [CrossRef]

26. Nair, V.V.; Dhar, H.; Kumar, S.; Thalla, A.K.; Mukherjee, S.; Wong, J.W.C. Artificial neural network based modeling to evaluate methane yield from biogas in alaboratory-scale anaerobic bioreactor. *Bioresour. Technol.* **2016**, *217*, 90–99. [CrossRef]

27. Fu, Z.M.; Mo, J.H. Springback prediction of high-strength sheet metal under air bending forming and tool design based on GA–BPNN. *Int. J. Adv. Manuf. Technol.* **2011**, *53*, 473–483. [CrossRef]

28. Liu, K.; Guo, W.; Shen, X.; Zhongfu, T. Research on the Forecast Model of Electricity Power Industry Loan Based on GA-BP Neural Network. *Energy Procedia* **2012**, *14*, 1918–1924.

29. Wang, S.; Na, Z.; Lei, W.; Wang, Y. Wind speed forecasting based on the hybrid ensemble empirical mode decomposition and GA-BP neural network method. *Renew. Energy* **2016**, *94*, 629–636. [CrossRef]

30. Liang, Y.; Chao, R.; Wang, H.; Huang, Y.B.; Zheng, Z.T. Research on soil moisture inversion method based on GA-BP neural network model. *Int. J. Remote Sens.* **2018**, 1–17. [CrossRef]

31. Zhu, Y.H.; Zhuang, D.Z. Application and Study of BP Neural Network and Genetic Algorithm for Optimizational Parameters Based on MATLAB. *Appl. Mech. Mater.* **2013**, *325–326*, 1726–1729. [CrossRef]

32. Dayhof, J.E.; Deleo, J.M. Artificial neural networks. *Cancer* **2001**, *91*, 1615–1634. [CrossRef]

33. Beigy, H.; Rezameybodi, M. A learning automata-based algorithm for determination of the number of hidden units for three-layer neural networks. *Int. J. Syst. Sci.* **2009**, *40*, 101–118. [CrossRef]

34. Murata, N.; Yoshizawa, S.; Amari, S. Network information criterion-determining the number of hidden units for an artificial neural network model. *IEEE Trans. Neural Netw.* **1994**, *5*, 865–872. [CrossRef]

Article

Accumulation Behaviors of Different Particles and Effects on the Breakdown Properties of Mineral Oil under DC Voltage

Min Dan [1,2], Jian Hao [1,*], Ruijin Liao [1], Lin Cheng [3,4], Jie Zhang [3,4] and Fei Li [3,4]

[1] State Key Laboratory of Power Transmission Equipment & System Security and New Technology, Chongqing University, Chongqing 400044, China; danmin@cqu.edu.cn (M.D.); rjliao@cqu.edu.cn (R.L.)

[2] State Grid Chongqing Nanan Power Supply Company, Chongqing 401223, China

[3] Najing NARI Group Corporation, State Grid Electric Power Research Institute, Nanjing 211000, China; chenglin@sgepri.sgcc.com.cn (L.C.); zhangjie3@sgepri.sgcc.com.cn (J.Z.); lifei6@sgepri.sgcc.com.cn (F.L.)

[4] Wuhan NARI Co. Ltd., State Grid Electric Power Research Institute, Wuhan 430077, China

* Correspondence: haojian2016@cqu.edu.cn; Tel.: +86-182-2301-0926

Received: 19 May 2019; Accepted: 12 June 2019; Published: 16 June 2019

Abstract: Particles in transformer oil are harmful to the operation of transformers, which can lead to the occurrence of partial discharge and even breakdown. More and more researchers are becoming interested in investigating the effects of particles on the performance of insulation oil. In this paper, a simulation method is provided to explore the motion mechanism and accumulation characteristics of different particles. This is utilized to explain the effects of particle properties on the breakdown strength of mineral oil. Experiments on particle accumulation under DC voltage as well as DC breakdown were carried out. The simulation results are in agreement with the experimental results. Having a DC electrical field with a sufficient accumulation time and initial concentration are advantageous for particle accumulation. Properties of impurities determine the bridge shape, conductivity characteristics, and variation law of DC breakdown voltages. Metal particles and mixed particles play more significant roles in the increase of current and electrical field distortion. It is noteworthy that cellulose particles along with metal particles cannot have superposition influences on changing conductivity characteristics and the electrical field distortion of mineral oil. The range of electrical field distortion is enlarged as the particle concentration increases. Changes in the electrical field distribution and an increase in conductivity collectively affect the DC breakdown strength of mineral oil.

Keywords: mineral oil; different particles; accumulation behavior; breakdown voltage; DC voltage

1. Introduction

In order to meet the urgent demand for energy delivery, China has vigorously developed large-capacity and long-distance ultra-high voltage transmission technology over the last 10 years. As a result, the market for large transformers has enlarged, which has strengthened the requirement for large volume and high-quality transformers. Mineral oil-paper insulation is widely used in large transformers, and this determines the safety, stability, and insulation performance of transformers. Transformer damage is mainly caused by insulation problems, among which particle pollution of insulating oil is a significant factor [1].

A large body of literature has expounded sources of particles in power equipment. For power transformers, metal and non-metal particles are the main components of solid particles. Metal particles mainly include copper and iron particles, while non-metallic particles are mainly made up of carbon particles and cellulose particles. Cellulose impurities are the main part of non-metallic impurities, accounting for more than 90% of these impurities [2–4]. Owing to the process of production, assembly, transportation, and incomplete oil filtering, impurities may be left in oil [5–9]. Meanwhile, impurities may be caused by the aging of oil–paper insulation, partial discharge, and part wearing due to long-term operation [5–9]. A few particles move to the joint position between different components with the oil flow. Others gather in the region where the flow rate is slower to form a path connecting conductors, which brings a severe challenge to the stable operation of transformers as well as leading to partial discharge or breakdown [10–12].

The CIGRE Working Group 12.17 conducted a statistical analysis of the fault causes of twenty-two transformers and forty bushings (765 kV) and pointed out that the source and hazard of particles must be paid attention to for transformers 400 kV and above [6]. In recent years, in view of the influence of particles on the insulation performance of oil, scholars at home and abroad have carried out research on the movement characteristics of particles in mineral oil as well as the influence of impurity bridges on insulation performance. Mahmud S. et al. studied the bridging phenomena of cellulose particles in mineral oil based on spherical electrodes and needle-plate electrodes under different DC and AC voltages. It was found that DC voltage was the main factor in bridge formation [13–17]. Yuan Li et al. studied the generation and development of cellulose bridges and their influences on partial discharge under DC voltage [18]. In [19,20], a millimeter-diameter metal particle's movement properties and its movement were studied using a theoretical model. In [21,22], the motion trajectory of a metal particle in flowing oil under AC voltage was provided. The movement process of a millimeter-diameter metal particle between electrodes has been investigated under various forms of voltage, and it has been stated that the voltage form and particle concentration have important impacts on the motion characteristics of metal particles [23,24]. The movement properties of millimeter-diameter metal particles under AC voltage have been investigated by numerous researchers. Nevertheless, studies of the motion characteristics of smaller metal particles and their DC breakdown features are lacking. The bridging processes of non-metallic and metal impurities has scarcely been researched. In addition, most simulation models used to analyze the impact of particles on the breakdown performance of insulating oil are still being devoted to the study of electrical field distortion of insulating oil caused by large spherical particles, and comprehensive analyses on the influence of current characteristics along with changes in the electrical field owing to particle accumulation in insulation oil are lacking.

In this paper, experiments on the motion characteristics of cellulose particles, copper particles, and mixed particles are carried out under DC voltage. In addition, DC breakdown voltages of mineral oil contaminated by different concentration particles are measured. A simulation model of particle accumulation is built through Comsol software. The computation model can be utilized to explain the differences in the motion and accumulation properties of different particles in mineral oil. Moreover, the variation in the DC breakdown strength of mineral oil containing different particles with different concentration is analyzed.

2. Simulation and Experiment

2.1. Motion Model of Metal Particles

Assuming that insulating oil cannot be compressed, suspended metal particles in oil withstand gravity and buoyancy in the vertical direction. In the horizontal direction, forces include the dielectrophoretic force (F_{DEP1}) [25], the viscous drag force (F_{drag1}) [25], and the Coulomb force (F_{C1}) [25]. Based on the microstructure of metal particles in Figure 1a, the dielectrophoretic force of spherical metal particles with the radius r can be calculated:

$$F_{DEP1} = 2\pi\varepsilon_m r^3 \frac{\delta_p - \delta_m}{\delta_p + 2\delta_m} \nabla|E|^2 \tag{1}$$

where ε_m represents the permittivity of insulation oil, and σ_m and σ_p stand for the conductivity of insulation liquid and the metal particle respectively.

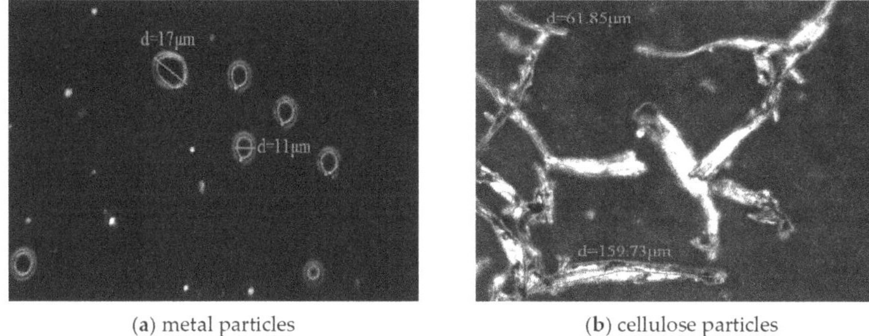

(a) metal particles   (b) cellulose particles

Figure 1. The microstructures of metal particles (**a**) and cellulose particles (**b**).

Viscous drag force (F_{drag1}) acts on the metal particle, as shown in Formula (2). Because of the nonlinearity of the viscous drag force, the correction factor (C_d) related to the Reynolds coefficient (Re) should be used to correct the viscous drag force.

$$F_{drag1} = 6\pi\eta r v_{p1} C_d \tag{2}$$

$$C_d = \begin{cases} 0.17 + 0.72\ln(\mathrm{Re}) & \mathrm{Re} \geq 2 \\ 1 & \mathrm{Re} < 2 \end{cases} \tag{3}$$

$$\mathrm{Re} = \frac{2r\rho_m |v_{p1}|}{\eta} \tag{4}$$

where η and ρ_m are the viscosity and density of the insulation liquid, and v_{p1} represents the velocity of the metal particle.

When the metal particle with the radius of r collides with one electrode, a certain amount of charge (q_1) is obtained, as calculated by Formula (5). Then, plugging Formula (5) into (6) yields the following expression:

$$q_1 = \frac{2}{3}\pi^3 \varepsilon_m \varepsilon_0 r^2 E \tag{5}$$

$$F_{C1} = 0.83 q_1 E = 0.553\pi^3 E^2 r^2 \varepsilon_m \varepsilon_0 \tag{6}$$

Owing to the dynamic model of metal particles represented by Equations (1) to (6), the velocity of the metal particle can be calculated by Formula (7):

$$v_{p1} = \frac{r^2 \varepsilon_m (\delta_p - \delta_m)}{3\eta(\delta_p + 2\delta_m)} \nabla E^2 \tag{7}$$

2.2. Motion Model of Cellulose Particles

As a result of the difference in micromorphology between metal particles and cellulose particles, their kinetic models are different. The micromorphology of cellulose particles is illustrated in Figure 1b. It is assumed that cellulose particles are regarded as ellipsoids to simplify the computational investigation. The kinetic model of cellulose particles is as follows:

$$F_{DEP2} = \frac{\pi\varepsilon_m\varepsilon_0}{12}d^2 l \left[\frac{\alpha}{\alpha-1} - f(\beta)\right]^{-1} \nabla E^2 \tag{8}$$

$$\alpha = \frac{\varepsilon_p}{\varepsilon_m} \tag{9}$$

$$f(\beta) = \xi[(1-\xi^2)\coth^{-1}\xi + \xi] \tag{10}$$

$$\xi = \beta(\beta^2 - 1)^{-1/2} \tag{11}$$

$$\beta = l/d \tag{12}$$

$$F_{drag2} = 3\pi \eta d v_{p2} g(\beta) \tag{13}$$

$$g(\beta) = \frac{8}{3}\left[\frac{-2\beta}{\beta^2 - 1} + \frac{2\beta^2 - 1}{(\beta^2 - 1)^{3/2}} \ln \frac{\beta + (\beta^2 - 1)^{1/2}}{\beta - (\beta^2 - 1)^{1/2}}\right]^{-1} \tag{14}$$

$$F_{C2} = 0.553\pi^3 E^2 r^2 \varepsilon_m \varepsilon_0 \tag{15}$$

where ε_m and ε_p represent the permittivity of insulation oil and cellulose impurities, and d and l stand for the diameter and length of the cellulose particle. Using Formulas (8)–(15), the velocity of the cellulose particle v_{p2} can be attained using

$$v_{p2} = \lim_{\beta \to \infty} \frac{\varepsilon_m \varepsilon_0}{24\eta} l^2 \frac{\ln 2\beta - 0.5}{\ln 2\beta - 1} \nabla E^2 \tag{16}$$

2.3. Accumulation Model of Particles

Based on the results of the impurity accumulation experiments, the velocity of the particles (v_p) is closely related to the concentration (c). There is a certain particle concentration (c_{crit}) that prevents particle movement, because a large number of particles are bound to form bridges [26]. Thus, the relationship between the velocity and concentration of particles is as follows [26]:

$$v'_p = \begin{cases} 0 & c \geq c_{crit} \\ v_p - v_p \frac{c}{c_{crit}} & c < c_{crit} \end{cases} \tag{17}$$

Fick's diffusion law, which contains the velocity of particles, is utilized to calculate the dynamic behavior of particle concentration. Because of no chemical reaction takes place, the reaction rate expression for species R_i is equal to 0. The flux vector (or molar flux) N is associated with the Fick equation and is used under boundary conditions and for flux computation. u_m (10^{-7} s*mol/kg) is the charge mobility, where c is the concentration of impurities, and D (10^{-11} m^2/s) is seen as the diffusion parameter:

$$\frac{\partial c_i}{\partial t} + \nabla \bullet N_i = R_i$$
$$N_i = -D_i \nabla c_i - z_i u_{m,i} F c_i \nabla V \tag{18}$$

The different physics interface involving only the scalar electric potential can be interpreted in terms of the charge relaxation process. The basic equation is Ohm's law:

$$J = \delta E + J_e$$
$$\delta = \delta_p^{2.3c} * \delta_{oil}^{1-2.3c} \tag{19}$$

where J_e is an externally generated current density, and $J_e = 0$. σ is the collective conductivity of insulation oil and impurities, which is derived according to the Looyenga Formula (31). The static form of the equation of current continuity then reads

$$\nabla \bullet J = 0 = -\nabla \bullet (\delta \nabla V - J_e) = -\nabla \bullet (\delta \nabla V) = 0 \tag{20}$$

Then, according to the fundamental equation of electrostatic field, the electric field distribution can be illustrated by

$$E = -\nabla V$$
$$\nabla^2 V = 0 \tag{21}$$

Based on motion models and accumulation model of particles, the results of the accumulation concentration together with the electrical field distribution variation for cellulose particles, metal particles, as well as mixed particles can be obtained.

2.4. Particle Accumulation and Oil Breakdown Measurement

A standard oil cup containing a sphere–sphere electrode made of copper material with a diameter of 13 mm was used in the experiments. In accordance with IEC 60156, an electrode distance of 2.5 mm was used in the DC breakdown experiments. To study the accumulation characteristics of the three kinds of particles, 7.5 mm and 12.5 kV were utilized. The oil cup was situated under a digital camera which recorded the process of particle accumulation. The current was measured by a Keithley electrometer (6517B). The experimental setup for the DC breakdown voltage testing is shown in Figure 2. HCDJC-100kV/5kVA was utilized to provide high DC voltage. The signal together with data was controlled and attained by the computer system.

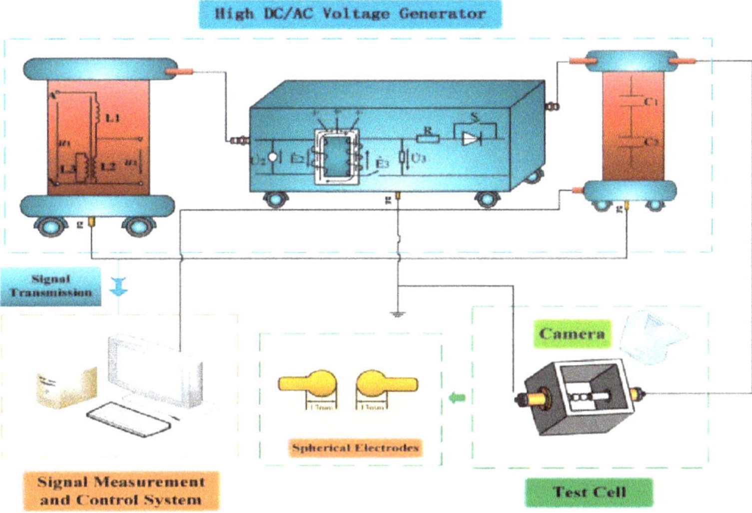

Figure 2. The experimental platform for the particle accumulation and DC breakdown test.

Three kinds of impurities were used in all experiments: cellulose particles, metal particles, and their mixture. Cellulose particles were produced by rubbing new insulation paper through metal files of different sizes. In this paper, cellulose particles with sizes of 63–150 μm and 150–250 μm were prepared. Spherical copper particles with 15 μm diameters came from Beijing Hongyu New Materials Co., Ltd. (Beijing, China). The contamination levels of particles can be seen in Table 1. Before the experiments, all samples were dried for 48 h using a vacuum box at 90 °C and 133 Pa. The parameters of clean mineral oil can be seen in Table 2. For the sake of simplicity, abbreviations are used in Table 3 to stand for different samples. After homogeneous mixing of particles and mineral oil, the samples were sealed in the vacuum box for 24 h.

Each particle accumulation experiment lasted 1500 s. To ensure even particle distribution, a stirrer was used to stir samples. Moreover, every sample was repeatedly broken down eight to ten times. The average value was treated as the final DC breakdown voltage for this sample. All tests were performed at room temperature.

Table 1. Parameters of mineral oil.

	Contamination Level				
Cellulose particles	0.001%	0.003%	0.006%	0.009%	0.012%
Metal particles	0.1 g/L	0.3 g/L	0.6 g/L	1 g/L	1.5 g/L
Mixed particles	0.003% + 0.1 g/L	0.003% + 0.3 g/L	0.003% + 0.6 g/L	0.003% + 1 g/L	—
	0.012% + 0.1 g/L	0.012% + 0.3 g/L	0.012% + 0.6 g/L	0.012% + 1g/L	—

Table 2. Parameters of mineral oil.

Parameters	Mineral Oil
Density in g/cm^3	0.89
Dynamic viscosity in mm^2/s (20 °C)	25.70
Permittivity (20 °C, 50 Hz)	2.20
Volume resistivity in Ω·m (20°)	4.68×10^{13}

Table 3. Abbreviations of samples analyzed in the experiments.

Samples	Sample Composition
DMCP	Dry mineral oil + cellulose particles
DMMP	Dry mineral oil + metal particles
DMCM	Dry mineral oil + mixed particles of cellulose and metal particles
MO	Pure mineral oil

3. Experimental Results and Discussion

3.1. Particle Accumulation Simulation Results

The parameters of the particle accumulation simulation model are described in Table 4. Three-dimensional simulation diagrams of the accumulation of cellulose particles, metal particles, and mixed particles in insulation oil at 10 s and 600 s are shown in Figure 3. With an increase in computational time, impurity accumulation between electrodes was more evident. Samples containing cellulose particles formed initial filamentous bridges faster than those only containing metal particles at the start stage of voltage application. At 600 s, the concentration of mixed particles was the largest, followed by cellulose particles and metal particles. The concentrations of cellulose particles and copper particles were, respectively, 1.2 and 3.7 times less than that of mixed particles.

Table 4. Parameters of the particle accumulation model.

Parameters	Cellulose Particles	Metal Particles
Density, g/cm^3 (20 °C)	1.2	8.6
Permittivity (20 °C, 50 Hz)	4.4	10^5
Size, μm	63–150	15
Volume resistivity, Ω·m (20 °C)	2×10^6	2×10^{-8}
Initial concentration (c_0), mol/m^3	0.0005	0.0005
DC voltage, kV	11.25	11.25
Distance, mm	7.5	7.5

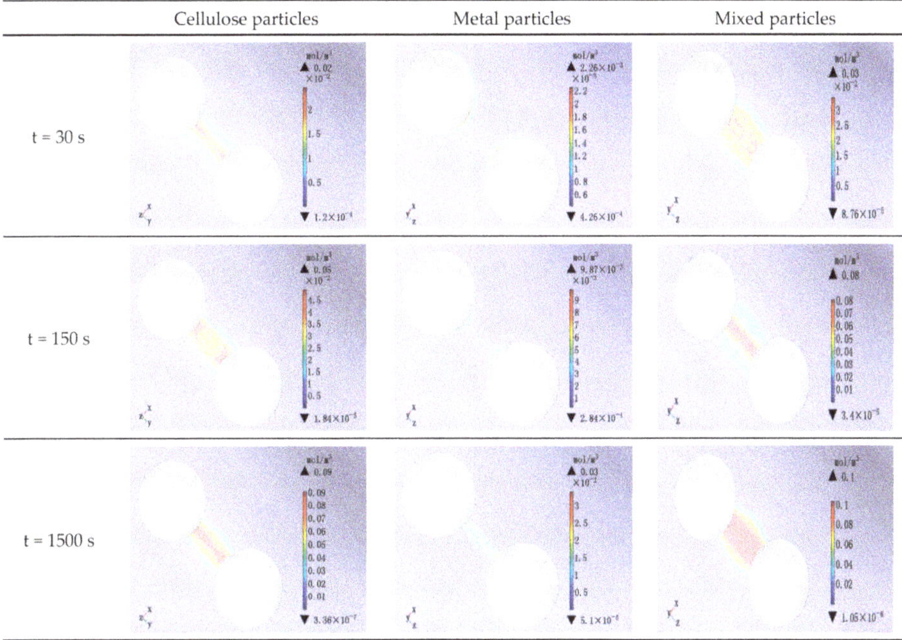

Figure 3. Simulation results of different kinds of particle accumulation in mineral oil at different times.

3.2. Experimental Results of Particle Accumulation.

Figure 4 illustrates the dynamic behavior of particle accumulation under DC voltage, in which the concentration levels of cellulose particles, metal particles, and mixed particles were respectively 0.009% by weight, 0.1 g/L, and the sum of both. The distance and DC voltage between the spherical electrodes were 7.5 mm and 11.25 kV, respectively. The left electrode was a cathode, and the right one was an anode.

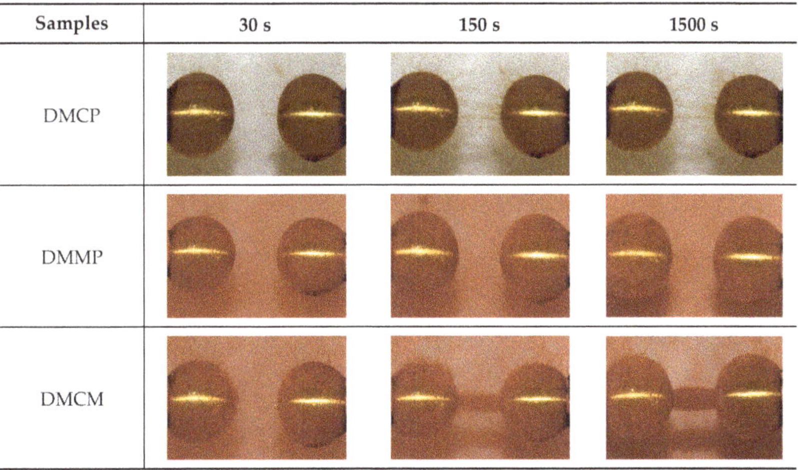

Figure 4. Particle accumulation status under the same DC voltage in mineral oil at different time.

It is obvious that impurity bridges form for the three kinds of particles, as shown in Figure 4. The non-charged particles move toward the higher electrical field area under the attraction of dielectrophoresis, i.e., the central position of the electrode. The particles are charged on the side of the adsorbed electrode. When the Coulomb force of the particles are enough to resist the dielectrophoresis force, the particle moves uniformly to the opposite electrode under combined forces including the Coulomb force, the viscous drag force, and dielectrophoresis force. A few of the particles move back and forth between the electrodes according to the above rule. It is noteworthy that large particles are more likely to move rapidly, because the electrical field force and dielectrophoresis force of the particle are proportional to the square of a particle's radius, and the viscous drag force is proportional to the particle size. However, as a result of lacking enough charge to resist the dielectrophoresis attraction in a certain time period, the rest of particles adhere to electrodes which form a tip on the surface of electrodes. The tip may enhance the local electrical field, attract surrounding particles to move there, and continuously lengthen the particle chain. The initial filamentous bridge is not completely parallel to the electrical field line, which results in parallel movement of the particle bridge under the action of force in the horizontal direction and merges with other small bridges; thus, its thickness increases.

As shown in Figure 4 (sample DMCP) a thin fiber bridge was observed after 30 s in mineral oil containing cellulose particles. After that, the thickness of cellulose bridges increased gradually. The complete bridge formed until approximately 600 s. Then, notable changes in the cellulose bridge were not seen. For mineral oil contaminated by metal particles, most metal particles sunk down, owing to their larger densities. A filamentous bridge formed at about 500 s (Figure 4, sample DMMP), which was so unstable that it migrated back and forth between electrodes, which may be attributed to the impact of the strong electrical field. The ultimate metal bridge appeared until 1200 s. As for mixed particles (Figure 4, sample DMCP), the complete bridge formed more quickly than those for single impurities. Furthermore, it was observed that the complete bridge was densest when the transverse force among thin bridges made up of mixed particles was larger. It can be inferred that the simulation model depicting the changing course of particles aggregation is in agreement with the actual process. Thus, the simulation model can accurately describe the impact of impurities on motion as well as on the accumulation characteristics of particles. Different attributes of particles determined the difference in the trajectory. The trajectories of cellulose particles and metal particles under DC electrical field observed from Figure 4 is shown in Figure 5.

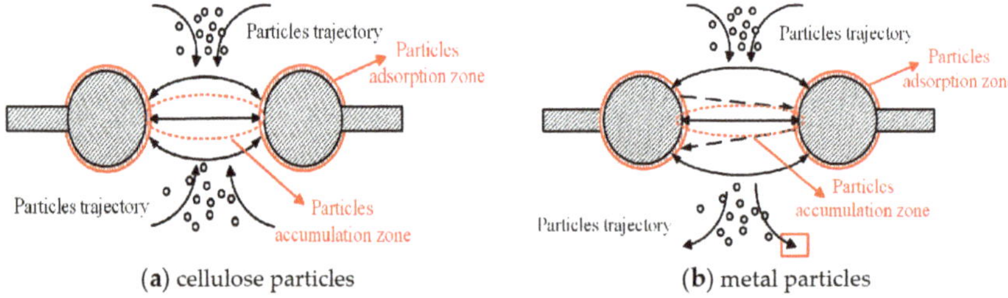

(**a**) cellulose particles   (**b**) metal particles

Figure 5. Motion trajectory of cellulose particles (**a**) and metal particles (**b**) under DC voltage in mineral oil.

3.3. Effect of Particles Accumulation on Oil DC Breakdown Voltage

The Weibull distribution model was utilized to analyze the DC breakdown voltages of samples. Its mathematical expression is shown by Equation (22), where t stands for the DC breakdown strength; α stands for the scale parameter, which describes the characteristic breakdown strength of oil; and β represents the shape parameter, which reflects the changing rate of breakdown probability with an increase in DC voltage. The data samples used in this study were complete, so the empirical distribution function $F_n(t_i)$ could be calculated by Formula (23), where i represents the order of test samples, and n is the number of samples. Figure 6 is the Weibull probability distribution plot of different samples under the DC breakdown voltage. With an increase in the concentration of particles, the Weibull curve moved to the left. Namely, the average breakdown values of samples reduced as the particle concentration increased. The relationship between the average breakdown voltages of mineral oil containing cellulose particles, metal particles, and mixed particles and the particles concentration are shown in Figure 7. The DC breakdown voltages of oil samples contaminated with metal particles and mixed particles were found to be lower than those of mineral oil containing cellulose particles. In summary, metal impurities and mixed particles have more significant impacts on the DC breakdown characteristics of mineral oil:

$$F(t; \alpha, \beta) = 1 - \exp[-(\frac{t}{\alpha})^{\beta}] \tag{22}$$

$$F_n(t_i) = \frac{i - 0.375}{n + 0.25} \times 100\% \tag{23}$$

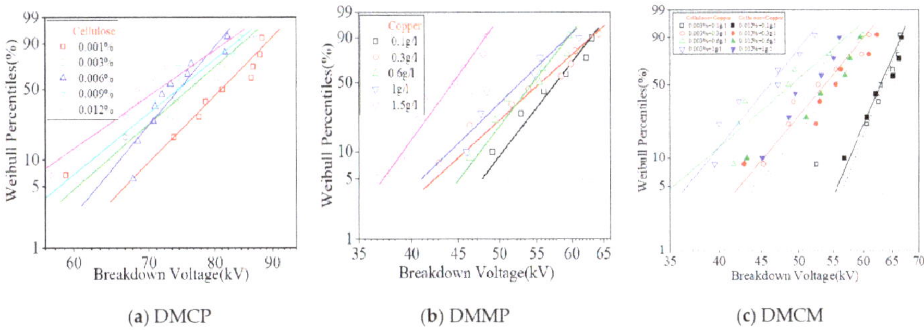

(**a**) DMCP   (**b**) DMMP   (**c**) DMCM

Figure 6. Weibull probability distribution plot of DC breakdown voltage for different samples DMCP (**a**), DMMP (**b**) and DMCM (**c**).

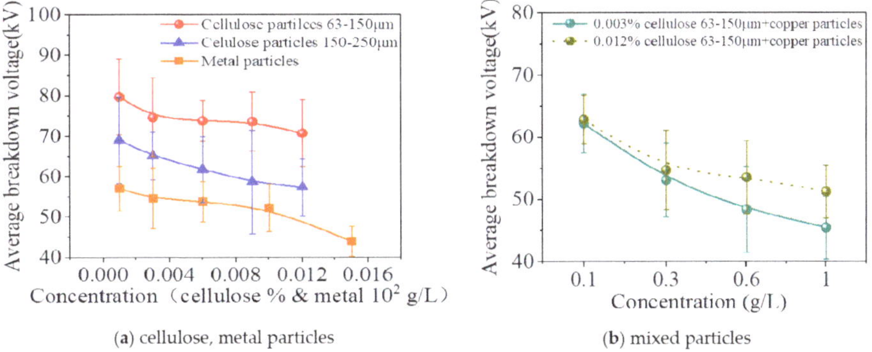

(**a**) cellulose, metal particles   (**b**) mixed particles

Figure 7. Average DC breakdown voltage of mineral oil with cellulose or metal particles (**a**), and mixed particles (**b**).

3.4. Difference Analysis of the Effects of Different Particles on the Oil DC Breakdown Voltage

First of all, impurities have a prominent effect on the conductivity of mineral oil. The changing current properties of mineral oil are displayed in Figure 8. Upon the application of DC voltage, instantaneous polarization current appears in both clean oil and contaminated oil. For clean oil, the polarization current decreases to a conduction current. Yet, due to the effect of particles' back and forward motion between electrodes, causing charging and discharging together with formation of bridges, the current in contaminated oil increases little by little until the saturation level is reached. The increasing degree of current saturation in contaminated oil compared to clean oil occurs in decreasing order for mixed particles, metal particles, and cellulose particles. The saturated current in mineral oil containing mixed particles is the largest, almost seven times larger than that of clean oil, followed by copper particles (5.5 times) and cellulose particles (2.8 times). As a result, metal particles or mixed particles more easily cause insulation oil to partially discharge and even break down. Since the sum of the saturated current of cellulose particles and metal particles is not equal to the saturated current of mixed particles, cellulose particles coupled with metal particles cannot have a superposition effect on the conductivity of oil. There is no doubt that the bridge formation of particles under a DC electrical field can significantly improve the conductivity of mineral oil, which is one of main reasons for the decrease in the breakdown strength of oil. Furthermore, there is a corresponding relation between current and particle aggregation state, which is shown in Figure 9.

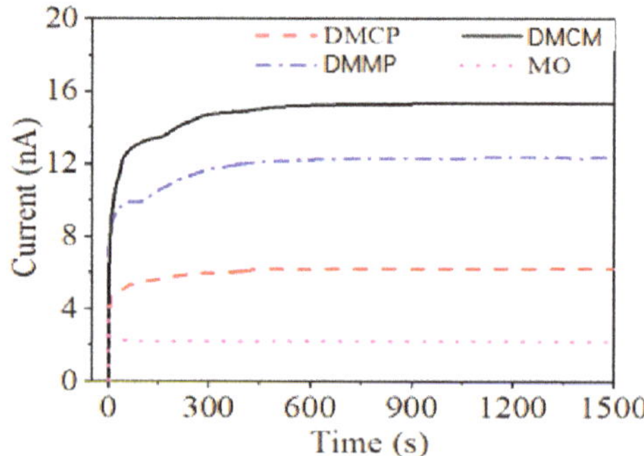

Figure 8. Changes in the current in mineral oil containing different particles under the same DC voltage.

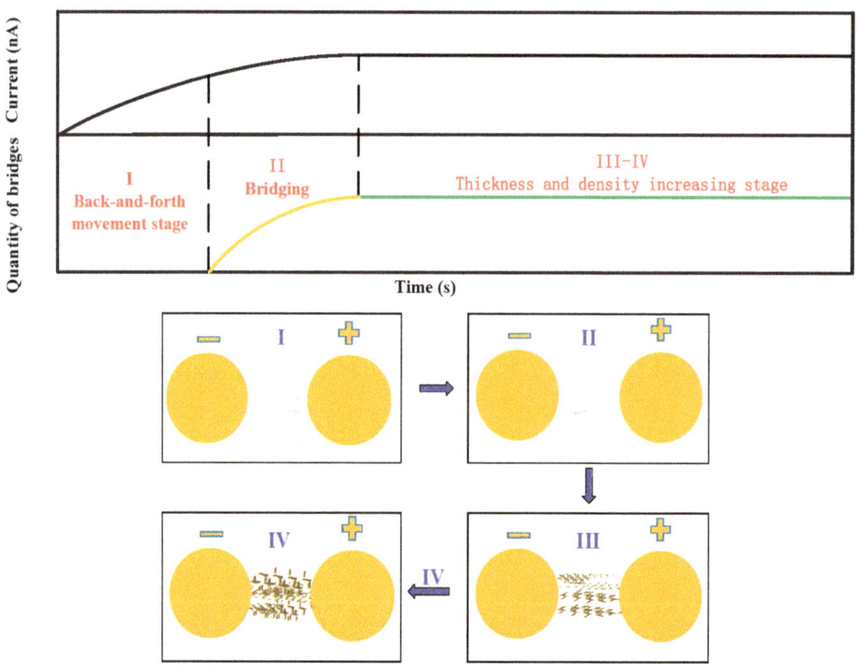

Figure 9. Relationship between oil conductivity current and particle aggregation.

Then, taking advantage of the effective simulation model shown in Section 3.1, relationships among concentrations, particles properties, and the variation in electrical field strength were analyzed in order to explain the DC breakdown results. Figures 10 and 11 display three dimensional particle accumulation patterns as well as the DC electric field distribution of insulation oil under three different initial concentrations at 600 s. It can be observed that firstly, accumulation degrees increase as the initial concentration increases. Secondly, the range of electrical field distortion caused by particle accumulation is also gradually enlarged.

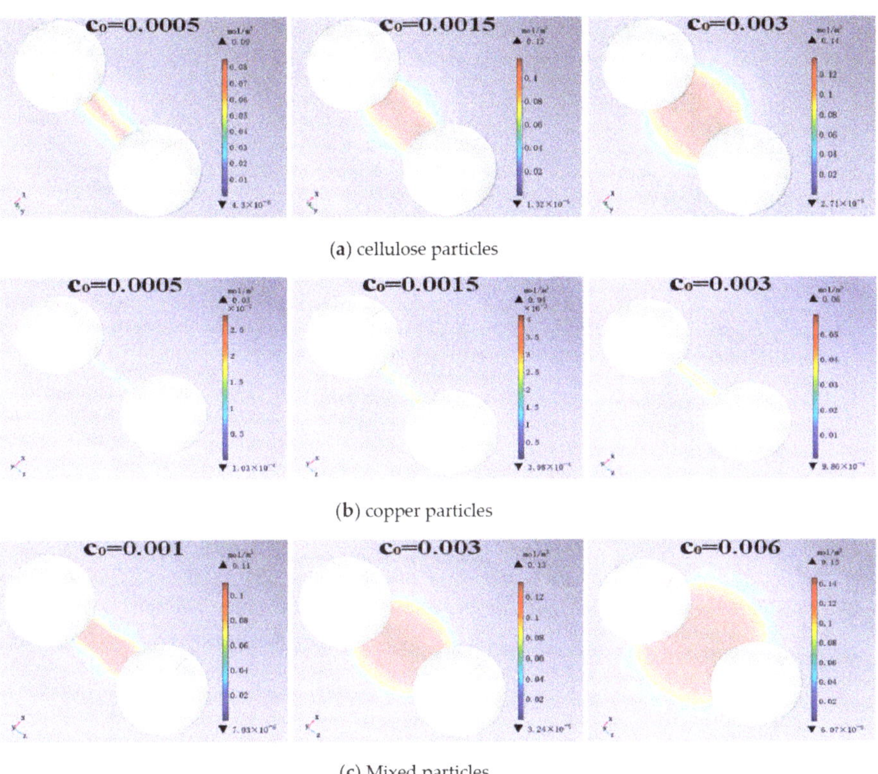

Figure 10. Particle accumulation patterns under three different initial concentrations for cellulose particles (**a**), copper particles (**b**) and mixed particles (**c**) at 600 s.

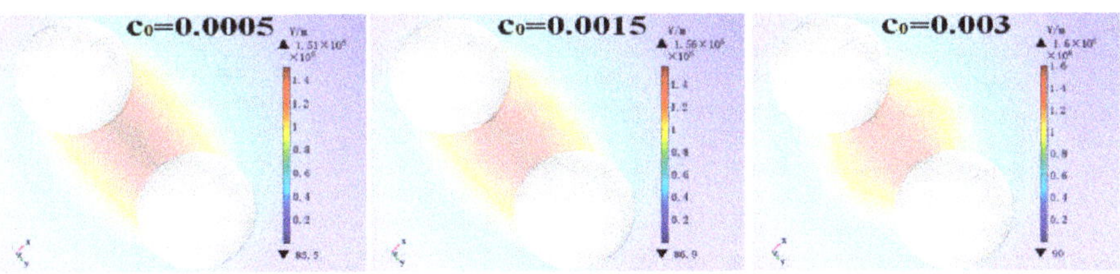

(**a**) cellulose particles

Figure 11. *Cont.*

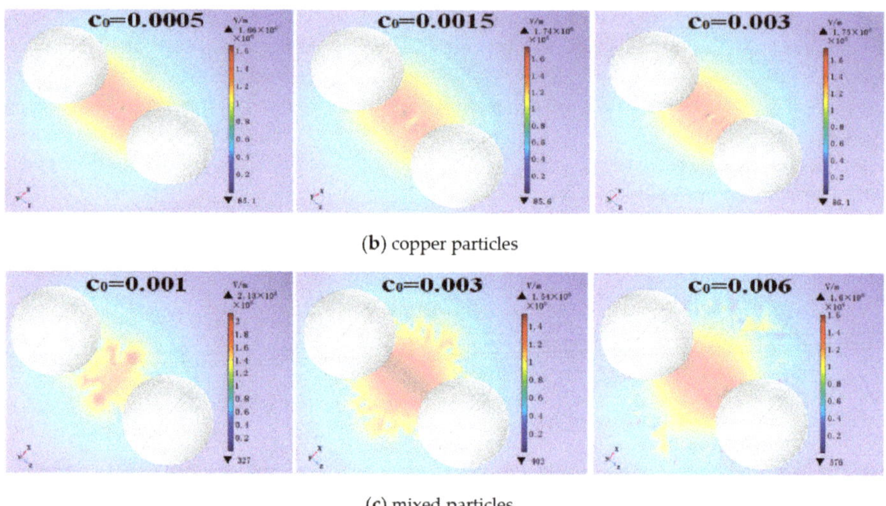

(b) copper particles

(c) mixed particles

Figure 11. DC electric field distribution of mineral oil contained cellulose particles (**a**), copper particles (**b**) and mixed particles (**c**) under three different initial concentrations at 600 s.

Figure 12 shows the influence of the particle concentration on the maximum electrical field strength (E_{max}) of contaminated mineral oil at 600 s. It can be seen that metal particles almost play a more prominent part in electrical field distortion than cellulose particles. Additionally, the saturated current of metal particles is larger than that for cellulose impurities, so the breakdown voltage of mineral oil polluted by metal particles is smaller than that contaminated by cellulose particles. Figures 10 and 11 also imply that cellulose particles as well as metal particles cannot have a superposition effect on the electric field distortion of insulation oil. For mineral oil contaminated by cellulose or metal particles, E_{max} increases as the particle concentration increases, which indicates that electrical field distortion combined with the conductivity variation leads to degradation of the insulation strength of mineral oil. As for mineral oil containing mixed particles, though electrical field distortion weakens with an increasing concentration, conductivity obviously increases because of the increasing accumulation degree. Hence, changes in the electrical field distribution together with an increase in conductivity collectively affects the DC breakdown characteristics of mineral oil.

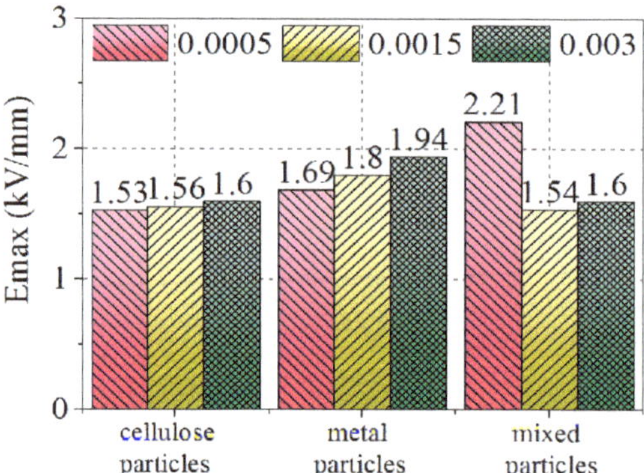

Figure 12. E_{max} of mineral oil and natural ester containing different particles under three different initial concentrations at 600 s.

The effect of pressure on breakdown voltage of insulation oil has been investigated, which indicates that the increasing of pressure can increase breakdown voltage of oil [27]. However, few researchers have studied the effects of temperature or pressure effect on particles' accumulation properties. This needs further study in the future.

4. Conclusions

In this paper, simulations and experiments were carried out to investigate the accumulation characteristics of cellulose, copper, and mixed particles in mineral oil under DC voltage. The DC breakdown voltages of mineral oil with different particle concentrations were measured. The conclusions obtained from this study are as follows.

The simulation model was able to reveal the process of particle bridging in mineral oil under DC voltage. The simulation results showed that as the experiment duration increased, particle accumulation became more evident. The accumulation concentration of mixed particles was the largest, followed by cellulose particles and metal particles, which is in agreement with the experimental results. In the particle accumulation experiments, it was also obvious that metal particles have difficulty forming stable bridges, while bridges of mixed particles were the densest and thickest among the tested compounds. Moreover, the larger the initial concentration of particles, the more obvious the accumulation phenomenon as well as the electrical field distortion.

The properties of particles determine the accumulation shapes, conductivity characteristics, and variation law of breakdown voltages. The increasing degrees of saturated current in contaminated oil compared to clean oil in decreasing order were found to be mixed particles, metal particles, and cellulose particles. Moreover, metal particles were also shown to play a more prominent part in electrical field distortion than cellulose particles. Therefore, the breakdown voltage of mineral oil contaminated by cellulose particles was larger than that of mineral oil containing copper particles. Nevertheless, it is noteworthy that cellulose particles along with metal particles cannot have a superposition effect on the conductivity characteristics and electrical field distortion of insulation oil.

The test and computational results indicated that changes in the electrical field distribution together with an increase in conductivity collectively affect the breakdown strength of mineral oil. The bridge formation of particles under DC electrical field was shown to significantly improve the conductivity of oil, which is one of main reasons for the decrease in breakdown strength of mineral oil containing particles. The simulation results of the particle accumulation model showed that as the duration of the experiment increased, the range of electrical field distortion caused by particles accumulation gradually expanded, which is also one of main reasons for the decrease in the DC breakdown strength of mineral oil containing particles.

Author Contributions: Designed the experiments, did the motion characteristics experiments and breakdown voltage measurement, wrote the paper, M.D.; improved the simulation method, analyzed the experiment data and revised the paper, J.H.; contributed discussion, R.L., L.C., J.Z., and F.L.

Funding: This research was funded by [National Key R&D Program of China] grant number [2017YFB0902704], [Science and Technology Project of State Grid Corporation-Research on Insulation Defect Analysis and Testing Technology of UHV Converter Transformer], [Joint Funds of the National Natural Science Foundation of China] grant number [U1866603], [Funds for Innovative Research Groups of China] grant number [51321063].

Conflicts of Interest: The authors declare no conflict of interest.

References

1. Wang, X.; Wang, Z. Particle effect on breakdown voltage of mineral and ester based transformer oils. In Proceedings of the IEEE Conference on Electrical Insulation and Dielectric Phenomena, Quebec City, QC, Canada, 26–29 October 2008; p. 598.
2. CIGRE. *Effect of Particles on Transformer Dielectric Strength*; WG 17/SC12; CIGRE: Paris, France, 2000.
3. Lu, W.; Liu, Q. Effect of cellulose particles on impulse breakdown in ester transformer liquids in uniform electric fields. *IEEE Trans. Dielectr. Electr. Insul.* **2015**, *22*, 2554–2564. [CrossRef]

4. Zhao, T. Research of the Effect of Bubbles and Cellulose Particles on Impulse Breakdown in Transformer Oil. Ph.D. Thesis, North China Electric Power University, Beijing, China, 2017.

5. Li, Z.; Ma, Q. Monitoring the particle pollution degree of 500 kV transformer's insulation oil. *Transformer* **1999**, *36*, 31–34.

6. Wu, R.; Liu, J.; Chen, J.; Zhang, Z. Analysis of insulating oil particulate pollution in 500 kV transformer and reactor. *Inn. Mongo. Electr. Power* **2013**, *31*, 35–37.

7. Xing, L.; Zhang, G. Test and analysis of graininess in 500kV transformer oil. *Transformer* **2009**, *46*, 40–43.

8. Wei, L.; Su, Z.; Qi, J. Study on particulate contamination detection of insulating oil for 1000 kV transformer and its influence factor. *Insul. Mater.* **2015**, *6*, 50–53.

9. Zhang, G.; Li, X.; Liu, H.; Lu, L. Study on particle size in 500 kV transformer insulation oil. In Proceedings of the Annual Academic Conference of Tianjin Electric Power Society, Tianjin, China, 20–22 October 2008.

10. Dan, M.; Hao, J.; Li, Y.; Liao, R.; Yang, L.; Wang, Q.; Zhang, S. Analysis of bridging phenomenon in mineral oil and natural ester contaminated with cellulose particles under different DC electrical field. In Proceedings of the 20th International Symposium on High Voltage Engineering, Buenos Aires, Argentina, 28 August 2017.

11. Dan, M.; Hao, J.; Qin, W.; Liao, R.; Zou, R.; Zhu, M.; Liang, S. Effect of different impurities on motion characteristics and breakdown properties of insulation oil under DC electrical field. In Proceedings of the 2018 IEEE International Conference on High Voltage Engineering and Application (ICHVE 2018), Athens, Greece, 10–13 September 2018; pp. 1–4.

12. Zhou, Y.; Hao, M.; Chen, G.; Wilson, G.; Jarman, P. Study of the charge dynamics in mineral oil under a non-homogeneous field. *IEEE Trans. Dielectr. Electr. Insul.* **2015**, *22*, 2473–2482. [CrossRef]

13. Hao, J.; Liao, R.; Dan, M.; Li, Y.; Li, J.; Liao, Q. Comparative study on the dynamic migration of cellulose particles and its effect on the conductivity in natural ester and mineral oil under DC electrical field. *IET Gener. Transm. Distrib.* **2017**, *11*, 2375–2383. [CrossRef]

14. Mahmud, S.; Chen, G.; Golosnoy, I.O.; Wilson, G. Bridging in contaminated transformer oil under AC, DC and DC biased AC electric field. In Proceedings of the 2013 IEEE Electrical Insulation and Dielectric Phenomena, Shenzhen, China, 20–23 October 2013; pp. 943–946.

15. Mahmud, S.; Chen, G.; Golosnoy, I.O.; Wilson, G.; Jarman, P. Experimental studies of influence of DC and AC electric fields on bridging in contaminated transformer oil. *IEEE Trans. Dielectr. Electr. Insul.* **2015**, *22*, 152–160. [CrossRef]

16. Mahmud, S.; Chen, G.; Golosnoy, I.O.; Wilson, G.; Jarman, P. Bridging phenomenon in contaminated transformer oil. In Proceedings of the International Conference on Condition Monitoring and Diagnosis, Piscataway, NJ, USA, 23–27 September 2012; pp. 180–183.

17. Li, J.; Zhang, Q.; Li, Y. Generation process of impurity bridges in oil-paper insulation under DC voltage. *High Volt. Eng.* **2016**, *12*, 211–218.

18. Li, Y.; Zhang, Q.; Li, J.; Wang, T.; Dong, W.; Ni, H. Study on micro bridge impurities in oil-paper insulation at DC voltage: Their generation, growth and interaction with partial discharge. *IEEE Trans. Dielectr. Electr. Insul.* **2016**, *23*, 2213–2222. [CrossRef]

19. Wang, S.; Shi, J.; Li, J. The effect of a macro-particle on the partial property of transformer oil. *High Volt. Eng.* **1994**, *20*, 26–29.

20. Fu, S. The acquired charge of macro-particle and its effect on the partial discharge of transformer oil. *High Volt. Eng.* **2000**, *26*, 49–50.

21. Tang, J.; Zhu, L.M.; Ma, S.X. Characteristics of suspended and mobile micro bubble partial discharge in insulation oil. *High Volt. Eng.* **2010**, *36*, 1341–1346.

22. Ma, S.X.; Tang, J.; Zhang, M.J. Simulation study on distribution and influence factors of metal particles in transaction transformer. *High Volt. Eng.* **2015**, *41*, 3628–3634.

23. Wang, Y.Y.; Li, Y.L.; Wei, C.; Zhang, J.; Li, X. Copper particle effect on the breakdown strength of insulating oil at combined AC and DC voltage. *J. Electr. Eng. Technol.* **2017**, *12*, 865–873. [CrossRef]

24. Wang, Y.; Li, X. Motion characteristic of copper particle in insulating oil under AC and DC voltages. In Proceedings of the 19th IEEE International Conference on Dielectric Liquid, Manchester, UK, 25–29 June 2017; pp. 25–29.

25. Dan, M.; Hao, J.; Liao, R.; Li, Y.; Yang, L. Different motion and bridging characteristics of fiber particles in mineral oil and natural ester under DC voltage. *Power Syst. Technol.* **2018**, *42*, 665–672.

26. Naciri, N. Finite Element Analysis for Power System Component: Dust Accumulation in Transformer Oil. Ph.D. Thesis, University of Southampton, Southampton, UK, 2011.
27. Butcher, M.; Neuber, A.; Krompholz, H.; Dickens, J. Effect of temperature and pressure on DC pre-breakdown current in transformer oil. In Proceedings of the 31st IEEE International Conference on Plasma Science, Baltimore, MD, USA, 28 June–1 July 2004; pp. 1–4.

Article

Dielectric Insulation Characteristics of Natural Ester Fluid Modified by Colloidal Iron Oxide Ions and Silica Nanoparticles

Vasilios P. Charalampakos [1], Georgios D. Peppas [2,*], Eleftheria C. Pyrgioti [2], Aristides Bakandritsos [3], Aikaterini D. Polykrati [4] and Ioannis F. Gonos [4]

[1] Department of Electrical and Computer Engineering, University of the Peloponnese, 26334 Patras, Greece
[2] Department of Electrical and Computer Engineering, University of Patras, 26500 Patras, Greece
[3] Regional Centre for Advanced Technologies and Materials, Department of Physical Chemistry, Faculty of Science, Palacký University Olomouc, 17. listopadu 1192/12, 771 46 Olomouc, Czech Republic
[4] School of Electrical and Computer Engineering, National Technical University of Athens, 15780 Athens, Greece
* Correspondence: peppas@ece.upatras.gr

Received: 29 June 2019; Accepted: 17 August 2019; Published: 23 August 2019

Abstract: In this study, the dielectric characteristics of two types of natural esters modified into nanofluids are studied. The AC breakdown voltage was investigated for colloidal Fe_2O_3 and SiO_2 nanoparticles effectively scattered in natural ester oil. The experimental results identify an increase in the breakdown voltage of the nanofluid with colloidal Fe_2O_3 conductive nanoparticles. In contrast, the breakdown voltage was reduced by adding SiO_2 nanoparticles in the same matrix. The potential well distribution of the two different types of nanoparticles was also calculated in order for the results of the experiment to be explained. The dielectric losses of the colloidal nanofluid are compared with the matrix oil and studied at 25 °C and 100 °C in the frequency regime of 10^{-1}–10^6 Hz. The experimental data and the theoretical study reveal that conductivity along with the permittivity of nanoparticles constitute a pivotal parameter in the performance of nanofluid. Specific concentrations of nanoparticles with different electrical conductivity and permittivity than those of matrix oil increase the breakdown voltage strength. Simultaneously, the addition of nanoparticles having electrical conductivity and permittivity comparable to the matrix oil results in reducing the breakdown voltage.

Keywords: nanofluids; nanoparticles; breakdown strength; transformer oils; permittivity; conductivity

1. Introduction

Insulating nanofluids have attracted the attention of researchers for the last twenty years. The dispersion of nanoparticles inside insulating oils can conditionally increase the breakdown voltage and thermal conductivity of matrix oil. Insulating fluids used in transformers exhibit high dielectric strength but their thermal conductivity is low and this introduces limitations on the power rating of the transformers, and, as a consequence, an increase in their size. Thus, the motivation for the insertion of nanoparticles within transformer oil was initially the increase of thermal conductivity. The idea of adding particles in order to increase thermal conductivity goes back to Maxwell in 1873 [1] and Choi et al. [2] were the first to add magnetic nanoparticles Fe_3O_4 to pure transformer oil and to develop the first nanofluid that demonstrated better thermal conductivity. However, several researchers soon realized that the dispersion of some nanoparticles increased the breakdown voltage at the same time [3–17] and their interest also turned to this direction.

Researchers in [3–10] studied iron oxide nanoparticles, Fe_3O_4 and Fe_2O_3, and found they enhanced dielectric and the cooling performance of the constitutive base fluids when they were dispersed in either mineral or natural ester oils. The dispersion of two-dimensional nanoparticles, such as boron nitride (BN) or graphene nanoparticles, also enhanced the cooling capacities of transformer oil [3,11]. After these promising results, different nanofluids were developed using nanoparticles such as TiO_2, SiO_2, nanodiamond-Ni nanoparticles, and boron nitride (BN) [7,11–19] and they were tested regarding their dielectric performance.

Semi-conductive TiO_2 nanoparticles were added to natural ester oil and mineral oil matrix [13,14] resulting in nanofluids with increased AC breakdown voltage (BDV) and lightning impulse withstand capability as compared to that of the matrix oils. In [15–17] high concentrations of SiO_2 nanoparticles in mineral oil resulted in increased AC breakdown voltage and positive lightning impulse withstand capability in comparison with the matrix dielectric liquid. On the other hand, the negative lightning impulse voltage withstand capability was decreased. In addition, they highlighted the strong influence of moisture on the performance of nanofluids. Particularly, the higher the presence of humidity in the aforementioned SiO_2 nanofluid the better its performance was. However, thermal conductivity seemed to be only slightly affected by the addition of SiO_2 nanoparticles.

Aluminum nitride (AIN), graphene oxide, and BN nanoparticles, as dispersants in mineral oil, were also studied [6,11,17,18]. The addition of AIN nanoparticles proved to increase positive lightning impulse voltage withstand capability and partial discharge ignition voltage, as well as the thermal conductivity of nanofluid as compared with pure oil, and simultaneously, AC breakdown voltage was decreased. Similar results were demonstrated with the addition of BN and graphene oxide nanoparticles where both the AC BDV and the thermal conductivity were improved as compared with the matrix oil. A satisfactory explanation for the higher AC BDV of insulating nanofluids with higher conductivity nanoparticles, with respect to the matrix oils, was given by the "electron traps" theory in [20]. According to this, electrons are very rapidly captured by conductive nanoparticles being transmuted into heavy negatively charged nanoparticles. As a consequence, the streamer speed was reduced resulting in an increased breakdown voltage. Experimental results concerning the space charge in both matrix oil and nanofluid oils [21] were in compliance with the proposed theory.

However, the aforementioned theory fails to adequately explain the superior performance of nanofluids with semi-conductive and non-conductive nanoparticles which have been observed with the conductive ones. This can be ascribed, according to a theory [22], to the different conductivity or permittivity between nanoparticles and their surrounding oil and, according to others [12,23,24], to the interfacial region shaped on the surface of the nanoparticles (NPs).

In this study the AC BDV of two completely different natural ester oil matrix nanofluids will be examined. Conductive in situ surface modified colloidal magnetic iron oxides NPs (MIONs) and SiO_2 nanoparticles were scattered into natural ester dielectric liquid (FR3®). Different nanoparticle concentrations were tested with respect to the optimal concentration in terms of AC dielectric strength performance. The theoretical model proposed in [22] was realized in order to interpolate the experimental results.

2. Materials and Methods

During the preparation of the nanofluids, natural ester dielectric liquid Envirotemp FR3® was used as base liquid which was selected due to its biodegrability and high temperature flash point.

At first, the matrix oil was triple filtered and dried as analytically described in [10]. In brief, low vacuum filtration was used by applying a 30 μm filter for the first level, a 1 μm glass microfiber filter for the second level, and a 0.8 μm membrane for the last one. Thereafter, the matrix liquid was dehumidified by means of a rotary evaporator in series with a vacuum pump, while it was positioned in a water bath heated at 80 °C for at least two days.

The preparation of the nanofluids (NFs) with oleate-coated colloidal magnetic iron oxide nanocrystals (colMIONs) required particular care because magnetic iron oxide nanoparticles, such as Fe_2O_3 and Fe_3O_4, have a tendency to indicate residual NPs after their dispersion in the matrix liquid, forming sedimentation within a few days. In an attempt to overcome agglomeration and increase the long-term stability of the nanofluid, oleate-coated colloidal magnetic iron oxide nanocrystals (colMIONs) were manufactured in the laboratory according to previous reports [9,25]. Initially, 3.62 g of iron-oleate complex and 3.4 g of oleic acid were dissolved in 1-octadence at a temperature of 25 °C. For 1 h, the mixture was magnetically agitated at room temperature, then it was kept under agitation (350 rpm) for 30 min at a temperature of 100 °C and, finally, it was further heated at 318 °C for another hour [26]. Afterwards, the commixture was cooled down and 8 mL of dichloromethane was added. Subsequently, acetone as a dissolver was introduced to decrease the solubility of the MIONs, as well as separate the reacting agents from them. This manufacturing method was repeated until the oleate-coated MIONs obtained a purity level higher than 80%. The colMIONs had a final diameter of approximately 10 nm with a very narrow dimensional distribution [9], and thereafter, they were introduced into the natural ester oil and the final liquid was ultrasonicated for at least 30 min. Six samples, having a concentration range from 0.004 to 0.014% w/w, in step 0.002% were prepared for further study.

Nanoparticles SiO_2 are well dispersed inside transformer oil and they do not indicate agglomeration effect. On the contrary, they may absorb humidity during the introduction of atmospheric air within 3–4 s. Therefore, the procedure took place in a shielded AtmosBag with dedicated gloves (Aldrich® AtmosBag) while inside N_2 was introduced. After the addition of 12 nm average diameter SiO_2 nanoparticles to the matrix oil, the final liquid was ultrasonicated for 30 min. Six different samples with concentrations ranging from 0.008–0.024% w/w in step 0.004% were also prepared.

The measurements of AC dielectric breakdown strength carried out for the samples of the two nanofluids according to IEC 60156 in [27] were enriched with additional ones. The measurement device used was a Baur DTA 100 C, measuring up to 100 kV, Rogowski electrodes based on IEC 60156 [28] with a gap distance of 2.5 mm, a voltage rise of 2 kV/s was adjusted at 50 Hz power system, the breakdown event is calculated based on the level of the current conduction (mA range). The distribution of the breakdown voltage of the experimental results is calculated based on the normal distribution [10]. Before each experimental set, the brass electrodes were polished and cleaned thoroughly. For the matrix oil, 150 breakdown tests were implemented for every sample, whereas, after every 50 successful breakdowns the sample under test was replaced in order to limit the degradation and its effect on the measurements.

Dielectric relaxation spectroscopy was studied by means of a Novocontrol Alpha analyzer (10^{-1}–10^6 Hz) monitored from a Novocontrol Quatro Cryosystem. The understudied dielectric liquid was studied in a custom-made cylindrical capacitor which consisted of two plane plates at 1–1.2 mm distance. The dielectric relaxation spectroscopy study was adopted at 20–100 °C with a 20 °C temperature increase rate at the frequency range of 10^{-1}–10^6 Hz.

Dynamic light scattering (DLS) was performed on oil dispersions of < 0.01% w/v in Fe_2O_3 using the viscosity of the oil 32.03 mm^2/s. A Malvern Instrument ZetaSizer Nano was used, equipped with a 4 mW He-Ne laser, operating at a wavelength of 633 nm. Scattered light was collected at a fixed angle of 173°. Diameter distribution was reported as number-based results. Transmission electron microscope images were collected with a JEOL 2100 on 200 kV. It should be noted that the DLS technique systematically provides a higher mean size of the crystallite size than the TEM technique, because in the former case the size is evaluated in the dispersion state and takes into account the organic coating of the magnetic crystallites, their solvation sphere, and possible formation of dyads between particles. In Figure 1 the size distribution of the colloidal MIONs in the oil matrix is depicted with a peak value of 23 nm, accordingly, in the same figure the colloidal stability of the nanoparticles in the matrix oil after 2 months of storage is depicted. In Figure 2 a detailed TEM micrograph from the oleate-coated MIONs, as manufactured through the thermolytic route [9], is shown, clearly demonstrating their proper dispersion in absence of agglomeration.

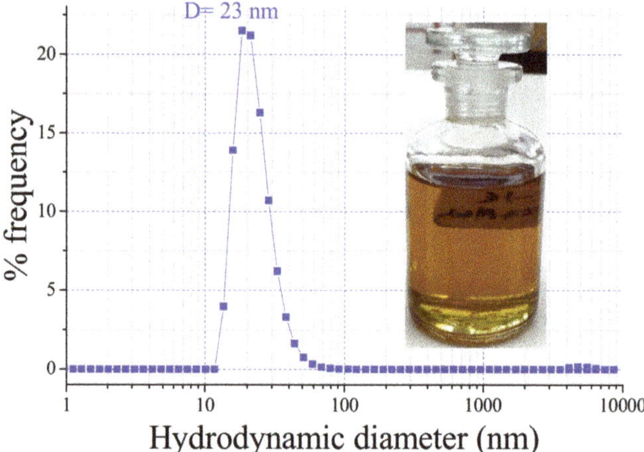

Figure 1. Size distribution diagram based on dynamic light scattering measurement of the colloidal magnetic iron oxides nanoparticles (MIONs) in the oil matrix and digital image colloidal nanofluid (colNF) after 2 months of storage (inset).

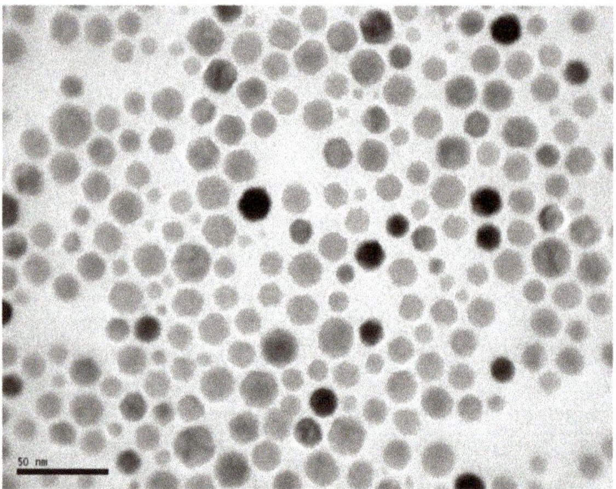

Figure 2. TEM micrograph from the oleate-coated MION colloids.

3. Results

Figure 3 shows the AC BDV of both measured nanofluids versus the concentration of nanoparticles as compared with the mean BDV for the natural ester matrix oil which is 64.5 kV.

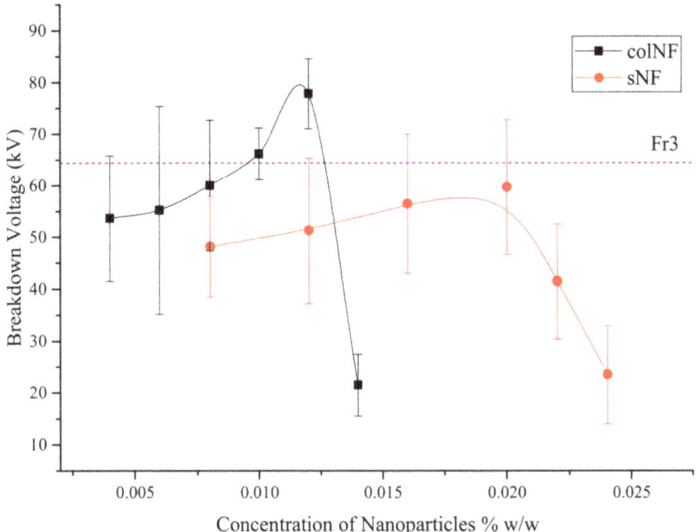

Figure 3. Mean AC breakdown voltage (BDV) versus nanoparticles concentration (colNF and silica NF).

As it is obvious from Figure 3, for the nanofluid with oleate-coated colloidal MIONs (colNF), the BDV increases as the concentration of nanoparticles increases until the 0.012% *w/w* concentration, at which point it reaches its maximum value of 77.8 kV, while at higher concentrations the BDV decreases sharply. However, colNF exhibits higher BDV than natural ester oil only at 0.008% *w/w* and 0.012% *w/w* concentrations.

As regards the BDV of nanofluid with SiO_2, nanoparticles (silica NF–sNF) exhibit similar behavior to the colNFs in terms of increasing nanoparticle concentration, although it always remains lower than the BDV of natural ester oil. The maximum BDV value of 59.7 kV is achieved at 0.02% *w/w*. Any further addition of silica nanoparticles results in an instant drop of the BDV.

Table 1 gives the probabilities 50%, 10%, and 1% for the three dielectric liquids, as well as their standard deviation. In addition, to enable a comparison, in Table 1 the results from two other nanofluids are given, i.e., the probabilities 50% and 10% for TiO_2 0.007% *w/w* in ester oil [14] and SiO_2 0.02% *w/w* in mineral oil [19].

Table 1. Breakdown voltage probabilities.

Probability	Breakdown Voltage (kV)			Standard Deviation (kV)
	$V_{50\%}$	$V_{10\%}$	$V_{1\%}$	
Natural Ester FR3	64.5	49.7	37.6	12
colNF 0.012%	77.8	69.1	62.1	6
sNF 0.02%	59.7	42.7	21.7	13
TiO_2 NF 0.007% [14]	30	22.7	N/A	N/A
SiO_2 NF 0.02% [19]	68.5	41.7	N/A	N/A

According to Table 1, the two nanofluids show differences not only in the mean value of BDV but also in standard deviation. The BDV values of sNF are more scattered at all ranges of concentrations and standard deviation varies from 9.5 kV up to 14.1 kV. The standard deviation of colNF is also large for a nanoparticle concentration of 0.008% or less, but above this value it declines sharply.

As it is known, the low probabilities of breakdown voltage play a pivotal role in designing electrical apparatus such as transformers [9]. The results from Table 1 show that the colNF appeared with the highest value of $V_{1\%}$ BDV which is 62.1 kV, while for natural ester oil the value is 37.6 kV and for sNF the value is 21.7 kV. This can be attributed to its extremely low standard deviation.

The Impact of Nanoparticles on Dielectric Behavior of Insulating Oil

A considerable amount of research has already been carried out to date to explain the enhanced dielectric behavior of nanofluids [12,20–25]. In [20,21], these are attributed to the capability of nanoparticles to trap electrons. This happens due to inductive charging for the conductive nanoparticles and by means of polarization for the non- and semi-conductive nanoparticles.

When an external electric field, E_0, is applied to the gap, the electrical charges of a conductive nanoparticle are redistributed inversely along the direction of the electric field within a relaxation time, τ_r. The relaxation time is expressed as follows [22]:

$$\tau_r = \frac{2\varepsilon_1 + \varepsilon_2}{2\sigma_1 + \sigma_2} \tag{1}$$

where, ε_1 and ε_2 are the permittivity of transformer oil and nanoparticle, respectively and σ_1 and σ_2 are the conductivities of transformer oil and nanoparticle respectively. Conductive NPs, such as Fe_3O_4 or ZnO, have a very small relaxation time of the order of 10^{-11}–10^{-14} s [20] as can be seen in Table 2. Non-conductive NPs on the other hand have large relaxation times. However, bound charges are formed on their surface due to polarization. Electronic and ionic displacement polarizations are generated very quickly in 10^{-15} s to 10^{-12} s. It should be noted that conductive NPs with large dielectric permittivity contain both induced and polarized charges on their surface.

Table 2. Relaxation time of indicative nanoparticles adapted from Hwang et al. [20].

Nanoparticles	Fe_3O_4	ZnO	Al_2O_3	SiO_2(Quartz)	SiO_2(Silica)
Relaxation Time (s)	7.47×10^{-14}	1.05×10^{-11}	12.2	36.3	5.12×10^{-2}

The potential well due to the redistributed charges on the surface of conductive nanoparticles is given for the direction of the applied field ($\theta = 0$) and for the opposite direction ($\theta = \pi$) by [22]:

$$\varphi_{cNP} = \begin{cases} \frac{\sigma_2 - \sigma_1}{2\sigma_1 + \sigma_2} R^3 E_0 \frac{1}{r^2} & \{\theta = 0, \ r \geq R \\ -\frac{\sigma_2 - \sigma_1}{2\sigma_1 + \sigma_2} R^3 E_0 \frac{1}{r^2} & \{\theta = \pi, \ r \geq R \end{cases} \tag{2}$$

where, σ_1 and σ_2 are the conductivities of transformer oil and nanoparticles respectively, R is the radius of nanoparticle, E_0 is the applied field, and r the distance from nanoparticles surface.

On the exterior of a non-conductive nanoparticle, the potential well is given by [22]:

$$\varphi_{ncNP} = \begin{cases} \frac{\varepsilon_2 - \varepsilon_1}{2\varepsilon_1 + \varepsilon_2} R^3 E_0 \frac{1}{r^2} & \{\theta = 0, \ r \geq R \\ -\frac{\varepsilon_2 - \varepsilon_1}{2\varepsilon_1 + \varepsilon_2} R^3 E_0 \frac{1}{r^2} & \{\theta = \pi, \ r \geq R \end{cases} \tag{3}$$

where, ε_1 and ε_2 are the permittivity of transformer oil and nanoparticle respectively. Fast moving electrons are captured by the potential well, forming negatively charged nanoparticles. The latter nanoparticles are introduced with slower mobility and charged negatively, as a consequence the streamer speed is reduced which leads to increased breakdown voltage. Equations (4) and (5) give the total amount of charges trapped by a conductive nanoparticle and non-conductive, respectively [22]:

$$Q_{cNP} = -12\pi\varepsilon_1 E_0 R^2 \tag{4}$$

$$Q_{ncNP} = -12\pi\varepsilon_1 E_0 R^2 \frac{\varepsilon_2}{2\varepsilon_1 + \varepsilon_2} \tag{5}$$

Table 3 depicts the conductivity and permittivity of the colMIONs, SiO_2, and natural ester oil FR3. As mentioned above, the colMIONs were synthesized in situ, thus their conductivity and permittivity are accurately considered equal to that of a commercial Fe_2O_3 NP.

Taking into account Equations (3) and (4), as well as the values of Table 3, the potential well of the suspended colMIONs and SiO_2 are given as [22]:

$$\varphi_{colMION} = 1.25 \cdot E_0 \cdot \frac{1}{r^2} \cdot 10^{-25} \ [V] \tag{6}$$

$$\varphi_{SiO2} = 1.9 \cdot E_0 \cdot \frac{1}{r^2} \cdot 10^{-26} \ [V] \tag{7}$$

where, E_0 is the average electric field and r is the distance from the surface of the nanoparticle.

Table 3. Breakdown voltage probabilities.

Dielectric Lquid	Conductivity σ (S/m)	Relative Permittivity ε (F/m)	Average Nanoparticle Diameter R (nm)
Natural Ester FR3	5×10^{-14}	3.2	-
colMION	1×10^4	80	10
SiO_2	1.4×10^{-11}–1×10^{-15}	3.7–4	12

The electric field, E_0, for colNF is calculated as 31.12×10^6 V/m and for sNF as 23.88×10^6 V/m considering that the breakdown voltage for colNF and sNF is 77.8 kV and 59.7 kV, respectively, for a gap of 2.5 mm. Substituting these values of electric field E_0, as well as the values of average nanoparticle diameter R from Table 3, into Equations (4) and (5) for the colMIONs and SiO_2 NPs, respectively, their saturation charges are:

$$Q_{colMION} = -0.83 \times 10^{-18} \ C \quad \text{and} \quad Q_{SiO_2} = -0.35 \times 10^{-18} \ C \tag{8}$$

Accordingly, substituting the above electric field values E_0 into Equations (6) and (7), the potential well as a function of the distance from the nanoparticles' surface before breakdown occurs is given in Figure 4 for both nanoparticles in question.

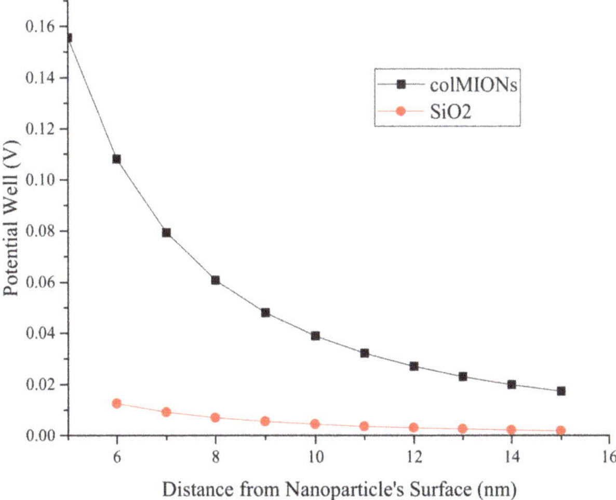

Figure 4. Potential well distribution of oleate-coated colloidal magnetic iron oxide nanocrystals (colMION) and SiO_2 nanoparticles versus the distance from the nanoparticle's surface.

Figure 4 shows the difference in the potential well between colMIONs and sNF and especially near the nanoparticle's surface. The saturation charges of nanoparticles Equation (8) with respect to the potential well of colMIONs (Figure 4), indicate their higher capability to trap electrons generated from ionization or injection in the bulk of dielectric liquid which can affect the streamer early development. The above is an indication of the better dielectric performance of colNF as compared to sNF which is due to the increased potential well of the colNF.

The results from dielectric relaxation spectroscopy study for natural ester oil and colNF 0.012% w/w at 20 °C and 100 °C for a frequency range of 10^{-1}–10^6 Hz are given in Figure 5.

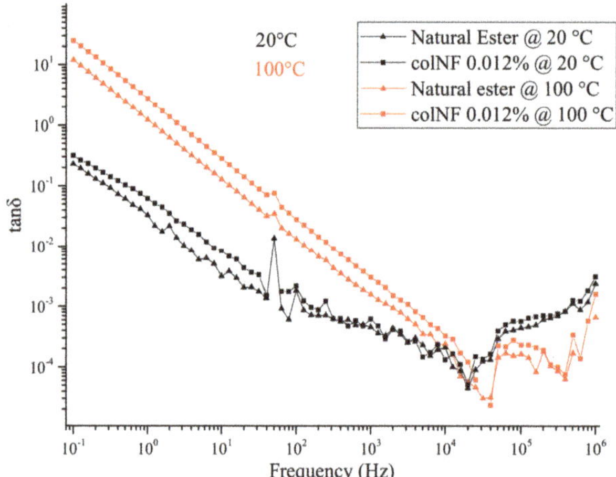

Figure 5. Dielectric losses (tanδ) for natural ester oil and colNF 0.012% at 20 °C and 100 °C.

From Figure 5 it is demonstrated that the dielectric losses (tanδ) are reduced by increasing the frequency and increased by increasing the temperature. A differentiation of the dielectric losses response is monitored at the frequency regime above 20 kHz associated with high frequency relaxations.

4. Discussion

It is well known that the presence of conductive and non-conductive solid particles inside insulating oils results in the decrease of breakdown voltage [29–32]. All the relevant studies that have been conducted concern solid particles with sizes of the order of several hundred μm. However, the basic theory can be extended to the nm scale. This effect is also observed in Figure 3, for low and high concentrations of both dispersed nanoparticles in the vicinity of the matrix oil.

However, the increase of BDV of nanofluids for specific concentrations of nanoparticles indicates the presence of an opposite physical mechanism which enhances the dielectric properties. This mechanism was explained satisfactorily in [20,22] and the experimental results of this work are in compliance with the proposed theory as analyzed in the previous section. Nanoparticles should have either high conductivity or higher permittivity than that of pure insulating oil, in order to act as electron scavengers and reduce the speed of streamer. The colMIONs succeeded due to their high conductivity while SiO_2 failed to improve the dielectric characteristics of insulating oil. This is also valid for TiO_2 nanofluid as its probabilities of BDV are much lower than the matrix oil [14], as can be seen in Table 1.

On the other hand, SiO_2 nanoparticles in [19] seem to have a better 50% probability (Table 1) just as other studies [15–17] reveal a better performance of silica nanofluids in contrast to the experimental results of this work. However, in all these studies, mineral oil was used as the base for the synthesis of nanofluids. Mineral oil exhibits lower permittivity ($e_r = 2$) as compared to natural ester oil ($e_r = 3.2$) and this fact enhances the ability of silica nanoparticles to capture electrons. Another issue seems to be the presence of moisture, since SiO_2 nanoparticles can absorb large amounts of moisture. Water absorption demonstrates higher conductivity and relative permittivity than SiO_2 and thus it enhances the electron trapping capability of SiO_2 nanoparticles.

In this study, nanofluid samples were tested shortly after their preparation, and therefore nanoparticles did not have the time to absorb moisture from the vicinity of the oil.

For lower concentrations than optimum, for both nanofluids tested, BDVs are lower than that of matrix oil but with an upward trend with the increase of nanoparticle concentration. For concentrations above the optimum, BDVs are also lower than that of matrix oil but with a steadily downward trend.

As was aforementioned, both conductive and non-conductive nanoparticles are charged when subjected to an external electric field. As the concentration of nanoparticles increases, they will form thin filaments that oscillate across the gap. When the filaments reach a critical length, a conductive bridge path is formed and a breakdown discharge will be triggered. As long as the enhancement mechanism overcomes the negative impact of solid impurities on transformer oil, BDV versus concentration will exhibit an upward trend. When the concentration of nanoparticles exceeds a critical value, the negative impact of the presence of solid particles dominates, and the BDV of the nanofluid rapidly decreases.

5. Conclusions

In this study, the dielectric insulation characteristics for two types of nanofluids based on a natural ester matrix were studied with respect to the influence of electrical conductivity and permittivity of the NPs on the AC dielectric strength of nanofluids. Two types of nanoparticles were studied, oleate-coated colloidal magnetic Fe_2O_3 nanocrystals (colMIONs) exhibiting high conductivity and permittivity and SiO_2 NPs exhibiting low conductivity and permittivity.

The experimental results of this study are in compliance with the proposed theories analyzing the impact of NPs on the dielectric strength of transformer oil-based nanofluids. NPs with higher electrical conductivity or permittivity can exhibit improved dielectric performance by adding them in the matrix oil. Particularly the introduction of colNF consisting of conductive colMIONs, demonstrates an improved AC breakdown voltage of 77.8 kV at 0.012% *w/w* concentration as compared with 64.5 kV AC BDV for the base matrix oil. Simultaneously, they exhibit improved field operation reliability taking into account that the U_1 breakdown voltage was considerably high, equal to 62.1 kV, which is remarkably higher than the U_1 of 37.2 kV of the matrix oil. The aforementioned performance is correlated to the substantially low standard deviation in the range of 6.7 kV. Additionally, the dielectric losses of the colNF studied at 25 °C and 100 °C in the frequency regime of 10^{-1}–10^6 Hz behaved similarly to the matrix oil, that is, they reduced as the frequency increased and increased as the temperature increased, but with slightly higher values than that of matrix oil.

In contrast, the AC breakdown voltage of silica NF with SiO_2 NPs is reduced as compared to that of the matrix oil for all the ranges of NPs concentrations that were investigated herein. The explanation of the abovementioned dielectric performance is correlated with the low electrical conductivity as well as the permittivity of silica NPs, along with their low capability of capturing free electrons. Furthermore, an in-depth research study is needed which takes into account the electrical conductivity and permittivity in order to discriminate the influence of moisture content on the dielectric strength performance of the nanofluids.

Author Contributions: Conceptualization: V.P.C., G.D.P., E.C.P.; Methodology: G.D.P.; Manufacturing of ColMIONs and characterization: A.B., G.D.P.; Preparation of NFs: A.B., G.D.P.; Measurements of BDV: G.D.P.; Measurements of dielectric losses: G.D.P., A.D.P.; Statistical analysis and Analysis of results: V.P.C., G.D.P., A.D.P.; Theoretical Analysis: V.P.C., G.D.P., A.D.P.; Writing: V.P.C., G.D.P., A.D.P.; Review: V.P.C., G.D.P., E.C.P., A.B., A.D.P., I.F.G.; Supervision: E.C.P., I.F.G.

Funding: This research received no external funding.

Acknowledgments: The authors warmly acknowledge Kevin Rapp Senior scientist of Cargill for the FR3 oil supply during this research.

Conflicts of Interest: The authors declare no conflict of interest.

References

1. Maxwell, J.C. *A Treatise on Electricity and Magnetism*, 2nd ed.; Clarendon Press: Oxford, UK, 1881; Volume 1, p. 435.
2. Choi, S.U.S.; Eastman, J.A. Enhancing Thermal Conductivity of Fluids with Nanoparticles. In Proceedings of the 1995 International Mechanical Engineering Congress Exposition, San Francisco, CA, USA, 12–17 November 1995; pp. 99–105.
3. Chiesa, M.; Sarit, D. Experimental investigation of the dielectric and cooling performance of colloidal suspension in insulating media. *Colloids Surf. A Physicochem. Eng. Asp.* **2009**, *335*, 88–97. [CrossRef]
4. Mergos, J.; Athanassopoulou, M.; Argyropoulos, T.; Vassiliou, P.; Dervos, C.T. Dielectric properties of nanopowder dispersions in paraffin oil. *IEEE Trans. Dielectr. Electr. Insul.* **2012**, *19*, 1502–1507. [CrossRef]
5. Nazari, M.; Rasoulifard, M.; Hosseini, H. Dielectric breakdown strength of magnetic nanofluid based on insulation oil after impulse test. *J. Magn. Magn. Mater.* **2016**, *339*, 1–4. [CrossRef]
6. Du, B.; Li, X.; Li, J. Thermal conductivity and dielectric characteristics of transformer oil filled with BN and Fe_3O_4 nanoparticles. *IEEE Trans. Dielectr. Electr. Insul.* **2015**, *22*, 2530–2536. [CrossRef]
7. Li, J.; Zhang, Z.; Zou, P.; Grzybowski, S.; Zahn, M. Preparation of vegetable oil based nanofluid and investigation of its breakdown and dielectric properties. *IEEE Electr. Insul. Mag.* **2012**, *28*, 43–50. [CrossRef]
8. Cavallini, A.; Negri, F. Behaviour of nanofluids under DC divergent fields. In Proceedings of the IEEE International Conference on Dielectrics, Montpellier, France, 3–7 July 2016. [CrossRef]
9. Peppas, G.D.; Bakandritsos, A.; Charalampakos, V.P.; Tucek, J.; Zboril, R.; Pyrgioti, E.C.; Gonos, I.F. Ultrastable natural ester based nanofluids for high voltage insulation applications. *ACS Appl. Mater. Interfaces* **2016**, *8*, 25202–25209. [CrossRef] [PubMed]
10. Peppas, G.D.; Charalampakos, V.P.; Pyrgioti, E.C.; Danikas, M.G.; Bakandritsos, A.; Gonos, I.F. Statistical investigation of AC breakdown voltage of nanofluids compared with mineral and natural ester oil. *IET Sci. Meas. Technol.* **2016**, *10*, 644–652. [CrossRef]
11. Taha-Tijerina, J.; Narayanan, T.N.; Gao, G.; Rohde, M.; Tsentalovich, D.A.; Pasquali, M.; Ajayan, P.M. Electrically insulating thermal nano-oils using 2D fillers. *ACS Nano* **2012**, *6*, 1214–1220. [CrossRef]
12. Atiya, E.G.; Mansour, D.E.A.; Khattab, R.M.; Azmy, A.M. Dispersion behavior and breakdown strength of transformer oil filled with TiO_2 nanoparticles. *IEEE Trans. Dielectr. Electr. Insul.* **2015**, *22*, 2463–2472. [CrossRef]
13. Du, Y.; Lv, Y.; Li, C.; Chen, M.; Zhou, J.; Li, X.; Zhou, Y.; Tu, Y. Effect of electron shallow trap on breakdown performance of transformer oil-based nanofluids. *J. Appl. Phys.* **2011**, *110*, 104104-1-4. [CrossRef]
14. Zhong, Y.; Lv, Y.; Li, C.; Du, Y.; Chen, M.; Zhang, S.; Zhou, Y.; Chen, L. Insulating properties and charge characteristics of natural ester fluid modified by TiO_2 semiconductive nanoparticles. *IEEE Trans. Dielectr. Electr. Insul.* **2013**, *20*, 135–140. [CrossRef]
15. Jin, H.; Andritsch, T.; Tsekmes, I.A.; Kochetov, R.; Morshuis, P.H.F.; Smit, J.J. Properties of Mineral Oil Based Silica Nanofluids. *IEEE Trans. Dielectr. Electr. Insul.* **2014**, *21*, 1100–1108. [CrossRef]
16. Rafiq, M.; Li, C.; Lv, Y.; Yi, K.; Arif, I. Breakdown characteristics of transformer oil based silica nanofluids. In Proceedings of the IEEE International Multi Topic Conference (INMIC), Islamabad, Pakistan, 5–6 December 2016; pp. 1–4. [CrossRef]
17. Cavallini, A.; Karthik, R.; Negri, F. The effect of Magnetite, grapheme oxide and silicone oxide nanoparticles on dielectric withstand characteristics of mineral oil. *IEEE Trans. Dielectr. Electr. Insul.* **2015**, *22*, 2592–2600. [CrossRef]
18. Liu, D.; Zhou, Y.; Yang, Y.; Zhang, L.; Jin, F. Characterization of high performance AlN nanoparticle-based transformer oil nanofluids. *IEEE Trans. Dielectr. Electr. Insul.* **2016**, *23*, 2757–2767. [CrossRef]
19. Lv, Y.Z.; Zhou, Y.; Li, C.R.; Wang, Q.; Qi, B. Recent progress in nanofluids based on transformer oil: Preparation and electrical insulation properties. *IEEE Electr. Insul. Mag.* **2014**, *30*, 23–32. [CrossRef]
20. Hwang, G.J.; Zahn, M.; O'Sullivan, F.M.; Pettersson, L.A.A.; Hjortstam, O.; Liu, R. Effects of nanoparticle charging on streamer development in transformer oil-based nanofluids. *J. Appl. Phys.* **2010**, *107*, 014310-1-17. [CrossRef]
21. Yang, Q.; Yu, F.; Sima, W.; Zahn, M. Space charge inhibition effect of nano-Fe_3O_4 on improvement of impulse breakdown voltage of transformer oil based on improved Kerr optic measurements. *AIP Adv.* **2015**, *5*, 097207-1. [CrossRef]

22. Sima, W.; Shi, J.; Yang, Q.; Huang, S.; Cao, X. Effects of conductivity and permittivity of nanoparticle on transformer oil insulation performance: Experiment and theory. *IEEE Trans. Dielectr. Electr. Insul.* **2015**, *22*, 380–390. [CrossRef]
23. Tanaka, T.; Kozako, M.; Fuse, N.; Ohki, Y. Proposal of a multi-core model for polymer nanocomposite dielectrics. *IEEE Trans. Dielectr. Electr. Insul.* **2005**, *12*, 669–681. [CrossRef]
24. Izzularab, M.; Ibrahim, M.; Abd-Elhaby, A. Effect on transformer oil breakdown strength: Experiment and theory. *IET Sci. Meas. Technol.* **2016**, *10*, 839–845. [CrossRef]
25. Liong, M.; Lu, J.; Kovochich, M.; Xia, T.; Ruehm, S.G.; Nel, A.E.; Tamanoi, F.; Zink, J.I. Multifunctional inorganic nanoparticles for imaging, targeting and drug delivery. *ACS Nano* **2008**, *2*, 889–896. [CrossRef]
26. Peppas, G.; Bakandritsos, A.; Pyrgioti, E. Method of Making and Synthesizing Dielectric Nanofluids. Patent WO2018020278A1, 12 July 2017.
27. Peppas, G.D.; Charalampakos, V.P.; Pyrgioti, E.C.; Bakandritsos, A.; Polykrati, A.D.; Gonos, I.F. A study on the Breakdown Characteristics of Natural Ester Based Nanofluids with Magnetic Iron Oxide and SiO2 Nanoparticles. In Proceedings of the IEEE International Conference on High Voltage Engineering and Application (ICHVE), Athens, Greece, 10–13 September 2018. [CrossRef]
28. IEC 60156:2016. *Insulating Liquids—Determination of the Breakdown Voltage at Power Frequency—Test Method*; International Electrotechnical Commission: Geneva, Swutzerland, 2016.
29. Chadband, W.G. The electrical breakdown of insulating oil. *Power Eng. J.* **1992**, *6*, 61–67. [CrossRef]
30. Chadband, W.G. Electrical breakdown of liquid insulation. In Proceedings of the IEE Colloquium on an Engineering Review of Liquid Insulation, London, UK, 7 January 1997. [CrossRef]
31. Mahmud, S.; Chen, G.; Golosnoy, I.O.; Wilson, G.; Jarman, P. Bridging phenomena in contaminated transformer oil. In Proceedings of the IEEE International Conference on Condition Monitoring and Diagnosis, Bali, Indonesia, 23–27 September 2012. [CrossRef]
32. Wang, X.; Wang, Z.D. Particle effect on breakdown voltage of mineral and ester based transformer oils. In Proceedings of the IEEE 2008 Annual Report Conference on Electrical Insulation Dielectric Phenomena, Quebec, QC, Canada, 26–29 October 2008. [CrossRef]

Article

Is the Dry-Band Characteristic a Function of Pollution and Insulator Design? [†]

Maurizio Albano *, A. Manu Haddad and Nathan Bungay

School of Engineering, Cardiff University, The Parade, Cardiff CF24 3AA, UK; Haddad@cardiff.ac.uk (A.M.H.);
BungayN@cardiff.ac.uk (N.B.)
* Correspondence: AlbanoM@cardiff.ac.uk; Tel.: +44-292-087-0672
† This paper is an extended version of our paper published in 2018 IEEE International Conference on High
Voltage Engineering (ICHVE 2018), Athens, Greece, 10–13 September 2018; 0-TM5-5.

Received: 22 July 2019; Accepted: 17 September 2019; Published: 21 September 2019

Abstract: This paper assesses the dry-band formation and location during artificial pollution tests performed on a 4-shed 11kV insulator with conventional and textured surface designs in a clean-fog chamber and with the application of a voltage ramp-shape source. The different designs present the same overall geometrical dimensions, but the textured ones are characterized by the application of a patented insulator surface design. Three pollution levels, extremely high, high and moderate, were considered. A newly developed MATLAB procedure is able to automatically recognize the perimeter of the insulator, the trunk and shed areas on infra-red recordings. In addition, using the vertical axis identification, all trunks are subdivided into zones and into left and right areas, significantly increasing the capability of abnormalities detection. Any temperature increase within these areas enables to detect the appearance and the extension of dry bands. The results of the analysis of the statistical location and extension development over time of the dry bands during these set of comparative tests show a clear distinction between designs and pollution levels. These results may offer interesting design guidelines for dry-band control.

Keywords: insulator design; dry band; pollution

---

1. Introduction

Pollution is a key aspect for high voltage insulator design and selection. Pollution deposition and wet conditions can initiate discharge activity and can lead to flashover and, consequently, the availability of a high-voltage transmission line is compromised. The adoption of silicone rubber insulators is justified by their improved performance under polluted environments in comparison with traditional ceramic and glass insulators, and particularly when the pollution level is severe. However, the superior performance of the composite material is compromised when discharge activity is established on the insulator area and, consequently, dry bands appear. This activity determines a hydrophobic reduction of the silicone rubber, and if it persists, no hydrophobicity recovery can be initiated. Moreover, further partial arcs can initiate localized erosion, permanently weakening the silicone rubber insulator surface, as presented in several research works [1–5].

Numerous works [6–11] have been published recently on dry-band formation and pollution estimation, showing the importance of this topic. The papers [6,7] propose an adaptive technique to predict the contamination level increase in order to warn the utility before reaching high levels, conditions that are favourable to extensive dry-band activities. In [8,9], dry-band formation is studied applying controlled non uniform areas of pollution on flat samples. In [10,11], the influence of multiple dry bands on flashover characteristics under various environmental and pollution conditions was investigated on rectangular silicone rubber test samples.

Previous research works [12,13] performed by the authors showed the valuable information obtained by the analysis of dry-band and infrared data. However, the dry-band location was not fully automatically detected along the full insulator, but always with user intervention requiring considerable amount of time.

In this research work, high voltage tests have been performed on 4-shed 11kV insulators with various surface designs in a clean-fog chamber test facility applying a wide range of artificial pollution levels [14]. The more common locations of the dry bands have been identified by analysing the infrared recordings. Further characteristics, such as extension and temperature profile, have been investigated during the time frame of the tests, applying a newly developed algorithm. The MATLAB procedure is able to identify the insulator physical boundaries, the vertical axis and trunk sub-areas, in order to evaluate, with good accuracy, the extension of each dry band and to monitor the variation of these hot zones continuously during the entire test. The series of clean-fog tests have been performed adopting conventional and three textured designs. In addition, after the application of the artificial pollution layer and a resting period of 24 hours, the equivalent salt deposit density (ESDD) and the Non-Soluble Deposit Density (NSDD) have been determined on all selected insulator designs. The ESDD and NSDD levels can contribute to the dry-band characteristics. These results confirm the interesting design advantage of texturing for dry-band control.

2. Experimental Set-Up

The experimental set-up adopted for this investigation was based on a clean-fog test chamber and a programmable voltage ramp-shape source to energise the insulators, previously artificially polluted. The high voltage source was provided by a Hipotronics 150 kVA transformer, and the primary circuit was fed via a Paschen transformer (0–960 V) providing a programmable voltage variation up to 75 kV. In this series of tests, the voltage control unit is set up to provide a constant rate of increase of 4 kV per minute to achieve the ramp shape until the flashover event; this permits to assess the insulator performance under increasing stress in a limited period of time. This methodology has been proposed by the Cardiff University research team, and it is described in detail in [15]. A schematic diagram of the measurement set-up used in the experiment is shown in Figure 1. Previous investigations focused on evaluating the impact of fog rate and pollution level applied as ESDD and NSDD on the insulator surface, and each test insulator was subjected to a ramp series (between 4 and 12 successive flashovers, with a 5 min interval). These comparative laboratory tests, based on voltage ramp shape and an artificially polluted insulator in a clean fog chamber, facilitated the evaluation of the impact on withstand levels of textured design compared with plain surface insulators. A clear indication of improved flashover withstand performance of textured insulators in comparison with conventional surfaces was demonstrated in the research investigation [16].

In this work, only the first ramp was selected in order to have a uniform temperature distribution over the insulator surface at the start of each test. Further tests have been performed using selected constant voltage levels to observe the stability of the dry-band extension over time and the most appropriate methodology to identify the dry-band formation.

The insulator designs selected for this investigation share an identical axial length (175 mm), trunk diameter (28 mm) and shed diameter (90 mm), as illustrated in Table 1. The increased creepage length offered by texturing the surface of the insulator (TT) is shown in the first column of Table 1. Additional increments of the creepage length is offered in the TTS design, where a logarithmic spiral double-ridge pattern is applied under the sheds. The insulator with an all plain surface is referred to in this work as the conventional design (CONV).

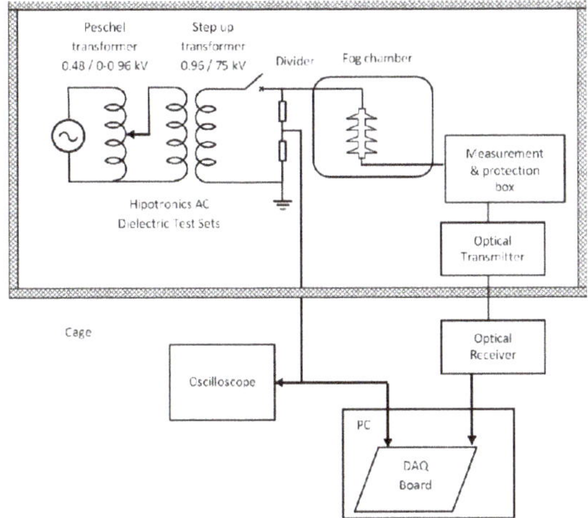

Figure 1. Schematic diagram of the measurement set-up used in the experiment [15].

Table 1. Insulator designs.

Design	Creepage Length (mm)	Dimple Size Radius (mm)	Under Shed Profile (mm)	Axial Length (mm)	Trunk Diameter (mm)
CONV	375	0	0	175	28
TT4	471	4	0	175	28
TTS4	503	4	4	175	28
TT6	471	6	0	175	28
TTS6	503	6	6	175	28

During the laboratory tests, video and images were recorded with a high-definition camcorder and a still camera to obtain signs of the discharge phenomena in the trunk areas. Unfortunately, the faint discharges were not clearly visible, and they could be observed only using long-time exposure techniques. Simultaneously, a FLIR A325 camera was adopted and fixed on the same tripod of the visual camera in order to achieve a close focal point between the two recordings. IR pictures were recorded using a 1 Hz frame rate to minimize memory storage requirement. The IR camera offers the capability to monitor the temperature at a frequency rate up to 60 Hz and with a precision of 0.5 °C. These capabilities are clearly not sufficient to record the instantaneous maximum temperature of the discharge. However, the IR records can show with good precision the maximum temperature of any area of the insulator. The camera does not offer any triggering signal, and only by using the time stamp associated with each IR frame recorded was it possible to synchronize the temperature with all the other data recorded during the test, such as electrical parameters and visual records, with a precision of 1 ms [14].

3. Experimental Results and Dry-Band Detection Procedure

A long series of laboratory tests were performed on the five insulator designs after the application of extremely high, high and moderate pollution levels, 1.15, 0.64 and 0.42 mg/cm^2, respectively, and a constant NSDD of 0.1 mg/cm^2.

In the initial tests, selected constant voltage levels were applied and some discharge activity was identified in visual recordings only using a long-exposure digital stills camera. These small discharges were not visible in the high-definition (HD) digital video camera recordings but only in the long-exposure photo. These discharges, of the corona/streamer type, are characterized by a purple-blue colour as shown in Figure 2b. The associated dry bands were localized only with the help of the IR camera, as shown in Figure 2c. The visual photos were post-processed in order to enhance the spark visibility. A second type of discharge, the streamer discharges, are shown by a red/orange colour crossing an extended area than that of the dry band.

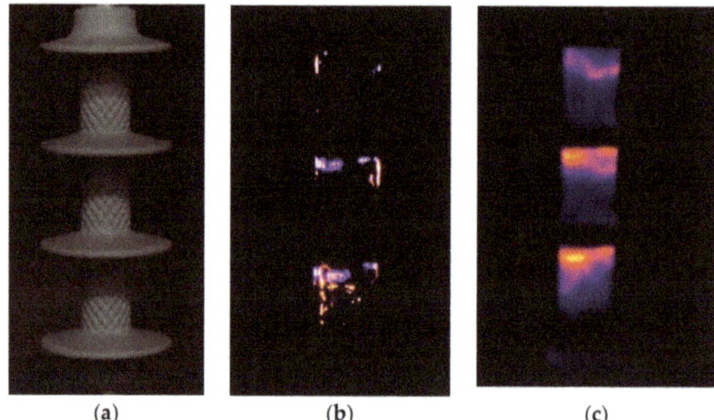

(a)   (b)   (c)

Figure 2. Visual frame before fog application (**a**), visual frame using long-exposure acquisition (**b**) and the IR frame (**c**) during the constant voltage test on sample TT4; equivalent salt deposit density (ESDD) was 0.64 mg/cm^2 with a fog spray rate of 3 L/h.

The synchronized IR images present areas of increased temperature (Figure 2c) where the visual images showed the presence of fully formed dry bands (Figure 2b). It is important to observe that no significant streamer activity is present in the upper trunk. However, the presence of a dry band in each trunk is confirmed by the IR photos, with a lower temperature magnitude on the upper one.

A differential temperature between the insulator surface and the surrounding fog along a vertical segment has been calculated for each test. After preliminary tests, using constant voltage, a fixed threshold of 5.5 °C appeared to be a quick identification method. However, if this methodology would be simply applied to the all extended ramp voltage tests the results would be not correct. In fact, a lower extension of dry bands would have been detected, with the temperature profile presented in Figure 3 as example. Moreover, the identification of the dry band is very time consuming if performed by the user by manually examining a significant number of frames and series of tests and this suggested the need for developing an automatic procedure.

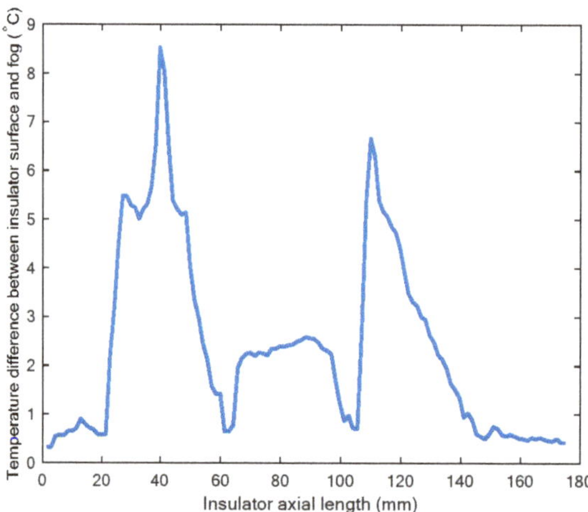

Figure 3. Temperature difference between the insulator surface and the surrounding fog.

The maximum average temperature on the whole insulator surface during the tests and the average values on identified areas can be evaluated from the thermal recordings. The extraction of the maximum temperature recorded during the test and the observation of its temporal variation can help to highlight only the timestamp of the spark/dry bands events but no information about their location, size or number. These limitations suggested the authors to develop a MATLAB computer procedure to acquire and process the temperature variation at different areas of the insulator. Preliminary investigations showed higher frequency of occurrence of dry-band and corona discharge events on the trunks in comparison with the shed areas. Consequently, increased surface temperature or spark events appeared mainly on these areas before flashover, suggesting limiting the analysis mainly on these areas.

3.1. Computer Technique for Identification of Dry-Band Boundaries

The precise identification of the insulator location and its perimeter on the thermal frame is a fundamental feature of the proposed analysis. The authors developed a MATLAB procedure to automatically detect the profile boundary of the insulator. This procedure is based on the analysis of the small variations between the ambient air surrounding the insulator and the insulator surface. Since the temperature difference between the insulator and the fog is quite small at the start of each test, the automatic identification is not a trivial task. In addition, possible reflections on the object can cause significant error in the outline identifications. A schematic diagram of the procedure adopted is shown in Figure 4.

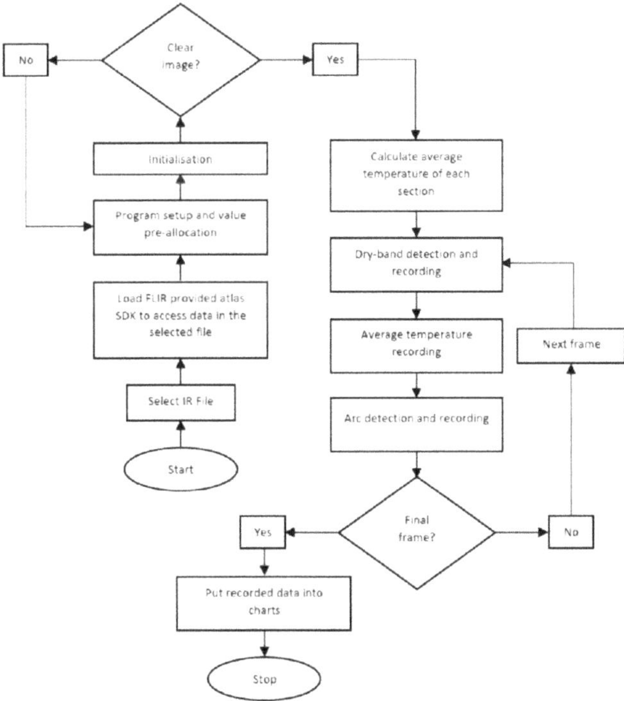

Figure 4. Schematic diagram of the procedure adopted.

The boundary detection procedure is based on Canny Edge detection, but other algorithms, such as Sobel, Prewitt and Roberts, were also tested. The Canny algorithm showed the best accuracy when applied to this environmental condition. The two threshold levels required by the Canny algorithm are automatically selected using two key areas of the frame, assuming the left corner as representation of the fog temperature and the hotter area indicating the insulator position within the frame. Once the perimeter is extracted, it is subdivided in two sections, the left- and right-hand boundaries, as can be seen in Figure 4 with the red and blue colours, respectively. Then, they are plotted over the frame for final acceptance by the user. As shown in Figure 5b, the boundaries are not always identified completely, and the user can decide to modify the thresholds or introduce some corrections. However, despite some uncertainty on sheds areas, the areas of interest, the trunks are correctly identified in most cases.

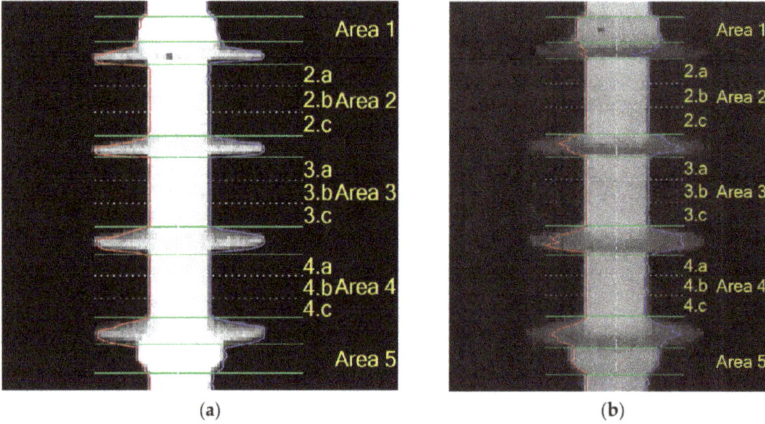

Figure 5. Automatic boundaries recognition and area selection applied to (**a**) a conventional and (**b**) a textured design.

A second step of the procedure aims to identify the shed and trunk areas, since only the trunk areas are processed on the temperature analysis. The trunk surfaces have been subdivided into three zones (a, b, c), identified by segmented white lines in Figure 5. The presence of the shed characterizes the type area a, meanwhile the type area b is located on the central area, more exposed to the fog and prone to a stronger wetting action. The remaining one, area type c, is located on the trunk near the top shed zone.

Assuming the insulator is always on an almost vertical position, the average of all the horizontal coordinates of the trunks permits to identify the position of the vertical axis of symmetry, drawn on Figure 3 as a dot–dash white line.

3.2. Average Temperature Calculation

The precise identification of the boundaries of the insulator enables the extraction only of the values related to each area from the IR temperature data, neglecting any measurement related to the fog. The procedure calculates the average temperature of each area for each frame. This permits to identify different trends related to the design. Figure 6 shows the two end-fitting areas during the whole test for the conventional and textured TT4 insulators.

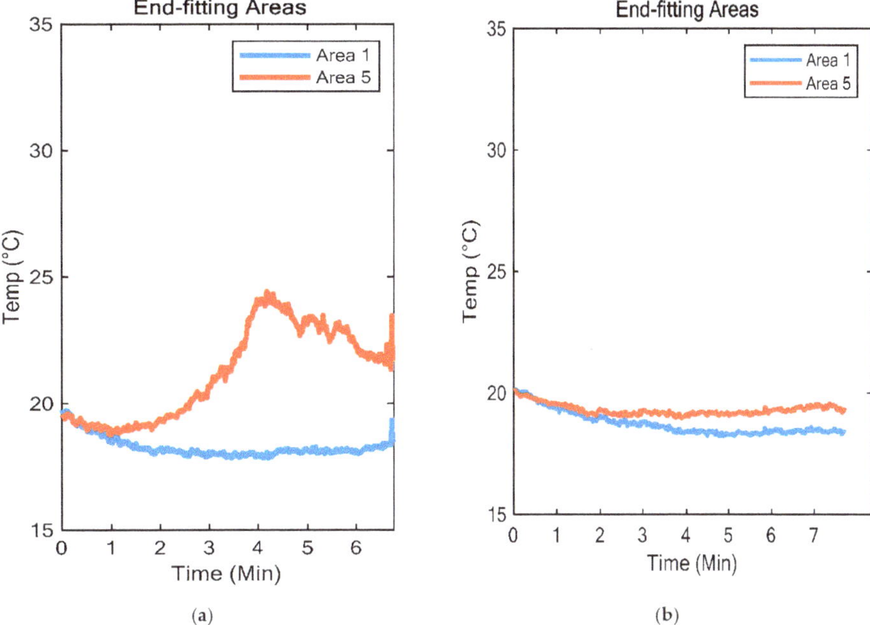

Figure 6. Average temperature of end-fitting areas during the whole test, on (**a**) conventional and (**b**) textured TT4 insulators.

In the first two minutes, the fog formation determines a cooling effect on the initial period. Afterwards, a steady increase in dry-band extension is observed as the voltage and current increase. The polluted insulator previously stored at laboratory temperature is placed in the fog chamber where, as the test begins, the fog cools the insulator surface until the discharge activity heats up the surface.

In particular, the temperature of area 5, which is the trunk close to the ground termination, is observed to increase in many tests of the conventional design. The increment of temperature in the graph confirms the presence of a dry band and discharge activity. This trend is not followed for the textured design, indicating no significant activity in this area. The area 1, which is the trunk area close to the high voltage termination, shows no discharge activity for both designs.

Figure 7 shows the average temperature of each zone of the three trunk areas for a TT4 design, with a pollution level of 0.64 mg/cm^2. The temperature trends are aggregated in two subplots to facilitate the comparison between the left- and right-hand areas. The comparison between left and right areas confirms that the artificial pollution has been applied correctly to achieve uniform ESDD on the surface and no significant variation is observed.

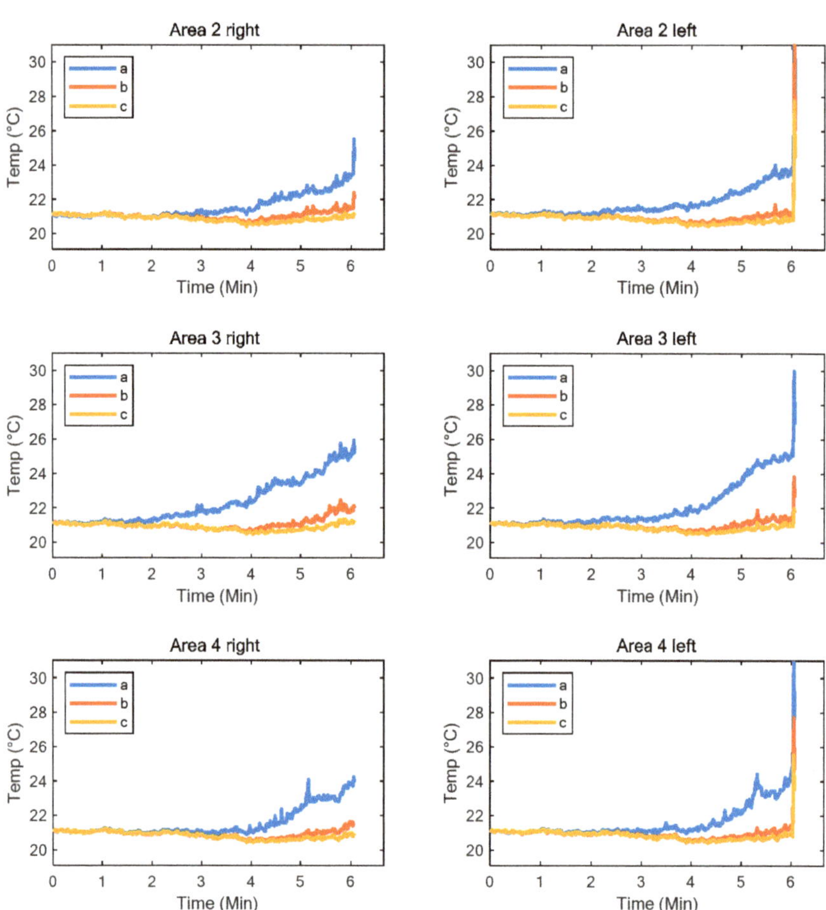

Figure 7. Average temperature of right and left trunk areas during the whole test.

From the plots, the different temperature trends are clearly visible for zone a, suggesting that the formation of dry bands is facilitated by the presence of the shed, shielding the wetting on this zone. Another clear indication from the average temperature of subareas is that subarea c always presents the lowest average temperature in each trunk. This suggests that the wetting is enhanced in the other two subareas, meanwhile in the area closer to the shed, the wetting is slightly reduced, creating better conditions for dry-band formation.

It is noted that the textured design tends to initiate dry bands on all the three areas, creating a more uniform distribution. This is not observed on the conventional design, exhibiting a more localized formation, as shown in the Appendix A in Figure A1. In fact, the middle trunk, area 3, does not show any significant increment, meanwhile the other two trunks show temperature increments in zone a. A rapid increase in temperature at around 5 to 6 min before the flashover event is observed.

3.3. Dry-Band Analysis

Another useful indication is the localization and progression of any dry band during the test. The identification of dry bands using a fixed selected threshold cannot be detected correctly, as described in the temperature profile reported in Figure 3. In addition, using a variable level function only on the first IR frame is still not always applicable because of the cooling trend by the fog, as described previously.

The proposed algorithm of dry-band detection is based on identifying the points/area where the minimum average value is exceeded within the trunk zone. This permits to take into account any cooling effect in the initial period or overall variation not caused by a localized event. Each dry-band width is then evaluated calculating the average temperature of each left- and right-hand horizontal row of data given by the individual IR pixels. If these values exceed the threshold, the row is flagged and counted. The total dry band extension along the vertical axis for each frame is calculated converting the number of pixels in mm, given the distance between the terminals of the insulator; and in this case, it is equal to 175 mm.

The sum of all dry-band extensions along the insulator axis for all the duration of the test (Design CONV, ESDD 0.64 mg/cm^2, fog rate 3 L/hr) is presented in Figure 8. An analogue test using a lower pollution level (ESDD 0.42 mg/cm^2) is shown in the Appendix A as Figure A3.

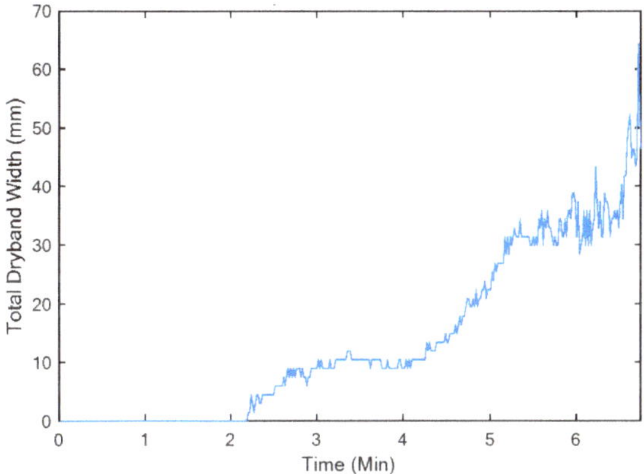

Figure 8. Cumulative dry-band extension on all the trunks. Ramp-test: Design CONV, ESDD = 0.64 mg/cm^2, and a fog rate of 3 L/hr.

In order to gain a deeper understanding of the growth of each dry band, the calculated dry-band location and duration were plotted on a single graph taking advantage of the contour facility introducing time as the x-axis value, the vertical position as the y-axis value and the temperature of the dry band as the colour level. In addition, the last frame of the IR image is automatically cropped according to the identified boundaries and rescaled along the vertical axis of symmetry, facilitating the user to localize the dry band on the insulator. The resulting plot is shown in Figure 9. This graph offers a valuable overview of the full test since each dry-band width is presented as function of time and as its dry-band temperature range. The graph allows the comparison of the dry-band temperature distributions for a specific design very readily. The individual dry-band extension function of time and the dry-band temperature range for the conventional design (Figure 9a) and textured TT4 insulator (Figure 9b) are also presented.

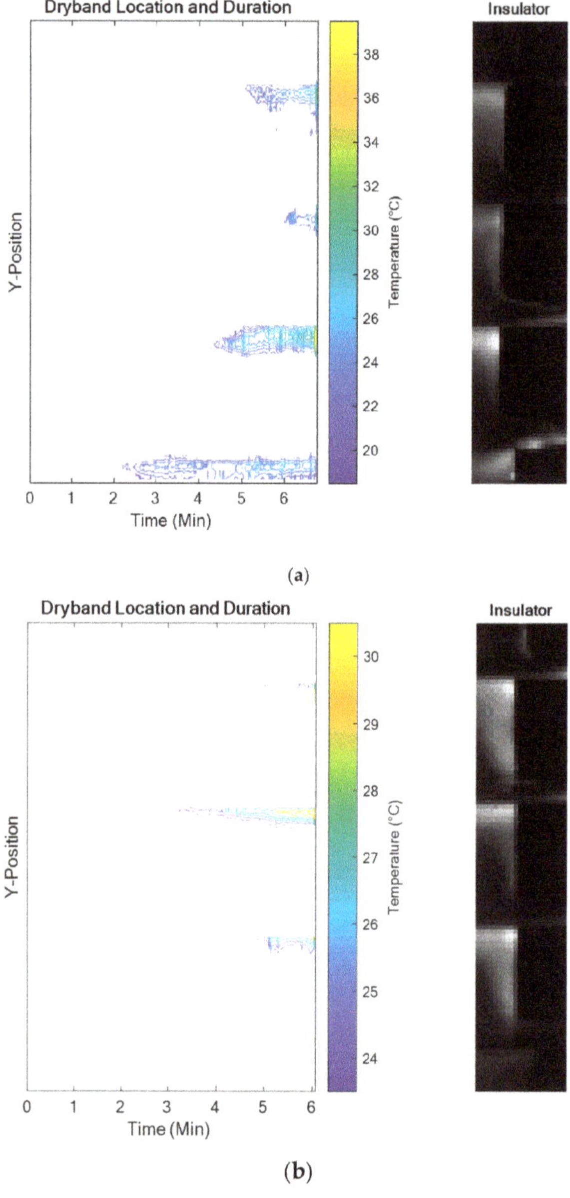

Figure 9. Dry-band extension as function of time and dry-band temperature range for the (**a**) conventional design and (**b**) textured TT4 insulator examples.

4. Discussion

In order to estimate the effect of the design on the dry band dimensions, the total dry-band extension of the selected designs for a given ESDD and fog spray rate were plotted on the same graph, as shown in Figure 10. The graph shows a selection of calculated dry band extensions for all four insulator designs with an applied pollutant on the surface equal to ESDD 0.64 mg/cm^2, and a fog spray rate equal to 3 L/h. It is clear that the conventional design exhibits dry-band formation much earlier than textured insulators. As early as after 1 min, the conventional (CONV) samples start to develop dry bands which continue to extend with increasing applied voltage. This trend is significantly delayed on textured (TT) samples, where the total dry-band length starts to be significant only after 4 min. The curves before flashover shows different trends. In fact, the dry-band dimensions on the TT6 design increases sharply, whereas for TT4 and TTS4 a delayed increase and a lower maximum width is observed before the flashover event. If the analysis was limited to only flashover voltage levels, the test results would not have highlighted any other details such as the clearly visible differences in this graph [10]. This suggests that the new procedure can offer a valuable tool to analyse and compare insulator performance under pollution.

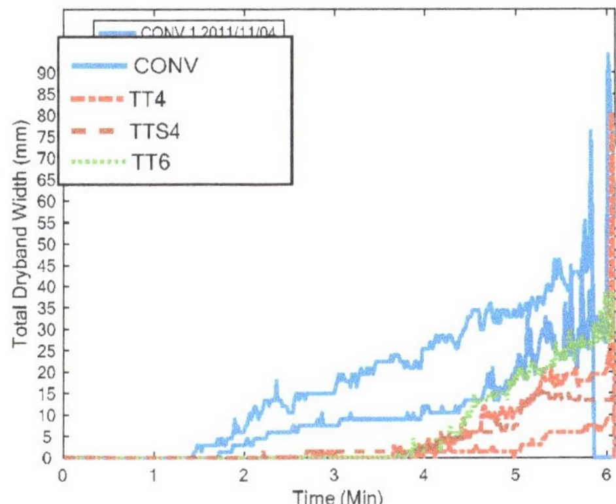

Figure 10. Comparison of the dry-band extension on all the trunks for the five designs, with ESDD = 0.64 mg/cm^2, and a fog spray rate of 3 L/h.

5. Conclusions

The paper presented a novel procedure to assess the formation, and to identify the location, of dry bands based on the automatic identification of selected areas and the calculation of the associated average temperature. In particular, the division of sub zones and left and right areas increase the detection capabilities from the IR recordings.

The proposed methodology has been applied on artificially polluted insulator surfaces under a ramp-shape voltage test in clean-fog chamber and the results of the analysis show that the two features are clearly a function of the insulator surface design and the pollution level applied to the insulator surface (ESDD). The automatic identification of the insulator boundaries allows an accurate estimation of the average local temperature, disregarding any area related to the surrounding environment. In addition, the choice of selected area on the trunks permits to take into account the presence of the sheds that can perturb the wetting action of the fog. The evaluation of local average temperature on symmetrical areas along the vertical axis permits to confirm the correct uniform application of a pollution layer and to warn about possible localized defects on the surface.

Pollution level is one of the most important parameters in insulator design selection. Any increased discharge activity caused by pollution on a specific insulator design affects the life expectancy, which is a key parameter that network operators have to asses in the adoption of a new design. Since one of the major causes of degradation is caused by continuous discharges on the polymeric insulator surfaces, this proposed new procedure based on the analysis of IR video recordings and on the spatial and time characterization of dry bands may provide an indication of the selection of the most appropriate surface design to maximize insulator-life extension.

The results show it is possible to estimate the location, extension and development-over-time of dry bands, and these features offer good indications to select the appropriate design for dry-band control.

Author Contributions: Conceptualization, M.A. and A.M.H.; writing—original draft preparation, M.A.; writing—review and editing, M.A. and A.M.H.; visualization, N.B.

Funding: This research received no external funding.

Appendix A

Appendix A shows an example of ramp test performed on insulator design CONV, with applied pollution of ESDD 0.42 mg/cm^2 and a fog rate 3 L/hr. Figures A1 and A2 shows the average temperature of the main trunk and top and lower trunk areas during the whole test respectively. Figure A3 shows the total dry-band length computed for the same selected test. Figure A4 shows the dry-band development on each trunk as a maximum temperature profile on a 3D plot versus time and vertical axis position.

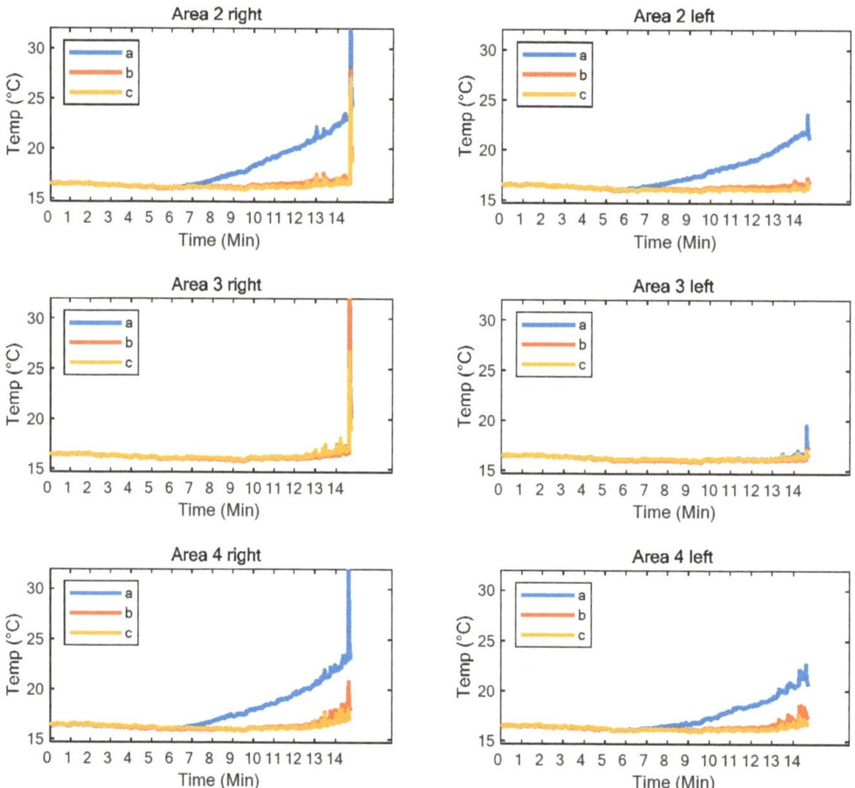

Figure A1. Average temperature of the main trunk areas during the whole test (left and right zones). Design CONV, ESDD = 0.42 mg/cm^2, and a fog rate of 3 L/hr.

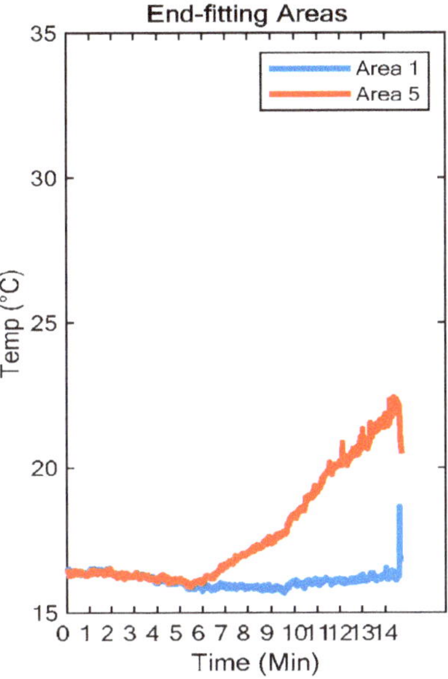

Figure A2. Average temperature of top and lower trunk areas during the whole test (left and right zones). Design CONV, ESDD = 0.42 mg/cm^2, and a fog rate of 3 L/hr.

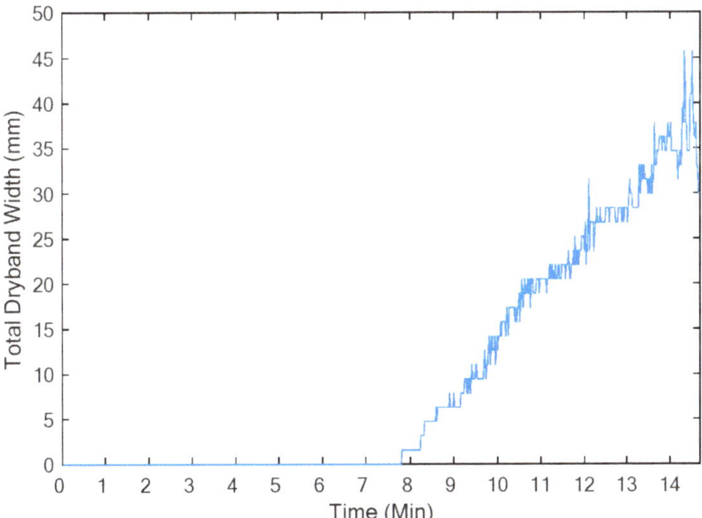

Figure A3. Total dry-band length. Ramp-test: Design CONV, ESDD = 0.42 mg/cm^2, and a fog rate of 3 L/hr.

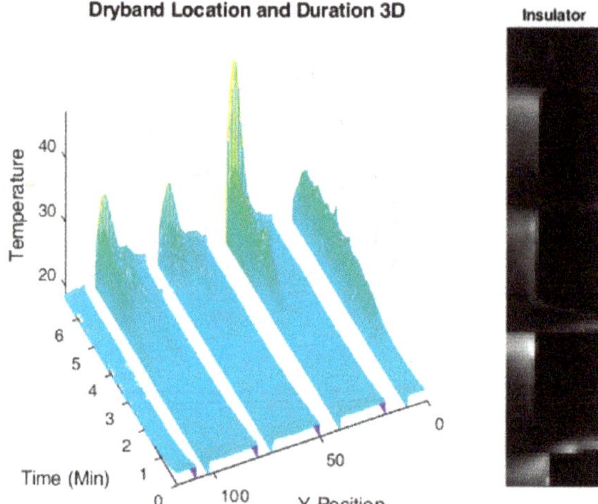

Figure A4. 3D representation of dry-band development on each trunk. Ramp-test: Design CONV, ESDD = 0.42 mg/cm^2, and a fog rate of 3 L/hr.

References

1. Gorur, R.S.; Cherney, E.A.; Hackam, R.; Orbeck, T. The Electrical Performance of Polymeric Insulating Materials Under Accelerated Aging in a Fog Chamber. *IEEE Trans. Power Deliv.* **1988**, *3*, 1157–1163. [CrossRef]
2. Moreno, V.M.; Gorur, R.S. Effect of long-term corona on non-ceramic outdoor insulator housing materials. *IEEE Trans. Dielectr. Electr. Insul.* **2001**, *8*, 117–128. [CrossRef]
3. Blackmore, P.; Birtwhistle, D. Urface discharges on polymeric insulator shed surfaces. *IEEE Trans Dielectr. Electr. Insul.* **1997**, *4*, 210–217. [CrossRef]
4. Moreno, V.M.; Gorur, R.S. Impact of corona on the long-term performance of nonceramic insulators. *IEEE Trans. Dielectr. Electr. Insul.* **2003**, *10*, 80–95. [CrossRef]
5. Meng, D.; Zhang, B.-Y.; Chen, J.; Lee, S.-C.; Jong-Yun Lim, J.-Y. Tracking and erosion properties evaluation of polymeric insulating materials. In Proceedings of the 2016 IEEE International Conference on High Voltage Engineering and Application (ICHVE), Chengdu, China, 19–22 September 2016; pp. 1–4.
6. Vita, V.; Ekonomou, L.; Chatzarakis, G.E. Design of artificial neural network models for the estimation of distribution system voltage insulators' contamination. In Proceedings of the 12th WSEAS International Conference on Mathematical Methods, Computational Techniques and Intelligent Systems (MAMECTIS '10), Kantaoui, Sousse, Tunisia, 3–6 May 2010; pp. 227–231.
7. Pappas, S.S.; Ekonomou, L.; Heraklion, N. Comparison of adaptive techniques for the prediction of the equivalent salt deposit density of medium voltage insulators. *WSEAS Trans. Power Syst.* **2017**, *12*, 220–224.
8. Nekahi, A.; McMeekin, S.G.; Farzaneh, M. Influence of Dry Band Width and Location on Flashover Characteristics of Silicone Rubber Insulators. In Proceedings of the 2016 Electrical Insulation Conference (EIC), Montréal, QC, Canada, 19–22 June 2016.
9. Nekahi, A.; McMeekin, S.G.; Farzaneh, M. Measurement of surface resistance of silicone rubber sheets under polluted and dry band conditions. *Electr. Eng.* **2017**. [CrossRef]
10. Dhahbi-Megriche, N.; Slama, M.E.A.; Beroual, A. Influence of dry bands on polluted insulator performance. In Proceedings of the 2017 International Conference on Engineering & MIS (ICEMIS), Monastir, Tunisia, 8–10 May 2017; pp. 1–4. [CrossRef]
11. Arshad; Mughal, M.; Nekahi, A.; Khan, M.; Umer, F. Influence of Single and Multiple Dry Bands on Critical Flashover Voltage of Silicone Rubber Outdoor Insulators: Simulation and Experimental Study. *Energies* **2018**, *11*, 1335. [CrossRef]
12. Albano, M.; Haddad, A.; Griffiths, H.; Waters, R.T. Dry-band characterisation using visual and IR data analysis. In Proceedings of the International Conference on High Voltage Engineering and Application—ICHVE2014, Poznan, Poland, 8–10 September 2014. [CrossRef]

13. Albano, M.; Waters, R.T.; Charalampidis, P.; Griffiths, H.; Haddad, A. Infrared Analysis of Dry-band Flashover of Silicone Rubber Insulators. *IEEE Trans. Dielectr. Electr. Insul.* **2016**, *23*, 304–310. [CrossRef]

14. Albano, M.; Haddad, A.; Bungay, N. Is the dry-band characteristic a function of pollution and insulator design? In Proceedings of the 2018 IEEE International Conference on High Voltage Engineering and Application (ICHVE), Athens, Greece, 10–13 September 2018; pp. 1–4. [CrossRef]

15. Charalampidis, P.; Albano, M.; Griffiths, H.; Haddad, A.M.; Waters, R.T. Silicone Rubber Insulators for Polluted Environments Part 1: Enhanced Artificial Pollution Tests. *IEEE Trans. Dielectr. Electr. Insul.* **2014**, *21*, 740–748. [CrossRef]

16. Albano, M.; Charalampidis, P.; Griffiths, H.; Haddad, A.M.; Waters, R.T. Silicone Rubber Insulators for Polluted Environments Part 2: Textured Insulators. *IEEE Trans. Dielectr. Electr. Insul.* **2014**, *21*, 749–757. [CrossRef]

Article

Two Dimensional Axisymmetric Simulation Analysis of Vegetation Combustion Particles Movement in Flame Gap under DC Voltage [†]

Ziheng Pu [1], Chenqu Zhou [1], Yuyao Xiong [1], Tian Wu [1,*], Guowei Zhao [2], Baodong Yang [2] and Peng Li [1]

[1]   College of Electrical Engineering and New Energy, China Three Gorges University, Yichang 443002, China; pzhdq@ctgu.edu.cn (Z.P.); zcq26@foxmail.com (C.Z.); xyy_ctgu@foxmail.com (Y.X.); lipeng_ctgu@163.com (P.L.)

[2]   State Grid Shanxi Province Datong Power Supply Company, Datong 443002, China; 13803422759@139.com (G.Z.); sx_toug@foxmail.com (B.Y.)

*   Correspondence: wutian_08@163.com; Tel.: +86-158-7171-4688

†   This paper is an extended version of our paper published in the 2018 IEEE International Conference on High Voltage Engineering and Application (ICHVE), Athens, Greece, 10–13 September 2018.

Received: 8 July 2019; Accepted: 16 September 2019; Published: 20 September 2019

Abstract: In recent years, extreme high temperature weather occurs frequently, which easily causes forest fires. The forest fire is prone to the trip accident of the transmission line. Previous studies show that charged combustion particles cause electric field distortion in the gap below the transmission line, and trigger discharges near the conductor area. The motion and distribution characteristics of combustion particles in the gap have an important influence on the discharge characteristics. Therefore, the size and morphology of combustion particles are analyzed through combustion experiments with typical vegetation. The combustion particles are mainly affected by the air drag force, electric field force and gravity. The interaction and influence of temperature, fluid, electric field and the multi-physical field of particle motion are comprehensively analyzed. A two dimensional (2D) axisymmetric simulation model is established by simplifying the flame region. According to the heat release rate of vegetation flame combustion, the fluid temperature and velocity are calculated. Combined with the fluid field and electric field, the forces on particles and movement are calculated. The results can provide a basis for the analysis of the electric field distortion, and further study the discharge mechanism of the gap under the condition of vegetation flame.

Keywords: combustion particle; electric field distortion; multi physical field; finite element method; particle movement characteristic

---

1. Introduction

In recent years, more and more extreme weather has led to frequent forest fires. China's energy center is in the central and western regions, while the load center is on the eastern coast. Therefore a large amount of electricity has to be transported over a long distance. Due to the limited transmission corridor, more and more extra high voltage/ultra-high voltage (EHV/UHV) transmission lines will inevitably pass through high forest fire risk areas. Tripping accidents of two-dimensional (2D) transmission line caused by forest fires are frequent [1–3]. There are also many reports in other countries, such as the forest fire in California in November 2018, which caused a considerable area of power supply interruption [4–7]. The influence of flame on gap breakdown mainly includes flame temperature, charged particles, ashes and so on [8–10]. High flame temperature reduces air density and promotes thermal ionization.

A large number of charged particles in flame are produced by a combustion reaction, thermal ionization and an electric field. They increase gap conductivity and charge combustion particles. The charged particles will distort the electric field. The trigger discharge is more likely to be produced when charged particles are near the electrode. It will cause the breakdown of the whole gap. These factors lead to transmission line tripping, which seriously affects the safe and stable operation of transmission lines.

Because of the difference in the corridor area, the vegetation of the transmission line is also quite different. The main vegetation that causes mountain fires and the trip of transmission lines are fir, eucalyptus, sugarcane, reed and straw, etc. The combustion of vegetation mainly produces carbon black particles. The particle size of carbon black varies from 0.2 to 5 microns, and it has a certain conductivity. They form granular chains and bridge parts intermittently, so that the gap insulation strength decreases rapidly [11]. In addition to small carbon black particles, vegetation combustion also produces large-scale ash. The maximum size of ash collected in relevant vegetation fire tests can reach 45 mm [8]. Materials with different dielectric constants and conductivities are used to simulate different vegetation particles. The effect of different vegetation particles on gap insulation strength was studied. The experimental results show that the higher the dielectric constant or conductivity, the greater the impact on the breakdown voltage [10]. But the test by Naidoo P. was not carried out in flames, and therefore cannot effectively reflect the impact of combustion particles on breakdown voltage under flame conditions. Typical vegetation in transmission line corridors burning experiments were conducted by some researchers. The direct current (DC) breakdown characteristics of clearance under flame conditions are studied. The effect of particles on the gap electric field distortion is preliminarily simulated and analyzed [12,13]. However, there is a lack of analysis of the motion and distribution of combustion particles under the action of electric field and hot air flow.

Experiments show that the spatial electric field distortion under positive and negative DC voltage is significantly different from that under the influence of particle charging and motion characteristics. The analysis of the distortion characteristics of the electric field plays an important role in studying the breakdown mechanism of the flame gap under DC voltage [14,15]. Distribution of uniform particles in the air gap is often used to analyze the electric field distortion in the existing literature, which is not in accord with the actual situation [16]. It is necessary to analyze the movement characteristics of vegetation combustion particles in a flame gap under DC voltage. Charged particles are affected by the drag force of the heat flow and electric field force. The motion and distribution of charged particles have great influence on the "distortion of space" electric field. In the previous research, the shape and size of different vegetation particles were observed, and the motion path of large particles was recorded by camera in some cases [17]. However, it is difficult to analyze the movement and distribution law of particles under the restriction of experimental conditions. It is necessary to establish a reasonable model for simulation.

Therefore, combined with the characteristics of vegetation combustion particles, temperature field, fluid field, electric field and fluid motion are comprehensively simulated and analyzed in this paper. Positive and negative DC voltages are applied to the electrode respectively, and then combustion experiments are carried out under the electrode using different typical vegetation. Both the size and morphology of typical combustion particles are statistically analyzed. The temperature distribution of the gap under different vegetation flames was measured. According to the measured thermal generation rate of vegetation flame, the temperature distribution and fluid characteristics are obtained by a coupling calculation of fluid and temperature fields. The drag force acting on the particles is further calculated. Considering the drag force, electric field force and gravity, the forces on particles and movement of the particles in the process of rising are calculated.

2. Combustion Experiment of Typical Vegetation

2.1. Experiment Arrangement and Steps

It is necessary to obtain the statistical law of the shape and size of vegetation combustion particles for simulation. The heat release rate and temperature distribution of vegetation combustion also need to be measured. The experimental arrangement of typical vegetation combustion and measurement is shown in Figure 1. The vegetation is placed on a plate electrode. A weight sensor is placed below the plate electrode to measure the rate of vegetation mass loss during combustion. Then the heat release rate can be calculated by combining the burning calorific value of different vegetation. The varied process of flame shape and height change is photographed and recorded, combined with a digital camera and ruler. An infrared thermal imager can be used to photograph the temperature distribution of the whole flame. At the same time, the thermocouple array can record the temperature values at different heights in the flame. Rod electrodes are suspended directly above the vegetation and applied with DC voltage.

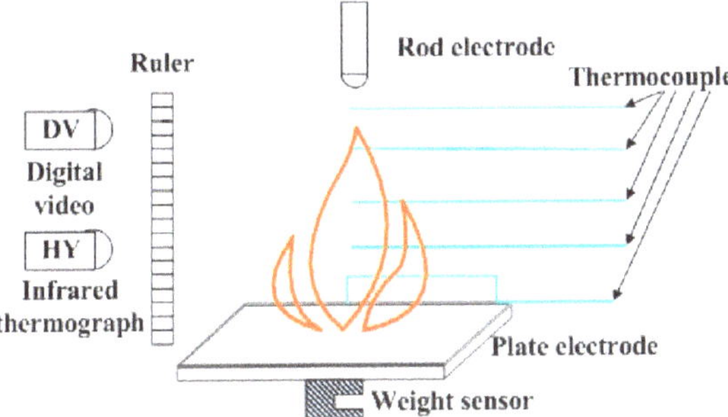

Figure 1. Experimental arrangement of typical vegetation combustion and measurement.

The arrangement of the wooden crib is simple and repeatable. Most of the previous studies for breakdown characteristics of clearance under flame conditions use this wooden crib. However, due to the difference of component content, the combustion effect of branches and leaves is quite different from that of the wooden crib. It shows that the breakdown voltage of clearance under the branch and leaf combustion is significantly lower than that of a wooden stack. Therefore, reed, straw and fir branches were selected as three typical plants to cause the trip of transmission lines due to a vegetation fire. At the same time, the wooden crib test was carried out as a comparison. As shown in Figure 2, the vegetation or wooden crib is arranged in a square stack of 21 cm × 21 cm × 10 cm. In order to ensure the same burning intensity each time, the typical vegetation was assigned with the same quality in repeated experiments. Typical vegetation is exposed to the sun for several days before the experiment to ensure dryness, as in nature. The effect of different moisture content on combustion is avoided.

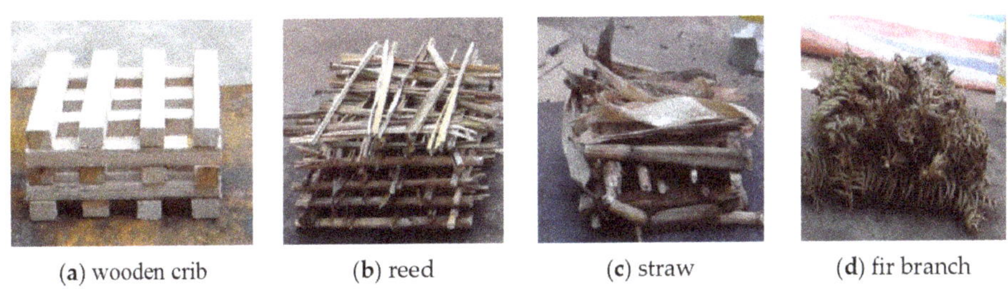

| (a) wooden crib | (b) reed | (c) straw | (d) fir branch |

Figure 2. Arrangement of typical vegetation.

The test steps are as follows: (1) Arrange the equipment according to Figure 1; (2) take the same weight of vegetation and arrange it into square stacks; (3) apply the specified DC voltage on the electrode; (4) ignite the vegetation and observe the temperature and shape of vegetation flame by digital camera and infrared camera; (5) measure and record the temperature of the thermocouple and measure the mass loss rate by weight sensor; (6) observe the triggered discharge of particles; (7) the size, shape and weight of particles are collected and measured after combustion. The experiment is carried out indoors to avoid the influence of wind, under dry conditions, and with typical values of T = 20~30 °C, RH = 50~75%. Tests for different vegetation are repeated three times.

2.2. Test Results: Temperature Distribution and Particle Size

Temperature variations at different heights on the axis of each planting flame are obtained by temperature sensors. The burning time of the wooden crib is the longest, being about 14 min. After ignition, the temperature of each thermocouple is recorded every minute. The burning time of reed, straw and fir branches is relatively short, so the thermocouple readings are recorded every 10 s after ignition. For instance, comparison of the flame temperature distribution between the wooden crib and straw is shown in Figure 3. With the increase of flame, the temperature increases gradually. When the maximum fire is reached, the combustion will remain stable for a period of time, and the temperature will reach the maximum. The maximum flame temperature of wood is 747 °C, while that of straw is 518 °C. The maximum temperatures of reed and fir branches are 306 °C and 273 °C, respectively. According to vegetation combustion, the gap can be divided into a continuous area, an oscillating area and a smoky area. The continuous area is about 0~30 cm above the vegetation. The highest temperature is found in the continuous flame area about 15 cm above the wood. The height of the oscillating area is about 30~55 cm. In the oscillating area, the flame will oscillate up and down, and the flame does not always exist. The smoky area is filled with particles and smoke generated by combustion, and the temperature is low.

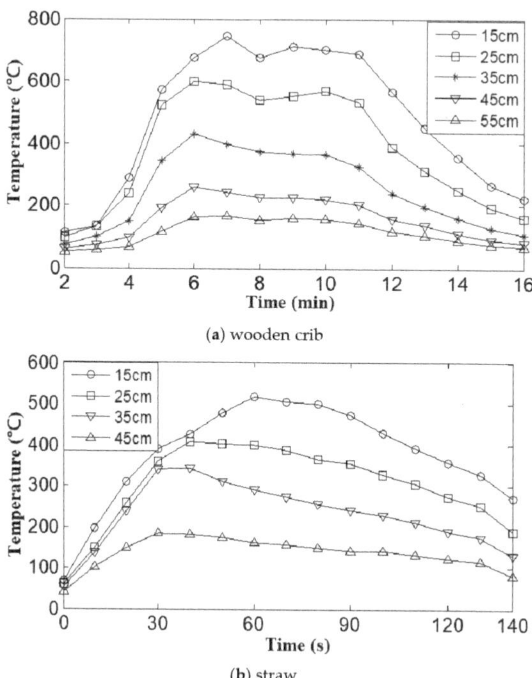

(a) wooden crib

(b) straw

Figure 3. Comparison of flame temperature distribution between the wooden crib and straw.

According to the experimental data, the mass loss rate is calculated. The maximum mass loss rates of wooden crib, reed, straw and fir branches are 2 g/s, 1.84 g/s, 1.34 g/s and 1.24 g/s, respectively, and it can be maintained at maximum flame for a period of time. The average heat release rates of the wooden crib, reed, straw and fir branches were 21.45 kJ/s, 16.89 kJ/s, 15.93 kJ/s and 15.19 kJ/s, respectively. According to the mass loss of the whole combustion process, the total heat generated by the four kinds of burners are 4.896×10^3 kJ, 2.197×10^3 kJ, 2.814×10^3 kJ and 2.066×10^3 kJ, respectively. The measurement results show that the heat generated is in good agreement with the temperature.

The stable combustion flames of the four combustible materials are shown in Figure 4. The smoke concentration produced during combustion is small. Because the burning material has been exposed to the sun for a week, the water content is low, and the burning is sufficient. The combustion products of vegetation in the experiment mainly consist of carbon black particles of micron magnitude and ashes of large size. A differential motion particle size meter (DMPS) was used to analyze the size of these carbon black particles. The particle size of carbon black is about 0.02–0.05 μm. These carbon black particles will aggregate into spherical or chain-like particles with a size of about 10–300 μm. For instance, the size of aggregated carbon black particles produced by the wooden crib combustion is from 0.01 μm to more than 43 μm. The two peaks of the contents of different sizes are 0.3 μm and 43 μm, respectively. These carbon black particles are charged, and may be adsorbed onto large size ashes. They further distort the space electric field and lead to trigger discharge.

(**a**) wooden crib (**b**) reed (**c**) straw (**d**) fir branch

Figure 4. The stable combustion flames of four combustible materials.

Large size ashes produced by the insufficient burning of vegetation have a greater impact on the gap discharge. Typical large-scale ashes from the burning of four vegetations are shown in Figure 5. The longest ash of reed can reach 12 cm, with its diameter up to 4mm. But the weight of reed ash is very light, about 0.2 g. The largest ash of the fir branch is about 10 cm in length, 2 mm in diameter and 0.7 g in weight. The largest ash of straw is about 8 cm in length, 6 mm in diameter and 0.5 g in weight. The maximum ash length and weight of wooden crib burning are 7 cm and 8 g, respectively. The measured data show that the density of ash in wooden stacks is high. The large size particles are heavy, and will not rise with the hot air flow, and only the smaller size of ash will rise. The ash density of other vegetation combustion is smaller, and the ash of larger length can also rise. Reed ashes have the largest size and the lightest weight. Under the combined influence of the electric field and thermal convection, they float into the gap, bridge the gap, and trigger the discharge.

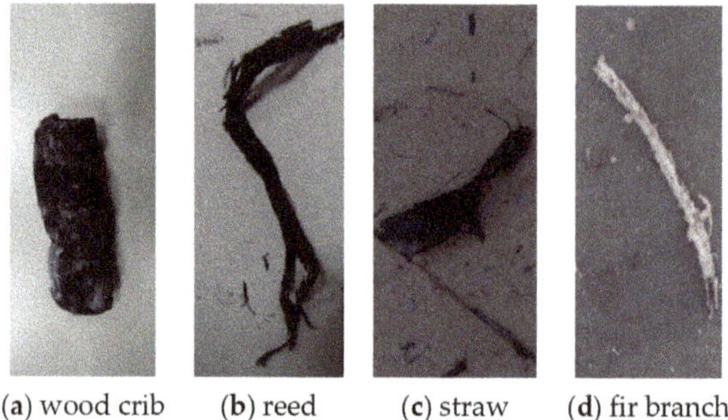

Figure 5. Typical ashes of four kinds of combustions.

3. Simulation Mechanism and Model Establishment

3.1. Mathematical Model of Multiple Physical Fields for the Combustion Particles

3.1.1. Charging Mechanism of Combustion Particles

The generation of charged particles in a vegetation fire mainly includes chemical reaction, collision ionization, thermal dissociation, etc. The main component of vegetation is cellulose, which is similar to hydrocarbons. The process of generating electrons and ions in the reaction of hydrocarbons is shown in Formula (1) [9]. During combustion, alkali metal and alkaline-earth metal salts can react with the combustion byproduct CO to produce electrons and CO_2, such as in Formula (2). Flame is a kind of plasma (ionized gas). The ion concentration in the reaction zone of mixed hydrocarbon and air flame is 10^9–10^{12}/cm^3, and the typical ion concentration of flame is 10^{10}/cm^3.

$$CH + O \rightarrow CHO^+ + e \tag{1}$$

$$CaO + CO \rightarrow Ca^+ + e + CO_2 \tag{2}$$

Vegetation combustion produces high temperatures. If the flame contains alkali metal salts (S(g)), ionization mainly comes from the thermal ionization of metals. Although the content is only in a ppm order of magnitude, it can produce a large number of electrons. For instance, the dry weight of vegetation contains up to 3.4% potassium salt. The ionization energy of potassium salt is relatively low, such as K_2CO_3 is 3.79 eV. When the burning efficiency of vegetation is 98%, 28% of potassium salt is dissolved. Thermal ionization of alkali metals can be expressed by Formula (3).

$$S(g) \rightarrow S(g)^+ + e \tag{3}$$

The high temperature of flame decreases the gas density and increases the average free travel of electrons. The ionization energy of the gas molecule does not change with the increase of temperature. Therefore, collision ionization is more likely to occur under flame conditions. Vegetation flames are filled with these positive and negative ions. Some small particles produced by combustion will continuously adsorb free ions or collide with charged particles when they float upward. Finally, small charged particles of the chain or agglomeration type are formed. The particles can reach the saturated charge in the flame. The saturated charge q_s of the particle can be calculated by Formula (4) [18].

$$q_s = 3\frac{\varepsilon_r}{\varepsilon_r + 2}\pi\varepsilon_0 d_p^2 E \tag{4}$$

where ε_r is relative permittivity, ε_0 is vacuum permittivity, d_p is particle diameter, and E is electric field strength.

3.1.2. Motion Control Equation of Combustion Particles

The fluid motion produced by a high temperature vegetation flame is typical turbulence. Compared with the standard model, the κ-ε model based on the renormalization group (RNG) takes turbulent vortices into account. The RNG κ-ε model is selected to describe the fluid field in this paper. The control equation is shown in Formulas (5) and (6).

$$\frac{\partial(\rho\kappa)}{\partial t} + \frac{\partial(\rho\kappa u_i)}{\partial x_i} = \frac{\partial}{\partial x_j}\left[\left(\mu + \frac{\mu_t}{\sigma_\kappa}\right)\frac{\partial\kappa}{\partial x_j}\right] + \mu_t\frac{\partial u_i}{\partial x_j}\left(\frac{\partial u_j}{\partial x_i} + \frac{\partial u_i}{\partial x_j}\right) \tag{5}$$

$$\frac{\partial(\rho\varepsilon)}{\partial t} + \frac{\partial(\rho\varepsilon u_i)}{\partial x_i} = \frac{\partial}{\partial x_j}\left[\left(\mu + \frac{\mu_t}{\sigma_\varepsilon}\right)\frac{\partial\varepsilon}{\partial x_j}\right] + C_{1\varepsilon}\frac{\varepsilon}{\kappa}\mu_t\frac{\partial u_i}{\partial x_j}\left(\frac{\partial u_j}{\partial x_i} + \frac{\partial u_i}{\partial x_j}\right) - C_{2\varepsilon}\rho\frac{\varepsilon^2}{\kappa} - R_\varepsilon \tag{6}$$

where κ is turbulent kinetic energy, ε is turbulent dissipation rate, σ_κ and σ_ε are the turbulent Plante numbers of κ and ε, respectively, and $C_{1\varepsilon}$ and $C_{2\varepsilon}$ are the computational constants for the models.

Due to the low concentration of particles, gas-solid flow in the dispersed phase can be used for analysis. The forces acting on moving particles in fluids usually include traction, gravity, buoyancy, saffman lift, thermophoresis, pressure gradient force, virtual mass force, etc. When the number of particles is small and the gas density is far less than the particle density, the forces acting on the particles are mainly the heading force, electric field force and the acceleration due to gravity. The other forces are several orders of magnitude smaller than the main forces. They can be neglected in simplifying calculation. The equation of motion of particles can be described as Formulas (7)–(9).

$$m_p d\vec{u}_p/dt = \vec{F}_D + \vec{F}_E + \vec{F}_g \tag{7}$$

$$\vec{F}_D = m_p\left(\vec{u} - \vec{v}\right)/\tau_p \\ \tau_p = \rho_p d_p^2/18\mu \tag{8}$$

$$\vec{F}_E = q\vec{E}/m_p \tag{9}$$

where m_p is the mass of particles, u_p is the velocity of particles, F_D, F_E and F_g are traction, electric field force and gravity, respectively, v is the velocity of fluid, τ_p is the influence factor, ρ_p is the particle density, d_p is the particle diameter, and finally μ is the dynamic viscosity.

3.2. Coupled Model of Multi-Physical Field

As shown in Figure 6, the test area can be divided into three areas: The vegetation area, the equivalent flame area and the smoky area. The total heat release rate is set in both the vegetation area and equivalent combustion area, respectively, in a certain proportion. The proper proportion is selected to make the temperature distribution in good agreement with the experiment. The particles are set to release from the surface of the vegetation areas. The equivalent flame area contains a large number of plasmas. Charged particles are subject to an electric field force and thermal airflow drag force in the gap, and float up above the flame to form the smoky area. Motion analysis of combustion particles involves the interaction of a temperature field, fluid field and electric field. It is complex and difficult to calculate a multi-physical field using a three dimensional (3D) model, especially involving the movement of fine particles. A 2D axisymmetric model is adopted for simplifying this numerical simulation. It can be considered that there is an approximate axisymmetric relationship between the flame center and its surroundings, and the electrostatic field produced by the electrodes has an approximate axisymmetric relationship. In this paper, the flame region is simplified as an axisymmetric model. Because the combustion material is stacked into a square shape, a simplified equivalent treatment is carried out. In order to ensure the same volume, the size of the vegetation area is set to 11.8 cm in radius and 10 cm in height. The flame area is set as a circular platform, and part of the heating power can be applied.

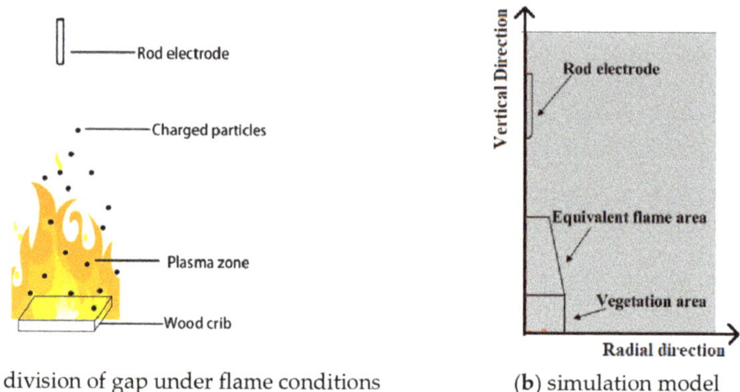

(**a**) Regional division of gap under flame conditions   (**b**) simulation model

Figure 6. Equivalent Model of the Vegetation Flame Gap under Direct Current (DC) Voltage.

The size of the flame area is set according to the experimental data of different vegetation. The radius of the rod electrode is 1.5 cm, and the length is 15 cm. The boundaries of the whole model are set to open boundaries.

Gas-solid two-phase motion is complex when the vegetation flame burns under DC voltage. If the physical field is considered to be closely related, the mathematical model considering all of the coupling terms will inevitably lead to extremely complex equations which are very difficult to solve. This paper mainly studies the movement of a small number of particles with different sizes. The simulation model is simplified as follows: (1) Considering the small volume ratio of particulate matter, the influence of particulate matter on the fluid motion is neglected; (2) the influence of particles on the background electric field is not considered.

The flow chart of the multi-physical field coupling analysis is shown in Figure 7. Firstly, the temperature field and fluid field are strongly coupled and simulated according to the heat release rate obtained from the experiment. The simulation model is improved by comparing it with the temperature distribution. Then, according to the statistical data of combustion particle size, the charge of particles is set. The maximum charge of particles does not exceed the saturated charge. Finally, the coupling simulation of particle motion, fluid field and electric field is carried out. The forces on particles and movement of the particles are simulated and analyzed.

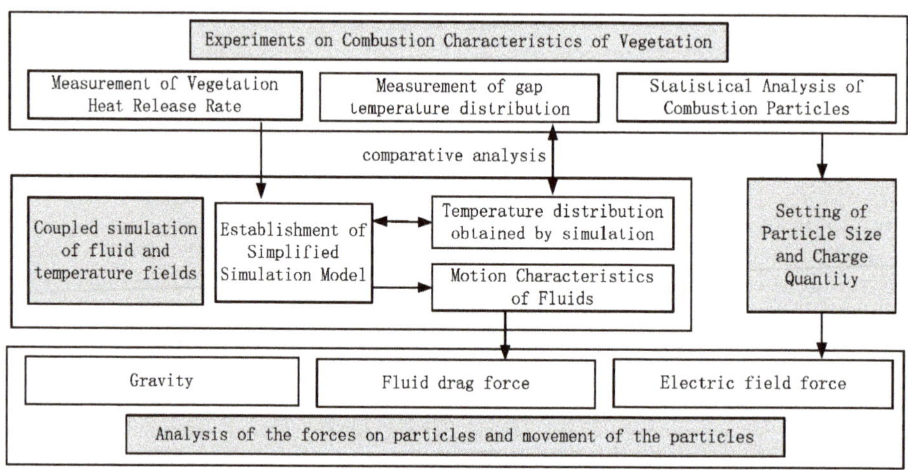

Figure 7. Flow chart of the multi-physical field coupling simulation.

4. Analysis of Simulation Results

4.1. Simulation Results of Temperature and Fluid Field

For instance, when straw burns steadily, the heat release rate is set at 15.93 kJ/s. The heat release rate is applied to the vegetation area and the equivalent flame area, correspondingly. By adjusting the equivalent flame area, the temperature distribution is closer to the experimental situation. The flame will oscillate in the process of steady combustion. The simulation results are shown in Figure 8.

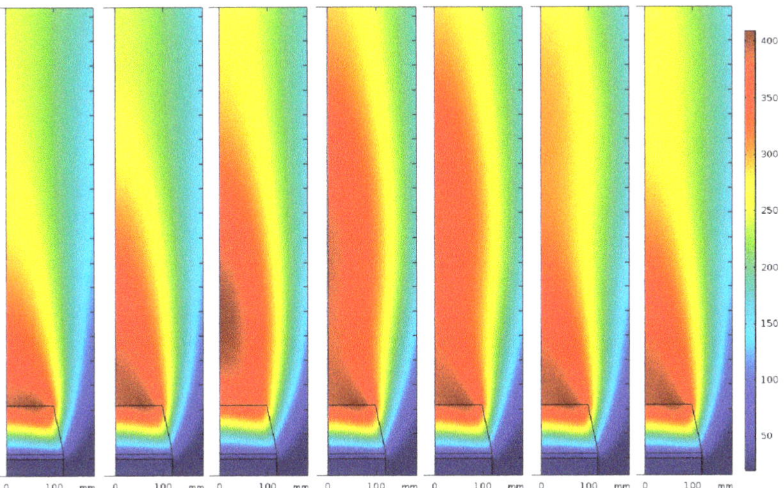

Figure 8. Coupled simulation results of temperature and fluid field in straw combustion.

The maximum temperature is about 500 °C and is close to the test results. The temperature near the stacking area is lower than that in the exterior flame area. The temperature decreases gradually with the increase in height from the outer flame area upward. These laws are in agreement with the experimental results. The average temperature of a period on the central axis is compared with the experimental measurements, as shown in Table 1. The deviation is about 10%. There are still some differences in the overall temperature distribution. The main reason is the uniform distribution of heat release rate in the flame zone, which is different from the actual situation. The initial velocity of flame fluid at the wood stack ranges from 0.25 m/s to 0.1 m/s. The velocity of fluid from the inside to the outside of the flame decreases gradually.

Table 1. Comparison of temperature data between experiment and simulation.

Height (cm)	Experimental Data (°C)	Simulation Data (°C)	Deviation (%)
15	465	421	9.03
25	387	365	5.68
35	306	318	3.92

4.2. Analysis of Force and Motion Characteristics of Combustion Particles

50 kV DC voltage with positive and negative polarities is applied to the electrodes respectively. Particles with different sizes and charges are set up to simulate and analyze. The gravity action is considered. The electrode is set to positive polarity. The combustion particles are set up with different mass, charge polarity and quantity. Firstly, five initial points are selected along the edge of the vegetation, and the overall trajectory of the particles is shown in Figure 9. When the polarity of the

particles is the same as that of the electrode, the particles are mainly affected by the drag force of the fluid in the initial drift stage.

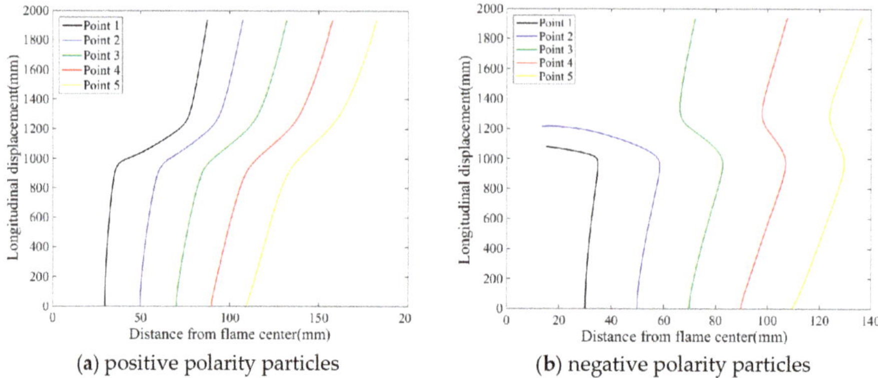

(a) positive polarity particles   (b) negative polarity particles

Figure 9. The overall trajectory of the combustion particles.

Particles rise with hot air flow, and diffuse slowly outward at the same time. When the particles approach the height near the electrode, the electric field increases and is just in the vertical direction. The particles move outward rapidly under the action of electric field force. The closer the electrode is, the stronger the electric field force is. When the polarity of the particles is opposite and approaching the height near the electrodes, the particles are attracted by the electric field and move towards the electrode. For a certain charge-mass ratio, there is a critical distance from the flame center. The particles will be adsorbed by the electrode when the distance between particle and electrode is less than the critical distance. Beyond the critical distance, the particles will re-move away from the electrode along with the hot air flow.

Then the force acting on the particles in the whole process is analyzed. Take a charged particle as an example, the variations of air drag force, electric field force and gravity are shown in Figure 10. Particles begin to rise from the initial position at a certain initial velocity. When the polarity of the particles is the same as that of the electrodes, the particle rises under the action of drag force to overcome the influence of gravity and of the electric field. As the electric field force increases near the electrode, the drag force increases rapidly. When away from the electrode, the drag force and electric field force in the x-axis direction decrease gradually. When particles move from below the electrode to above it, the electric field force in the y-axis direction will be reversed, and the corresponding air drag force direction will also be reversed, and then gradually reduce. When the polarity of particles is opposite to that of electrodes, there are two situations: One is that the particles are adsorbed by the electrodes, the other is not adsorbed. In initial period, particles are subjected to electric field force and drag force to overcome gravity and rise. When the particle rises near the electrode height, the x-axis electric field force increases rapidly, which makes the particle move toward the electrode. The Y-axis electric field force increases slowly and then decreases as it approaches the electrode. If the particles are adsorbed by the electrodes, the y-axis electric field force is approximately 0, and the x-axis electric field force reaches the maximum. If the particles are not adsorbed by the electrodes, the particles are first close to the electrodes and then far away. The Y-axis electric field force and air drag force will be reversed, and because it is closer to the electrode, the reverse peak will be larger, and then gradually decrease.

As shown in Figure 11a, when the particle size is the same, the larger the charge-mass ratio, the greater the longitudinal drag force on the particle. When the particle is near the center of the flame, the drag force on the particle will increase gradually. As shown in Figure 11b, when the charge of particles is constant, the larger the particle size, the larger the longitudinal drag force. With the increase in size, the drag force of fluid mainly balances gravity, and the electric field force is relatively small.

When the particle size is large, the distance between the particle and the center has little effect on the drag force.

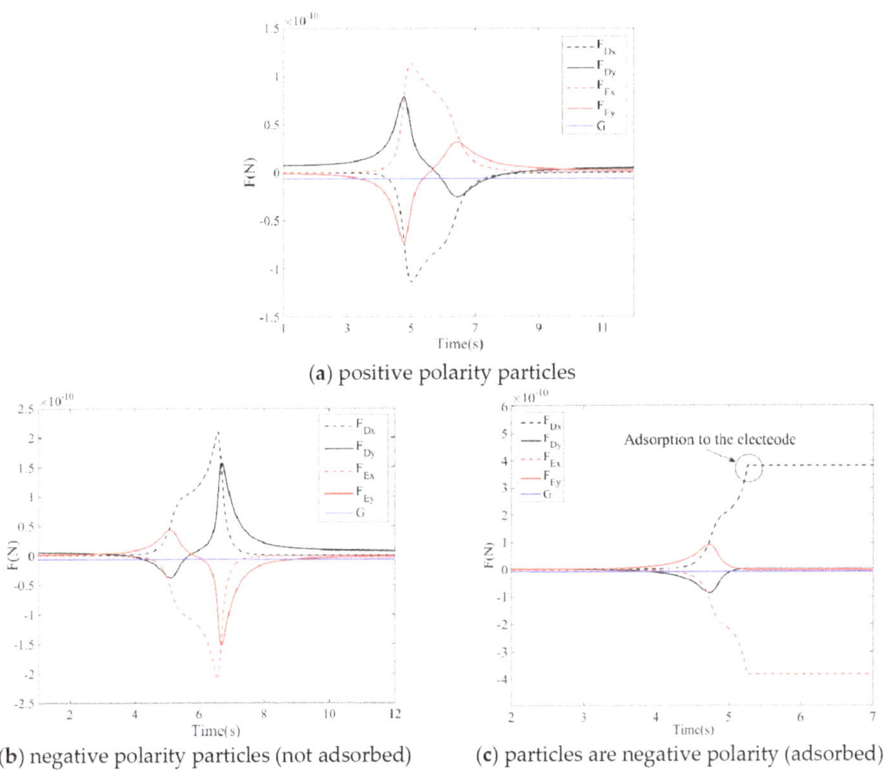

(**a**) positive polarity particles

(**b**) negative polarity particles (not adsorbed)   (**c**) particles are negative polarity (adsorbed)

Figure 10. Force analysis of combustion particles.

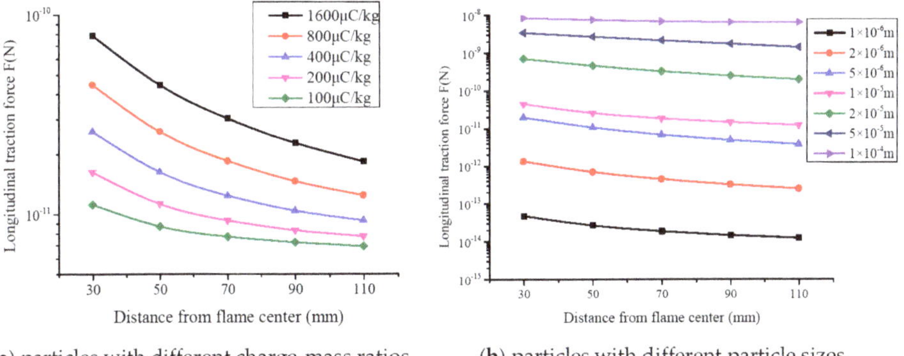

(**a**) particles with different charge-mass ratios   (**b**) particles with different particle sizes

Figure 11. Force analysis of particles under the same polarity.

As shown in Figure 12a, the particles are easily adsorbed by electrodes when the charge of particles is large. Although the drag force decreases with the distance from the electrode, the electric field force decreases more. The farther away from the electrode, the harder the charge is to be absorbed. As shown in Figure 12b, the farther away from the electrode, the larger the critical charge-mass ratio required for particles to be adsorbed by the electrode, and the critical charge-mass ratio decreases with the increase of particle size. Under the same charge-mass ratio, larger particles are easier to be adsorbed by the electrode.

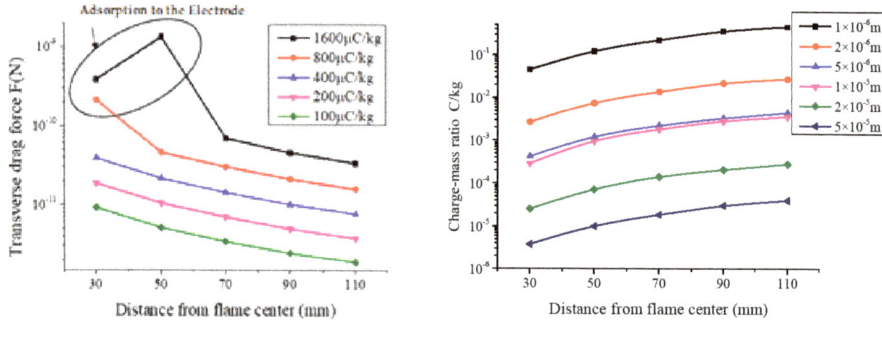

(**a**) particles with different charge-mass ratios (**b**) critical charge-mass ratio (adsorbed)

Figure 12. Force analysis of particles under different polarities.

5. Conclusions

In this paper, combustion experiments under DC voltage are carried out using typical vegetation. The heat release rate and spatial temperature distribution of vegetation combustion were measured. The shape and size of combustion particles were collected and analyzed. Through the coupled simulation of the temperature field, fluid field, electric field and particle motion, the forces on particles and the movement of the particles are analyzed. Combining with the actual vegetation combustion heat release rate, the simulation analysis is carried out. The model is modified by comparing with the actual temperature distribution. The simulation results are in good agreement with the experimental results. The situation of ash and carbon black particles in flame was obtained by experiments. In this paper, the motion of carbon black particles with different sizes is mainly analyzed. The force acting on the combustion particles is constantly changing during the rising process. When the charge of particles is the same as the polarity of the electrodes, the electric field force is always opposite to the drag force. When the charge of the particle is opposite to the polarity of the electrode, the particle will approach the electrode. The larger the charge-mass ratio, the easier it is to be captured and adsorbed by the electrodes, and the more charged are the particles farther away from the electrodes. Under the same charge-mass ratio, the larger the particle size, the easier it is to be captured by the electrode. The number of particles with a positive polar charge in flame is larger than that with a negative polar charge. The distribution characteristics of particles are different under positive and negative polarity DC voltage. The influence of charge statistical distribution on the space electric field will be considered for further analysis.

Author Contributions: Z.P. and T.W. propose the main idea of the paper. G.Z. and B.Y. conceived and designed the experiments. P.L. and C.Z. performed the experiments. Z.P. and C.Z. analyzed the data. The paper is written by Z.P. and is revised by T.W. and Y.X.

Funding: This work has been supported by the Natural Natural Science Foundation of China project (51607103).

Conflicts of Interest: The authors declare no conflict of interest.

References

1. Hu, Y.; Liu, K.; Wu, T.; Liu, Y. Analysis of influential factors on operation safety of transmission line and countermeasures. *High Volt. Eng.* **2014**, *40*, 3491–3499.
2. Zhou, Z.Y.; Ai, X.; Lu, J.Z. A real-time analysis approach and its application for transmission-line trip risk due to wildfire disaste. *Proc. CSEE* **2017**, *37*, 5321–5330.

3.   Huang, D.; Shu, Y.; Ruan, J.; Hu, Y. Ultra-high voltage transmission in China: Developments, current status and future prospects. *Proc. IEEE* **2009**, *97*, 555–583. [CrossRef]

4.   Maabong, K.E.; Mphale, K.; Letsholathebe, D.; Chimidza, S. Measurement of Breakdown Electric Field Strength for Vegetation and Hydrocarbon Flames. *J. Electromagn. Anal. Appl.* **2018**, *3*, 53–66. [CrossRef]

5.   Sukhnandan, A.; Hoch, D.A. Fire induced flashovers of transmission lines: Theoretical model. In Proceedings of the IEEE Africon 6th Africon Conference in Africa, George, South Africa, 2–4 October 2002.

6.   Wu, T.; Ruan, J.; Hu, Y.; Liu, B.; Chen, C. Study on forest fire induced breakdown characteristics and mechanism of 500kV transmission line. *Proc. CSEE* **2011**, *31*, 163–170.

7.   Peng, L.; Jiangjun, R.; Daochun, H. Study on Breakdown Characteristic and Discharge Model of Conductor-plane Gap Under Typical Vegetation Flame. *Proc. CSEE* **2016**, *36*, 4001–4010.

8.   Robledo-Martinez, A.; Guzman, E.; Hernandez, J.L. Dielectric characteristics of a model transmission line in the presence of fire. *IEEE Trans. Electr. Insul.* **1991**, *26*, 776–782. [CrossRef]

9.   Mphale, K.M.; Heron, M.; Ketlhwaafetse, R.; Letsholathebe, D.; Casey, R. Interferometric measurement of ionization in a grassfire. *Meteorol. Atmos. Phys.* **2010**, *106*, 191–203. [CrossRef]

10.   Naidoo, P.; Swift, D.A. Large particle initiated breakdown of an atmospheric air gap: Relating to AC power line faults caused by sugar cane fires. In Proceedings of the 8 th International Symposium on High Voltage Engineering, Yokohama, Japan, 23–27 August 1993.

11.   Fonseca, J.R.; Tan, A.L.; Silva, R.P.; Monassi, V.; Assuncao, L.A.R.; Junqueira, W.S.; Melo, M.O.C. Effects of agricultural fires on the performance of overhead transmission lines. *IEEE Trans. Power Deliv.* **1990**, *5*, 687–694. [CrossRef]

12.   Daochun, H.; Peng, L.; Jiangjun, R.; Yafei, Z.; Tian, W.U. Review on discharge mechanism and breakdown characteristics of transmission line gap under forest fire condition. *High Volt. Eng.* **2015**, *41*, 622–632.

13.   Hexun, X.I.; Guangfu, T.A.N.G.; Junzheng, C.A.O.; Jie, L.I.U.; Xiaoguang, W.E.I. Research porgress of electromagnetic field and electromagnetic compatibility of UHVDC converter valves. *Proc. CSEE* **2012**, *32*, 1–6. (In Chinese)

14.   Hu, Y. Analysis on operation faults of transmission line and countermeasure. *High Volt. Eng.* **2007**, *33*, 1–8. (In Chinese)

15.   Wu, T.; Hu, Y.; Ruan, J.J.; Liu, K.; Liu, T.; Chen, C. Air gap breakdown mechanism of model AC transmission line under forest fire. *High Volt. Technol.* **2011**, *37*, 1115–1122. (In Chinese)

16.   Hu, X.; Lu, J.Z.; Zeng, X.J.; Zhang, H.X. Ayalysis on transmission line trip caused by mountion fire and discussion on tripping preventing measure. *J. Electr. Power Sci. Technol.* **2010**, *25*, 73–78. (In Chinese)

17.   You, F.; Chen, H.; Zhang, L.; Zhang, Y.; Zhou, J.; Zhu, J. Experimental study on flashover of high-voltage transmission lines induced by wooden crib fire. *Proc. CSEE* **2011**, *31*, 192–197. (In Chinese)

18.   Pu, Z.; Xiong, Y.; Wu, T.; Lu, Z.; Fang, C. Simulation Analysis on Influence of Combustion Particles on the Gap Electric Field under DC Voltage. In Proceedings of the 2018 IEEE International Conference on High Voltage Engineering and Application, Athens, Greece, 10–13 September 2018.

Article

Computation of Transient Profiles along Nonuniform Transmission Lines Including Time-Varying and Nonlinear Elements Using the Numerical Laplace Transform [†]

Rodrigo Nuricumbo-Guillén [1,*], Fermín P. Espino Cortés [1], Pablo Gómez [2] and Carlos Tejada Martínez [1]

[1]   Departamento de Ingeniería Eléctrica SEPI-ESIME ZAC, Instituto Politécnico Nacional, Mexico City 07738, Mexico

[2]   Electrical and Computer Engineering Department, Western Michigan University, Kalamazoo, MI 49008, USA

*   Correspondence: rodrigo.ng.85@gmail.com

†   This paper is an extended version of our paper presented in the 2018 IEEE International Conference on High Voltage Engineering and Application (ICHVE), Athens, Greece, 10–13 September 2018.

Received: 30 June 2019; Accepted: 19 August 2019; Published: 21 August 2019

Abstract: Electromagnetic transients are responsible for overvoltages and overcurrents that can have a negative impact on the insulating elements of the electrical transmission system. In order to reduce the damage caused by these phenomena, it is essential to accurately simulate the effect of transients along transmission lines. Nonuniformities of transmission line parameters can affect the magnitude of voltage transients, thus it is important to include such nonuniformities correctly. In this paper, a frequency domain method to compute transient voltage and current profiles along nonuniform multiconductor transmission lines is described, including the effect of time-varying and nonlinear elements. The model described here utilizes the cascade connection of chain matrices in order to take into consideration the nonuniformities along the line. This technique incorporates the change of parameters along the line by subdividing the transmission line into several line segments, where each one can have different electrical parameters. The proposed method can include the effect of time-dependent elements by means of the principle of superposition. The numerical Laplace transform is applied to the frequency-domain solution in order to transform it to the corresponding time-domain response. The results obtained with the proposed method were validated by means of comparisons with results computed with ATP (Alternative Transients Program) simulations, presenting a high level of agreement.

Keywords: electromagnetic transients; nonuniform transmission line; numerical Laplace transform; time-dependent elements; transmission line modeling

---

1. Introduction

Electromagnetic transients can produce overvoltages and overcurrents that can have a negative impact on electric power systems. In order to reduce the potential deterioration or damage due to this condition, accurate transient simulations are needed [1]. Typically, the transient analysis of electrical systems is performed by means of two-port models of uniform transmission lines, thus the voltage/current measurements are available at certain nodes of the network. However, the maximum transient overvoltages and overcurrents may appear at interior points of the transmission line [2]. In such cases, the traditional simulation methods may not be well suited to correctly analyze this kind of phenomena. Additionally, nonuniformities can be present along the transmission line in the form of parameter variations such as the height of the line or the properties of the terrain; if these nonuniformities are too prominent, the distribution and magnitude of the voltage and currents can drastically be affected along the transmission line in comparison with uniform transmission lines (lines with space-independent per-unit-length parameters) [3,4].

There has been a considerable interest in developing methods to accurately model nonuniform transmission lines during electromagnetic transients. Previous works have presented line models applying different techniques such as the numerical Laplace transform, the method of characteristics, rational approximations, among others, with good results [5–12]. However, in general, these methods are only able to provide voltage and current information at the ends of the line, which in some cases may not be enough to correctly analyze the transient behavior of a transmission line [2], such as for insulation design or protection purposes.

A frequency domain method for the computation of transient voltage and current profiles along transmission lines was reported in [13,14], and later extended to include time-varying and nonlinear elements [15]. However, this method cannot be applied to nonuniform transmission lines. In [16], Laplace-domain and time-domain methods were tested in the computation of transient profiles on a nonuniform electronic system. It was found that the line model that used the inverse numerical Laplace transform (INLT) provided the most accurate results. However, the method described in [16] can only be applied to time-invariant linear systems.

Expanding upon the aforementioned publications, the main contribution of the present paper is the complete description and verification of a method for the computation of transient voltage and current profiles along nonuniform transmission lines. This method utilizes a modeling approach defined in the frequency domain and based on the cascaded connection of chain matrices. Furthermore, using the superposition technique, the proposed method can include time-dependent and nonlinear elements (such as switching devices and surge arresters) in the computation of transient profiles along nonuniform lines, something that has not been conducted in any previously reported research. Since the line model is defined in the frequency domain, it can take into account the frequency dependence of electrical parameters in a straightforward manner, providing more accurate results in comparison with existing methods defined in the time domain.

The method presented here makes use of the inverse numerical Laplace transform [17,18] to convert the computed frequency domain solution to a time-domain transient response. This method has strong potential for application in fault location and insulation coordination, with particular accuracy benefits for lines with prominent nonuniformities, such as river crossings, hilly terrains, and other substantial sagging conditions, which are commonly encountered in large countries such as China, Canada, India, Russia, and Brazil. For example, very challenging river crossings are found in Brazil for overhead lines constructed to connect the power generation in the Amazon Basin to the main load centers of the country. These river crossings are in the order of 2 km leading to very tall towers and wide line spans [19]. Accurate fault location under these circumstances requires an appropriate consideration of wave propagation along nonuniform lines, which can be achieved with the method described here. Additionally, the proposed method can be expanded to the modeling of other power system nonuniform elements, such as transmission towers [20] and rotating machines [21].

In order to validate the accuracy of the proposed method, the results obtained are compared with those obtained from simulations performed with ATP (Alternative Transients Program) [22]. In the ATP simulations, the J. Marti line model [23] was used, and the transmission line was subdivided into several line segments to allow the connection of measuring probes at internal points of the line, as well the inclusion of nonuniformities.

The computation of transient profiles along nonuniform transmission lines including nonlinear and time-varying conditions has not been previously reported, providing an original contribution to the current state of the art of the topic.

2. Nonuniform Transmission Line Model for the Transients Profiles Computation

This section describes the transmission line model used to compute the transient profiles as well as the technique used to include time-varying and nonlinear elements. Additionally, a brief explanation for the implementation of the INLT algorithm is presented.

2.1. Nonuniform Transmission Line Model

This work introduces the nonuniformities along transmission lines by means of the technique of cascade connection of chain matrices, as it has been previously shown to be an effective technique for the simulation of electromagnetic transients when nonuniform transmission lines are considered [11,24].

Initially, the uniform transmission line of Figure 1 is considered. This line can be represented by a two-port model known as transfer or **ABCD** matrix model:

$$\begin{bmatrix} V_L(s) \\ I_L(s) \end{bmatrix} = \begin{bmatrix} A & B \\ C & D \end{bmatrix} \begin{bmatrix} V_0(s) \\ I_0(s) \end{bmatrix} \tag{1}$$

where

$$A = \cosh(\gamma L)$$
$$B = -\mathbf{Z}_0 \sinh(\gamma L)$$
$$C = \mathbf{Y}_0 \sinh(\gamma L) \tag{2}$$
$$D = -\cosh(\gamma L)$$
$$\gamma = \sqrt{\mathbf{ZY}}$$

In Equation (2) $\mathbf{Z}_0$, $\mathbf{Y}_0$, $\mathbf{Z}$, $\mathbf{Y}$, and L are the line's characteristic impedance, characteristic admittance, series impedance, shunt admittance and length, respectively. Additionally, the Laplace variable is defined as $s = c + j\omega$, where ω is given by $2\pi f$.

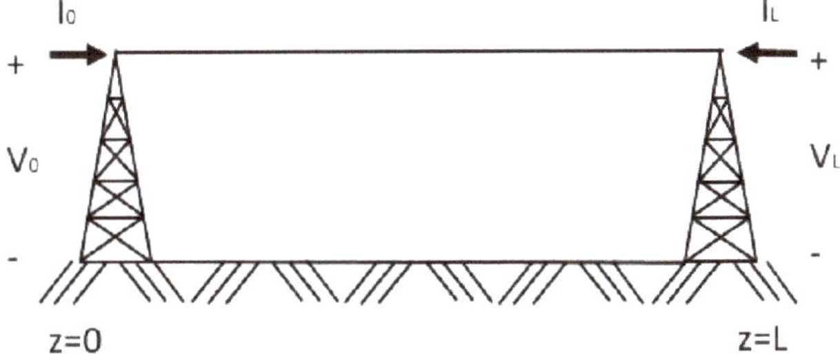

Figure 1. Uniform transmission line representation. V_0 and V_L are the voltages of the sending and receiving node, and I_0 and I_L are the injected currents at the sending and receiving node, respectively, and L is the line's length.

By changing the direction of I_L in Figure 1, Equation (1) is modified in the following manner:

$$\begin{bmatrix} V_L(s) \\ I_L(s) \end{bmatrix} = \begin{bmatrix} A & B \\ -C & -D \end{bmatrix} \begin{bmatrix} V_0(s) \\ I_0(s) \end{bmatrix} \tag{3}$$

or in a compact form:

$$\begin{bmatrix} V_L(s) \\ I_L(s) \end{bmatrix} = \Phi \begin{bmatrix} V_0(s) \\ I_0(s) \end{bmatrix} \tag{4}$$

Matrix Φ in Equation (4) is the chain matrix of the transmission line. Due to the fact that the currents at both ends of the line have the same direction, multiple transmission lines can be cascade-connected by using their corresponding chain matrices, as it can be seen in Figure 2.

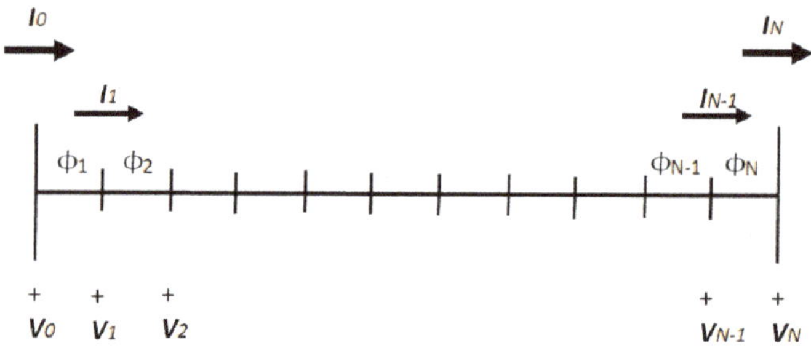

Figure 2. Cascaded connection of chain matrices, where Φ_n, V_n, and I_n are the chain matrix, sending voltage, and sending current of the *n*-th cascade-connected line, respectively.

Figure 2 can also be interpreted as one transmission line subdivided into several smaller line segments, where each segment is represented by a unique chain matrix with its own independent electrical properties. Using this approach, it is possible to build a transmission line model that includes nonuniformities. With this consideration in mind, it can be deduced from Figure 2 that the voltage and current at the beginning of each line segment and the chain matrix of the same segment can be used to compute the voltage and current of the next segment:

$$
\begin{bmatrix} V_1(s) \\ I_1(s) \end{bmatrix} = \Phi_1 \begin{bmatrix} V_0(s) \\ I_0(s) \end{bmatrix}
$$
$$
\begin{bmatrix} V_2(s) \\ I_2(s) \end{bmatrix} = \Phi_2 \begin{bmatrix} V_1(s) \\ I_1(s) \end{bmatrix} = \Phi_2 \Phi_1 \begin{bmatrix} V_0(s) \\ I_0(s) \end{bmatrix} \tag{5}
$$

or in a general way:

$$
\begin{bmatrix} V_N(s) \\ I_N(s) \end{bmatrix} = \Phi_N \Phi_{N-1} \cdots \Phi_3 \Phi_2 \Phi_1 \begin{bmatrix} V_0(s) \\ I_0(s) \end{bmatrix} \tag{6}
$$

Equation (6) can be used to compute the voltage and current profiles along a nonuniform transmission line. The transient profiles are computed in a sequential manner, obtaining the voltage and current at the end of the first line segment from the chain matrix and from the voltages and currents at the beginning of the same segment; this process is repeated until the voltage and current along the whole line have been computed. It can also be observed in Equation (6) that V_0 and I_0 are required as initial values of the algorithm; in order to compute such values, a two-port admittance representation of the complete transmission line from the chain matrix Φ_{FL} is defined as:

$$
\Phi_{FL} = \Phi_N \Phi_{N-1} \cdots \Phi_3 \Phi_2 \Phi_1 = \begin{bmatrix} \Phi_{FL11} & \Phi_{FL12} \\ \Phi_{FL21} & \Phi_{FL22} \end{bmatrix} \tag{7}
$$

$$
\begin{bmatrix} I_0(s) \\ I_L(s) \end{bmatrix} = \begin{bmatrix} Y_{ss} & -Y_{sr} \\ -Y_{rs} & Y_{rr} \end{bmatrix} \begin{bmatrix} V_0(s) \\ V_L(s) \end{bmatrix} \tag{8}
$$

where

$$
\begin{aligned}
Y_{ss} &= -\Phi_{FL12}^{-1} \Phi_{FL11} \\
Y_{sr} &= -\Phi_{FL12}^{-1} \\
Y_{rs} &= \Phi_{FL21} - \Phi_{FL22} \Phi_{FL12}^{-1} \Phi_{FL11} \\
Y_{rr} &= -\Phi_{FL22} \Phi_{FL12}^{-1}
\end{aligned} \tag{9}
$$

$V_0(s)$ is obtained by solving (8) for the voltages vector. $I_0(s)$ is computed as follows:

$$
I_0(s) = Y_{ss} V_0(s) - Y_{rr} V_L(s) \tag{10}
$$

2.2. Modeling of Time-Varying Elements

It can be difficult to include time-varying conditions, such as switching maneuvers, when methods defined in the frequency domain are used to simulate electromagnetic transients. However, the principle of superposition has demonstrated to be an efficient method to include such conditions in the frequency domain [25], as explained below.

The closing of a switch can be computed using the circuits presented in Figure 3, which represent the state of the circuit before and after the closing maneuver. V_E, V_S, and V_R are the voltage at the source side, at the line's sending node, and at the line's receiving node, respectively. The nodal voltage vector $V_{NL}(s)$ is formed by the three subvectors in Figure 3 as shown below:

$$V_{NL}(s) = \begin{bmatrix} V_E(s) & V_0(s) & V_L(s) \end{bmatrix}^T \tag{11}$$

and can be obtained from the following expression [11]:

$$V_{NL}(s) = Ybus_0^{-1}I_{N0} + Ybus_1^{-1}I_{N1} \tag{12}$$

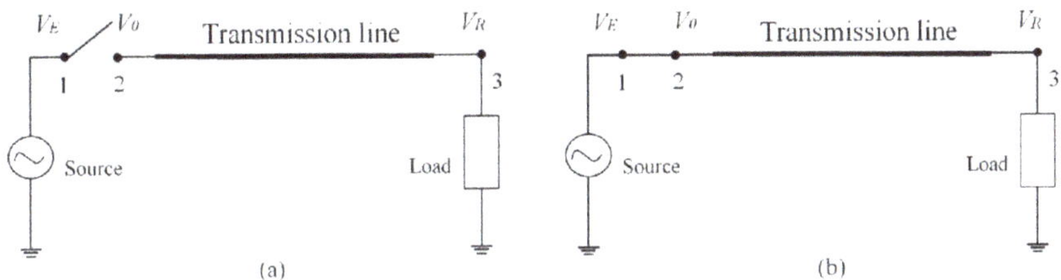

Figure 3. (**a**) Transmission line before the closing operation of a switch connected at the sending node. (**b**) Transmission line after the closing operation of a switch connected at the sending node.

In Equation (12), $Ybus_0$ is the nodal admittance matrix before the switch closing (Figure 3a), $Ybus_1$ is the admittance matrix modified by the closing operation (Figure 3b), I_{N0} contains the initially injected currents, and I_{N1} is the injection current vector due to the switch operation. An example of the construction of the nodal admittance matrices $Ybus_0$ and $Ybus_1$ for the transmission line in Figure 3 is presented below:

$$Ybus_{0,1} \begin{bmatrix} Y_{11} & -Y_{12} & -Y_{13} \\ -Y_{21} & Y_{22} & -Y_{23} \\ -Y_{31} & -Y_{32} & Y_{33} \end{bmatrix} \tag{13}$$

If $Ybus_0$ (Figure 3a) is to be built using (13), the elements Y_{11}, Y_{12}, Y_{21}, and Y_{22} would assume the following values: $Y_{12} = Y_{21} = Y_{switcho}$, where $Y_{switcho}$ is the open switch's admittance between its poles, ideally $Y_{switcho} = 0$. $Y_{11} = Y_s + Y_{swtcho}$, with Y_s being the admittance connected at the sending node, and $Y_{22} = Y_{LL} + Y_{switcho}$ where Y_{LL} represents the self-admittance of the transmission line. The matrix $Ybus_1$ (Figure 3b) can be built in a similar way, but replacing $Y_{switcho}$ by $Y_{switchc}$, that is, the admittance of the switch when closed.

The procedure presented above can be extended to any number of changes in the circuit topology using the following expression:

$$V_{NL}(s) = Ybus_0^{-1}I_{N0} + \sum_{i=1}^{m} Ybus_i^{-1}I_{Ni} \tag{14}$$

where $Ybus_i$ and I_{Ni} are the modified admittance matrix and the injection current vector corresponding to the *n*-th switch operation, respectively. A comprehensive explanation of this procedure can be found in [11].

$\mathbf{V}_0(s)$ in (11) is the voltage at the sending node of the line needed in (6) to compute the transient voltage and current profiles. Meanwhile, the current at the beginning of the line is computed with (10) using $\mathbf{V}_0(s)$ and $\mathbf{V}_L(s)$ from (11).

2.3. Inclusion of Nonlinear Elements

It has been demonstrated that the principle of superposition can be used to overcome the difficulties of the inclusion of nonlinear components in the frequency domain [25]. This is achieved by means of a sequence of switching operations.

In order to include a nonlinear element in a simulation performed using a method based in the frequency domain, first, its nonlinear characteristic curve must be approximated by a piece-wise curve, as illustrated in Figure 4.

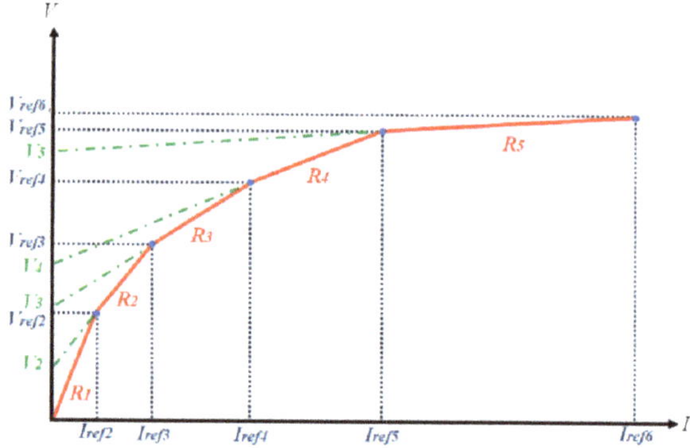

Figure 4. Five-segment piece-wise approximation of a nonlinear *v-i* characteristic curve [11].

The element R_n of the n-th linear element from Figure 4 and the voltage V_n are computed as follows:

$$R_n = \frac{V_{refn+1} - V_{refn}}{I_{refn+1} - I_{refn}} \tag{15}$$

$$V_n = \left(-R_{n+1}I_{refn}\right) + V_{refn+1} \tag{16}$$

By approximating the *v-i* characteristic of a nonlinear element as shown above, such an element can be represented as a network of N parallel-connected branches, as shown in Figure 5.

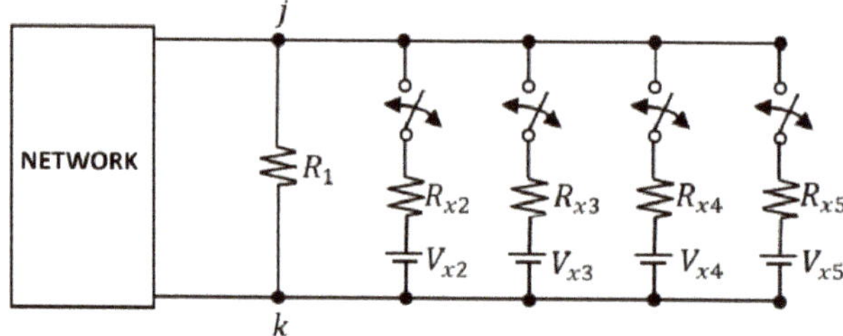

Figure 5. Circuital representation of the piecewise approximation for the characteristic curve of a nonlinear element presented in Figure 5 [11].

In the circuit presented in Figure 5, the switch in the n-th branch operates according to the reference voltage; it closes when the voltage across nodes j and k goes above V_{refn} and it opens when the voltage drops below V_{refn}. It is important to mention that the switches must operate in a successive manner, meaning that a switch can only close when the one in the preceding branch is closed, and a switch can only open when the one in the following branch is open.

When the switch connected to the n-th branch is closed, the equivalent Thevenin resistance of the circuit in Figure 5 must be equal to the value of R_n, that is, the slope of the n-th linear segment in the approximation of Figure 4:

$$R_n = \frac{R_{n-1}R_{xn}}{R_{n-1} + R_{xn}} \tag{17}$$

Finally, by solving (17) for R_{xn}, the value of such resistance can be computed:

$$R_{x_n} = \frac{R_{n-1}R_n}{R_{n-1} - R_n} \tag{18}$$

V_{xn} is computed as:

$$V_{x_n} = \frac{R_{n-1}V_n - V_{n-1}R_n}{R_{n-1} - R_n} \tag{19}$$

With this approach, it is possible to compute the transient profiles along nonuniform transmission lines with nonlinear and/or time-varying elements using (6), calculating the voltage and current at the sending node with (10) and (14), and implementing the nonlinear element model presented in this section.

2.4. Inverse Numerical Laplace Transform

As the last step, the INLT is applied to the computed transient profiles in the frequency domain with (6); this is done in order to transform the results to the time domain. This method has proven to be very accurate for the study of electromagnetic transients [11,24,25]. The implementation of the INLT algorithm is described below in a brief manner; references [17,18] can be consulted for a thorough explanation. Considering a time function $f(t)$ as a real and causal function, the inverse Laplace transform can be written as:

$$f(t) = Re\left\{ \frac{e^{ct}}{2\pi} \int_0^\infty F(s)e^{j\omega t}d\omega \right\} \tag{20}$$

In this work, the numerical evaluation of (20) is done considering an odd sampling in the frequency spectrum (using a spacing of $2\Delta\omega$), and normal time steps Δt in the time domain. With these considerations in mind, the following definitions for the discrete functions in the time and frequency domain for N equally spaced samples are made:

$$f_n = f(n\Delta t) \qquad F_m = F(c + j(2m+1)\Delta\omega) \tag{21}$$

where $n, m = 0, 1, 2, \ldots, N-1$ and c hat reduces the aliasing errors of the algorithm; in this work, c is defined as $c = 2\Delta\omega$.

Additionally, it is necessary to establish finite integration limits for the numerical evaluation of (20), that is, the maximum frequency Ω and the observation time T. The observation time is computed from:

$$T = \frac{\pi}{\Delta\omega} \tag{22}$$

and the following relations can be established:

$$\Delta t = \frac{T}{N}$$
$$\Delta\omega = \frac{\Omega}{2N} = \frac{\pi}{T} \tag{23}$$

Finally, by implementing the odd sampling presented in (21) and including a window function σ_m, the Laplace transform in (20) can be numerically approximated by:

$$f_n = Re\left\{C_n\left[\sum_{m=0}^{N-1} F_m\sigma_m exp\left(\frac{j2\pi m}{N}\right)\right]\right\} \text{for } n, m = 0,1,2,\ldots,N-1 \tag{24}$$

where:

$$C_n = \frac{2\Delta\omega}{\pi}exp\left(cn\Delta t + \frac{j\pi n}{N}\right) \tag{25}$$

In Equation (24), the term inside the brackets corresponds to the fast Fourier transform algorithm; this allows computing time savings if N is equal to an integer power of two.

3. Test Cases

Three test cases are presented in order to validate the proposed method. The frequency dependence of the line's electrical parameters was taken into account by means of the application of the complex image method to introduce the earth-return impedance, as well as the complex penetration depth to include the skin effect in the line conductors [26]. The accuracy of the method was validated through comparisons with measurements from simulations performed with ATP, where the frequency-dependent J. Marti line model was discretized in several smaller segments to allow the introduction of nonuniformities and the connection of measuring probes at interior points. The length of the simulated lines in the test cases was kept short in order to reduce the implementation burden in ATP; with the proposed method, the length and discretization of the line did not represent a problem.

3.1. Highly Nonuniform Transmission Line

A highly nonuniform three-phase transmission line was considered. The nonuniformity was included by the sagging of the conductors as shown in Figure 6. The line was excited by an ideal unit step voltage source (1 p.u.) connected at phase A of the sending node, while all of the other nodes were left open.

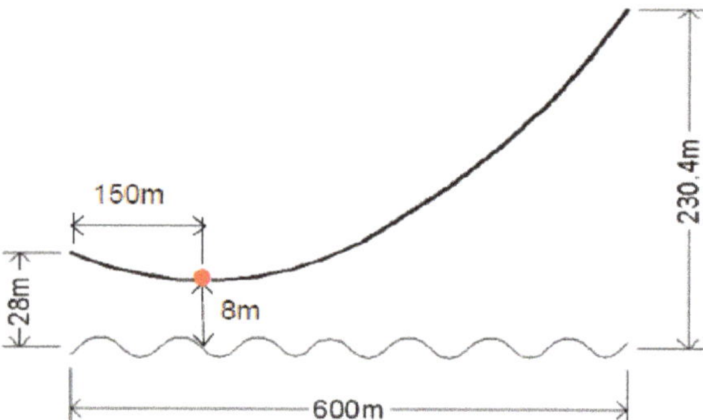

Figure 6. Three-phase non-uniform transmission line used in the example in Section 3.1. A second-degree polynomial equation was utilized to approximate the conductors' height, as presented in [23]. There was a 10 m separation between the conductors.

Figure 7 presents the transient voltage profile along phase A computed with the proposed method; in order to include the nonuniformities along the transmission line, it was subdivided into 80 chain matrices. This figure illustrates the propagation of the traveling waves along the line and how these waves were reflected when they reached the receiving end. It can also be observed in the transient profiles that there were periods where negative voltages appeared at some points along the line, which is not a commonly observed phenomenon when similar simulations with uniform lines are performed.

Additionally, a comparison at the middle of the line between the transient voltage waveforms obtained with the proposed method and results from ATP simulations is presented (Figure 8). The simulations performed with ATP required a time step 10 times smaller than the proposed method to achieve similar results.

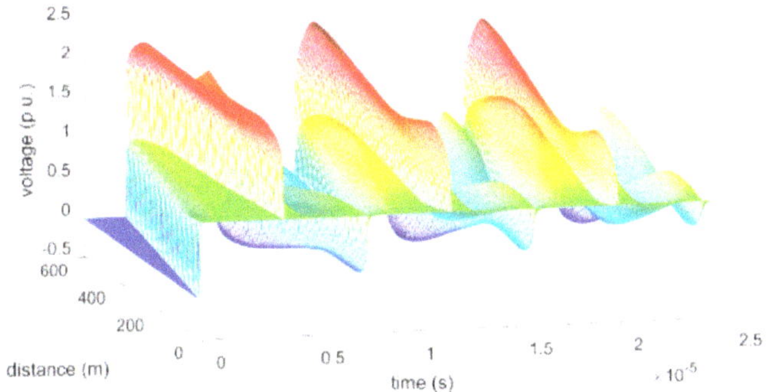

Figure 7. Computed voltage profile along phase A (example 3.1).

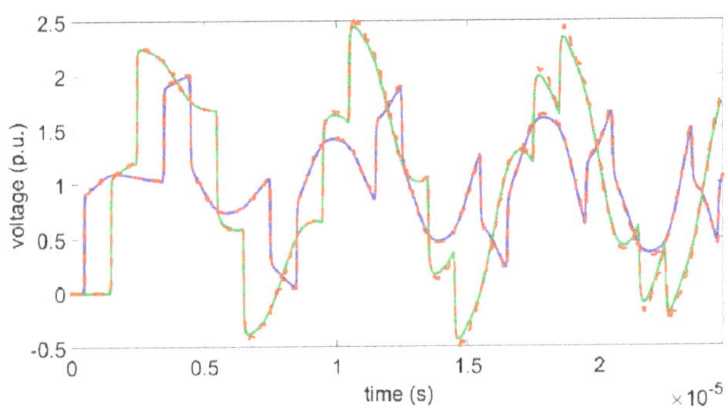

Figure 8. Comparison of transient voltage waveforms at two specific distances between the computed results with the proposed method (solid lines) and results obtained with ATP simulations (dashed lines). The solid blue and dashed red lines correspond to voltage measurements at 150 m from the sending node; the solid green and the dashed orange lines were obtained at 450 m. The oscillations observed in the results from the ATP simulations are attributed to the error accumulation due to the discretization of the transmission line [2].

This example was simulated with the proposed method considering 20 and 40 chain matrices. Figure 9 presents a comparison of voltage measurements at the middle point of the transmission line from simulations considering 20, 40, and 80 chain matrices. From this comparison, it can be observed that the curves from the three simulations, in general, have the same shape (curves for 40 and 80 chain matrices are overlapped). However, the curve corresponding to the simulation performed with 20 chain matrices presents some oscillations in comparison with the other two. As it can be seen with the curves corresponding to 40 and 80 chain matrices, with an increase in the number of chain matrices used in the simulations, the curves become smoother; however, a further increase in the number of chain matrices used is barely noticeable. This indicates that although the method's accuracy is dependent on the number of chain matrices considered, it does not require a large number of chain matrices to achieve good results.

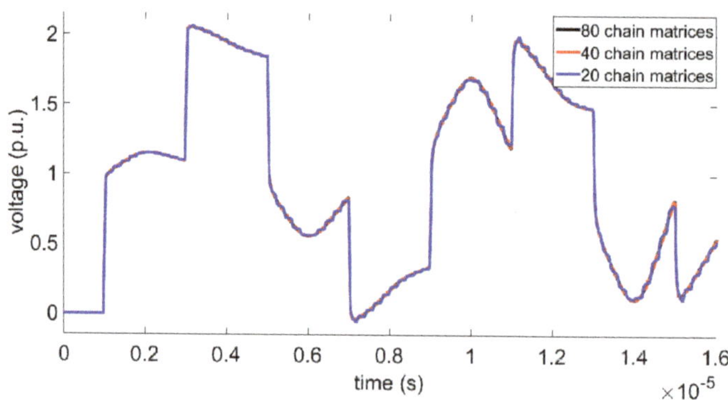

Figure 9. Comparison of transient voltage waveforms at the middle point of the line computed with different numbers of chain matrices. The curve computed with 20 chain matrices is not as smooth as the ones computed with 40 and 80.

3.2. Sequential Energization

A 5 km transmission line was used in this case. The nonuniformity was introduced by means of the line's sag: there was a transmission tower every 500 m, the conductors' height was maximum at the towers (20 m), and the height was minimum at the midspan between towers (15 m). The consideration for the height's variation as well as the horizontal separation between conductors was the same as in the previous example. The line was connected to an AC voltage source (1 p.u.) at the sending node through a three-phase switch. The switch's poles operated in a sequential manner with closing times of 3, 6, and 9 ms in an ABC sequence. The line's receiving node was left open.

The transient voltage profiles of phases A and B are presented in Figures 10 and 11, respectively. The comparison with ATP simulations is shown in Figure 12.

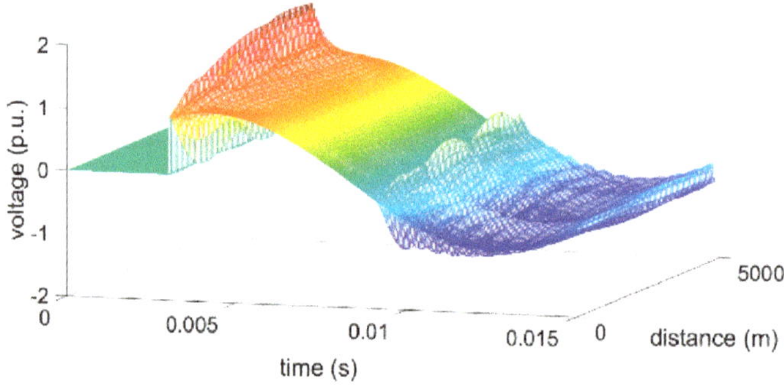

Figure 10. Voltage profile computed along phase A (example 3.2).

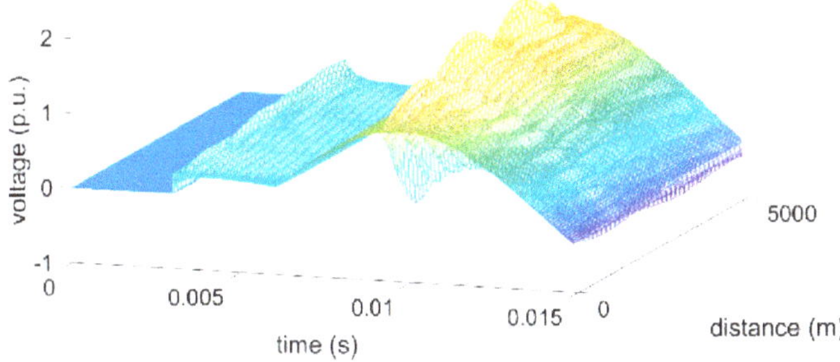

Figure 11. Transient voltage profile along phase B in example 3.2. The complete voltage profile induced in phase B due to the energization of phase A can be observed. This type of detailed information cannot be easily achieved using ATP-like simulation software.

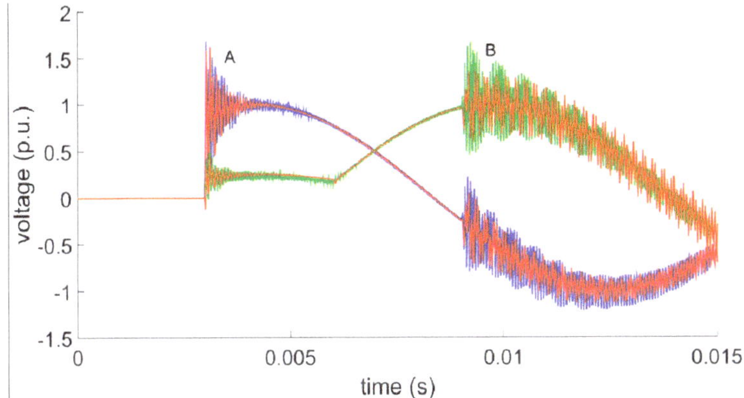

Figure 12. Comparison between the transient voltage computed with the proposed method (blue and green lines) and those obtained with ATP simulations (red and orange lines) in example 3.2. The comparisons are made at the middle of the transmission line (2500 m from the sending node) in phases A and B. A good level of agreement can be observed, but the ATP simulations needed a time step 50 times smaller than the proposed method in order to achieve such results.

3.3. Surge Arrester Operation

This example presents the operation of surge arresters during a direct lightning strike; due to their *v-i* characteristic, the arresters were modeled as nonlinear elements as described in Section 2.3. The same line configuration presented in Section 3.1 was considered. The line was excited by a direct lightning strike (1.2/50 μs), the impact point was at the phase A of the sending node, and the line was impedance-matched at both ends in order to avoid reflections. The injected current to the line was approximated by a double exponential current source defined as:

$$i(t) = I_0\left(e^{-\frac{t}{\tau_1}} - e^{-\frac{t}{\tau_2}}\right) \tag{26}$$

where $\tau_1 = 68.199$ μs, $\tau_2 = 0.405$ μs, and $I_0 = 10.37$ kA.

Surge arresters were connected to the three phases at the receiving node. The nonlinear characteristic of the arresters was simulated by means of a five-segment piecewise-linear approximation. Table 1 presents the *v-i* coordinates of this representation.

Table 1. Nonlinear *v-i* characteristic of the arresters [27].

Vref (kV)	Iref (kA)
0	0
484	0.1760
616	0.3226
748	0.7626
836	1.6426
880	12.6426

First, a simulation was performed without surge arresters connected to the line. Figure 13 presents the transient voltage profile along phase A computed in this simulation. On the other hand, Figure 14 shows the voltage profile along phase A when the arresters were connected at the receiving node of the line. By comparing Figures 13 and 14, the influence that the surge arresters' operation had on the magnitude of the voltages along the line is easily observed. The voltage along the line in Figure 14 was considerably lower in comparison with the transient profile in Figure 13. Additionally, a comparison is presented with results obtained from ATP simulations (Figure 15). In a similar way to the example in Section 3.1, the time step required by ATP was 10 times smaller than the proposed method in order to obtain similar results.

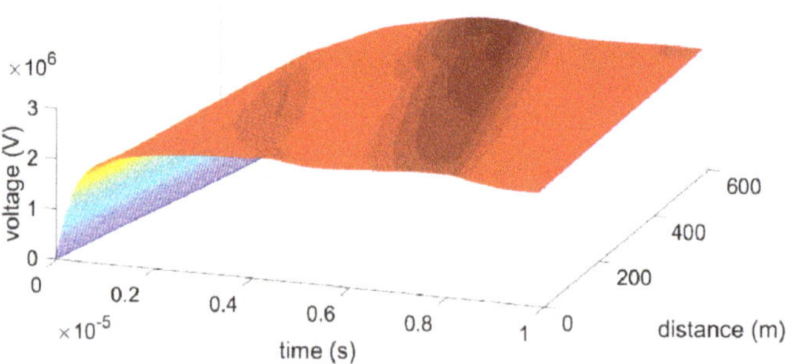

Figure 13. Computed transient voltage profile along phase A when there were no surge arresters connected at the receiving end of the line (example 3.3).

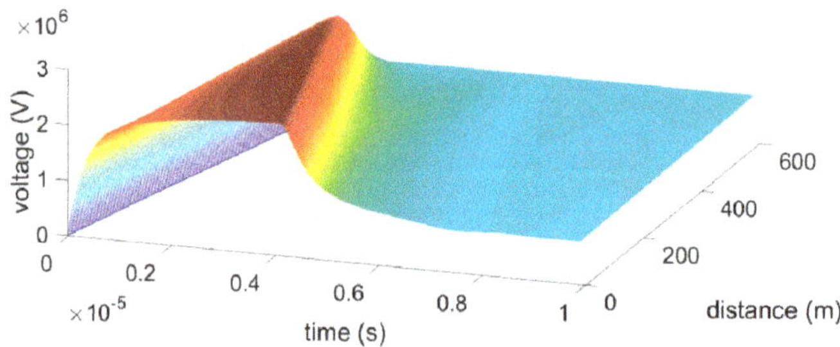

Figure 14. Computed transient voltage profile along phase A with surge arresters connected at the receiving node (example 3.3). The voltage magnitude significantly decreased along the transmission line compared to the transient profile presented in Figure 13.

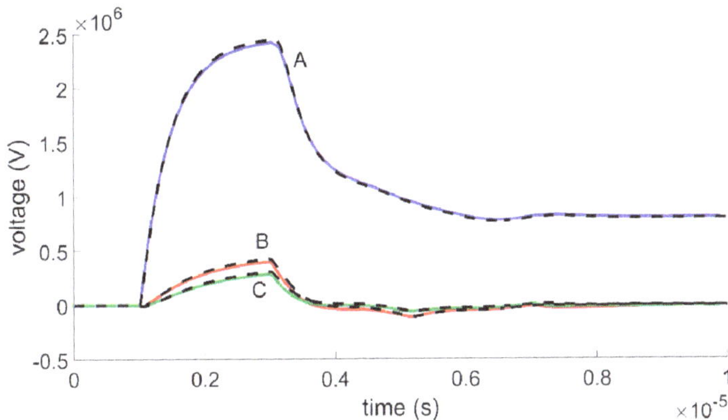

Figure 15. Voltage comparison between the results from the proposed method (solid lines) and those obtained from ATP simulations (dashed lines) for example 3.3. The comparisons were made at the middle point of the line (300 m from the sending node) for phases A, B, and C.

4. Discussion

Transient simulation of nonuniform transmission lines using traditional software (such as ATP in this paper) is a challenging task. This difficulty is due to the fact that the nonuniformities are typically approximated by cascade-connecting several small line segments. Such approximation requires the use of a very small time step in the simulation (at least five times smaller than the traveling time of the line segments [6]), which may translate into excessive simulation times and the saturation of the software's available memory, as was the case in this work. Additionally, the connection of several line segments can introduce errors in the simulation results [2].

In contrast, the proposed method is able to accurately compute transient profiles along nonuniform transmission lines with substantially larger time steps in comparison to ATP, resulting in a better memory usage as expected from the findings reported in [16] with regards to the use of the INLT algorithm. As it can be observed in Figures 7, 10, 11 and 14, the main advantage of the proposed method is the fact that it allows visualizing the voltage and current transient behavior along the line and not only at its ends, which can be advantageous when designing lines with a high level of nonuniformities and cannot be easily done using traditional simulation software.

The precision of the proposed method was validated by comparisons with the results obtained with ATP simulations. In general, there was a very good level of agreement between the results from both methods, as it can be seen by the mean relative difference presented in Tables 2–4. There is a slight difference in the comparison of results presented in Figure 12 corresponding to the simulation of sequential pole closure. This is attributed to the previously mentioned ATP limitations and the difference in the time step used in each method, which can result in deviations in the operation time of the switch model, leading to the observed variations.

Table 2. Mean relative difference of transient voltage in example 3.1.

Measuring Point	Mean Relative Difference (%)		
	Phase A	Phase B	Phase C
150 m	0.0708	0.1837	0.2893
450 m	0.3839	0.3351	0.6832

Table 3. Mean relative difference of transient voltage in example 3.2.

Measuring Point	Mean Relative Difference (%)		
	Phase A	Phase B	Phase C
2.5 km	0.3889	0.9861	1.2761

Table 4. Mean relative difference of transient voltage in example 3.3.

Measuring Point	Mean Relative Difference (%)		
	Phase A	Phase B	Phase C
300 m	0.0374	0.7738	1.9183

5. Conclusions

This paper describes a frequency domain method to compute transient voltage and current profiles along nonuniform multiconductor transmission lines, where the nonuniformities along the line are introduced in the model by means of the cascaded connection of chain matrices. The method can incorporate nonlinear and time-dependent elements by using the superposition principle. The profiles obtained provide useful information to locate possible overvoltages at interior points along the transmission line, in contrast to traditional methods that only provide information at the line's ends. This information can be used as a helpful instrument in the insulation coordination design of transmission lines, as well as an educational tool in electrical engineering courses.

The results computed with the described method were compared with time-domain simulations using the well-known software ATP. In all of the comparisons, a high level of agreement was observed, demonstrating that the proposed method has a high level of accuracy.

It is worth mentioning that, although the results from both methods were very similar, the ATP simulations required substantially more time samples (10 to 50 times more samples) to achieve such results, which can result in a large computational burden. Additionally, in order to obtain measurements at interior points of the transmission line and to include nonuniformities along the transmission line in the ATP simulations, it is necessary to subdivide the line into many line segments, which can be a time-consuming process and can lead to error accumulation. Because of these issues, the proposed frequency domain method is a superior alternative to analyze the transient behavior of a transmission line when the voltage at interior points of the line is of interest.

The relevance of the proposed method lies in its strong potential for application in the accurate fault location and insulation coordination of transmission systems with prominent nonuniformities, which are commonly encountered in many power systems around the world.

Author Contributions: Conceptualization, R.N.-G. and F.P.E.C.; investigation, P.G. and F.P.E.C.; methodology, R.N.-G.; funding acquisition, P.G. and C.T.M.; software, R.N.-G.; validation, R.N.-G. and F.P.E.C.; visualization, C.T.M.; writing—original draft, R.N.-G.; writing—review & editing, R.N.-G., F.P.E.C. and P.G.

Funding: This research received no external funding.

References

1. Martínez-Velasco, J.A. Parameter determination for electromagnetic transient analysis in power systems. In *Power System Transients: Parameter Determination*, 1st ed.; CRC Press: Boca Raton, FL, USA, 2010; pp. 1–16.
2. Marti, L.; Dommel, H.W. Calculation of voltage profiles along transmission lines. *IEEE Trans. Power Deliv.* **1997**, *12*, 993–998. [CrossRef]
3. Pinto, A.J.G.; Costa, E.C.M.; Kurokawa, S.; Monteiro, J.H.A.; de Franco, J.L.; Pissolato, J. Analysis of the electrical characteristics and surge protection of EHV transmission lines supported by tall towers. *Int. J. Electr. Power Energy Syst.* **2014**, *57*, 358–365. [CrossRef]

4. Pinto, A.J.G.; Costa, E.C.M.; Kurokawa, S.; Pissolato, J. Analysis of the electrical characteristics of an alternative solution for the Brazilian-Amazon transmission system. *Electr. Power Comp. Syst.* **2011**, *39*, 1424–1436. [CrossRef]

5. Martínez, D.; Moreno, P.; Loo-Yau, R. A new model for non-uniform transmission lines. In Proceedings of the 12th International Conference on Electrical Engineering, Computing Science and Automatic Control (CCE), Mexico City, Mexico, 28–30 October 2015.

6. Lima, A.C.S.; Moura, R.A.R.; Gustavsen, B.; Schroeder, M.A.O. Modelling of non-uniform lines using rational approximation and mode revealing transformation. *IET Gener. Transm. Distrib.* **2017**, *11*, 2050–2055. [CrossRef]

7. Saied, M.M.; Al-Fuhaid, A.S.; El-Shandwily, M.E. S-domain analysis of electromagnetic transients on non-uniform lines. *IEEE Trans. Power Deliv.* **1990**, *5*, 2072–2081. [CrossRef]

8. Oufi, E.A.; Al-Fuhaid, A.S.; Saied, M.M. Transient analysis of loss-less single phase nonuniform transmission lines. *IEEE Trans. Power Deliv.* **1994**, *9*, 1694–1700. [CrossRef]

9. Correia de Barros, M.T.; Almeida, M.E. Computation of electromagnetic transients on nonuniform transmission lines. *IEEE Trans. Power Deliv.* **1996**, *11*, 1082–1087. [CrossRef]

10. Semlyen, A. Some frequency domain aspects of wave propagation on nonuniform lines. *IEEE Trans. Power Deliv.* **2003**, *18*, 315–322. [CrossRef]

11. Gómez, P.; Escamilla, J.C. Frequency domain modeling of nonuniform multiconductor lines excited by indirect lightning. *Int. J. Electr. Power Energy Syst.* **2013**, *45*, 420–426. [CrossRef]

12. Nuricumbo-Guillén, R.; Espino-Cortés, F.P.; Gómez, P.; Tejada-Martínez, C. Computation of transient profiles along non-uniform trasmission lines using the numerical Laplace transform. In Proceedings of the IEEE International Conference on High Voltage Engineering and Applications, Athens, Greece, 10–13 September 2018.

13. Nuricumbo-Guillén, R.; Gómez, P.; Espino-Cortés, F.P.; Uribe, F.A. Accurate computation of transient profiles along multiconductor transmission lines by means of the numerical Laplace transform. *IEEE Trans. Power Deliv.* **2014**, *29*, 2385–2395. [CrossRef]

14. Gómez, P.; Vergara, L.; Nuricumbo-Guillén, R.; Espino-Cortés, F.P. Two dimensional definition of the numerical Laplace transform for fast computation of transient profiles along power transmission lines. *IEEE Trans. Power Deliv.* **2016**, *31*, 412–414. [CrossRef]

15. Nuricumbo-Guillén, R.; Vergara, L.; Gómez, P.; Espino-Cortés, F.P. Laplace-based computation of transient profiles along transmission lines including time-varying and non-linear elements. *Int. J. Electr. Power Energy Syst.* **2019**, *106*, 138–145. [CrossRef]

16. Brancik, L. Time and Laplace-domain methods for MTL transient sensitivity analysis. *Int. J. Comput. Math. Electr. Electron. Eng.* **2011**, *30*, 1205–1223. [CrossRef]

17. Gómez, P.; Uribe, F.A. The numerical Laplace transform: An accurate tool for analyzing electromagnetic transients on power system devices. *Int. J. Electr. Power Energy Syst.* **2009**, *31*, 116–123. [CrossRef]

18. Moreno, P.; Ramírez, A. Implementation of the numerical Laplace transform: A review. *IEEE Trans. Power Deliv.* **2008**, *23*, 2599–2609. [CrossRef]

19. Souza, L.; Lima, A.C.S.; Carneiro, S., Jr. Modeling Overhead Transmission Line with Large Asymmetrical Spans. In Proceedings of the 2011 International Conference on Power Systems Transients, Delft, The Netherlands, 14–17 June 2011.

20. Guo, X.; Fu, Y.; Yu, J.; Xu, Z. A non-uniform transmission line model of the ±1100 kV UHV tower. *Energies* **2019**, *12*, 445. [CrossRef]

21. Hussain, M.K.; Gómez, P. Optimal filter tuning to minimize the transient overvoltages on machine windings fed by PWM inverters. In Proceedings of the 2017 North American Power Symposium (NAPS), Morgantown, VA, USA, 17–19 September 2017.

22. Dommel, H.W. *Electromagnetic Transient Program Theory Book*; Bonneville Power Administration: Portland, OR, USA, 1986.

23. Marti, J. Accurate modelling of frequency-dependent transmission lines in electromagnetic transient simulations. *IEEE Trans Power App. Syst.* **1982**, *PAS-101*, 147–157. [CrossRef]

24. Gómez, P.; Moreno, P.; Naredo, J.L. Frequency domain transient analysis of nonuniform lines with incident field excitation. *IEEE Trans. Power Deliv.* **2005**, *20*, 2273–2280. [CrossRef]

25. Moreno, P.; Gómez, P.; Naredo, J.L.; Guardado, J.L. Frequency domain transient analysis of electrical networks including non-linear conditions. *Int. J. Electr. Power Energy Syst.* **2005**, *27*, 139–146. [CrossRef]

26. Gary, C. Approche complète de la propagation multifilaire en haute fréquence par utilisation des matrices complexes. *EDF Bull. Dir. des Études et Rech.* **1976**, *3*, 4.
27. Gómez, P.; Moreno, P.; Naredo, J.L.; Guardado, L. Frequency domain transient analysis of transmission networks including non-linear conditions. In Proceedings of the 2003 IEEE Bologna Power Tech Conference, Bologna, Italy, 23–26 June 2003.

Article

Optimization of Radio Interference Levels for 500 and 600 kV Bipolar HVDC Transmission Lines [†]

Carlos Tejada-Martinez [1,*], Fermin P. Espino-Cortes [1], Suat Ilhan [2] and Aydogan Ozdemir [2]

[1] Departamento de Ingeniería Eléctrica SEPI ESIME Zacatenco, Instituto Politécnico Nacional, Mexico City 7738, Mexico

[2] Department of Electrical Engineering, Istanbul Technical University, 34467 Istanbul, Turkey

* Correspondence: ctejadam@ipn.mx

† This paper is an extended version of our paper presented in the 2018 IEEE International Conference on High Voltage Engineering and Application (ICHVE 2018), Athens, Greece, 10–13 September 2018.

Received: 14 June 2019; Accepted: 2 August 2019; Published: 20 August 2019

Abstract: In this work, a method to compute the radio interference (RI) lateral profiles generated by corona discharge in high voltage direct current (HVDC) transmission lines is presented. The method is based on a transmission line model that considers the skin effect, through the concept of complex penetration depth, in the conductors and in the ground plane. The attenuation constants are determined from the line parameters and the bipolar system is decoupled by using modal decomposition theory. As application cases, ±500 and ±600 kV bipolar transmission lines were analyzed. Afterwards, parametric sweeps of five variables that affect the RI levels are presented. Both the RI and the maximum electric field were calculated as a function of sub-conductor radius, bundle spacing, and the number of sub-conductors in the bundle. Additionally, the RI levels were also calculated as a function of the soil resistivity, and the RIV (radio interference voltage) frequency. Following this, vector optimization was applied to minimize the RI levels produced by the HVDC lines and differences between the designs with nominal and optimal values are discussed.

Keywords: bundle electric field; corona; HVDC transmission lines; optimization; radio interference

1. Introduction

High voltage direct current (HVDC) transmission systems are being studied and developed in several countries around the world. In Mexico and Turkey, the future installation of HVDC transmission lines is planned because of the advantages this kind of technology presents over HVAC transmission lines, mostly in energy transmission over long distances [1]. During the design stage of an HVDC transmission system, it is necessary to carry out studies like insulation coordination, protections, and transient stability, among others [2–4].

Corona discharge appears at the surface of the conductors when a certain critical value of the electric field is reached, causing the ionization of the air surrounding the conductors. Some of the main consequences of corona are power loss, audible noise, and radio interference [5]. Radio interference (RI) can be any effect on the reception of a wanted radio signal due to an unwanted disturbance within the radio frequency spectrum [6]. Corona performance and radio interference are key issues during the design stage of both HVAC and HVDC transmission lines, usually as a part of electromagnetic compatibility studies (EMC) [7,8].

Concerning bipolar DC lines, corona in the positive pole is the dominant source of electromagnetic interference since the current pulses produced by this polarity have much higher amplitude than those from the negative pole. There is also a noteworthy difference between RI performance in AC and DC lines; in AC lines, the RI levels increase significantly in heavy rain weather while the RI levels in bipolar DC lines, by contrast, decrease in rain or wet snow [9,10].

There are still only a few analytical methods for calculating RI levels from DC lines, mainly due to difficulty in defining the RI excitation function through experimental studies. Based on measurements made on test lines, empirical formulas for predicting the RI levels for bipolar HVDC lines have been developed by different researching groups, some of which are described in [5,9,10]. In general, these kinds of formulas are defined for fair weather conditions, and they depend on the maximum voltage gradient, conductor diameter, and the distance between the conductor and the measuring point. Both the analytical methods presented in [5] and the empirical formulas mentioned above are only defined for horizontal configurations of bipolar lines. Also, they do not consider the soil resistivity which is a variable that affects the RI levels produced by corona in DC lines, as will be discussed later in this work.

In this paper, the computation of RI lateral profiles produced by corona discharge in HVDC transmission lines is presented. The proposed method has already been applied for calculating RI levels in HVAC lines in a previous work, where the results were very close to the measurements made on AC lines [11]. The proposed method, described in Section 2, considers a transmission line model where the skin effect is taken into consideration by using the concept of complex penetration depth in the pole conductors and in the ground plane. The soil resistivity is considered in the calculation of the complex penetration depth which, in turn, affects the series impedance of the line and the horizontal component of the magnetic field. Moreover, the attenuation constants are calculated from the line parameters and the multiconductor system is decoupled by using modal decomposition theory, which makes it possible to analyze not only horizontal bipolar lines but also any other configuration even hybrid transmission lines.

In Section 3, a comparison of computed values and measurements of RI reported in a previous publication for a DC test line of ±800 kV is also presented. In Section 4, parametric sweeps are presented in order to investigate the impacts of five important variables on RI levels and on the maximum bundle electric field. The variables are sub-conductor radius, bundle spacing, the number of sub-conductors in a bundle, soil resistivity, and the RIV (radio interference voltage) frequency.

Afterwards, Section 5 describes how vector optimization is applied in order to minimize the RI levels produced by two bipolar DC lines, one of ±500 kV and the other of ±600 kV. In the optimization study, three independent variables are considered, namely, sub-conductor radius, bundle spacing, and the number of sub-conductors in the bundle. Finally, RI lateral profiles computed with the optimal values are compared with those obtained with the nominal values.

2. Transmission Line Model for Corona Propagation Analysis

The line model used to simulate the propagation of the corona current along a multiconductor transmission line is derived from the per unit length equivalent circuit for a Δz section, presented in Figure 1. The transmission line is considered to be infinitely long with uniform corona current density injections per unit length (J).

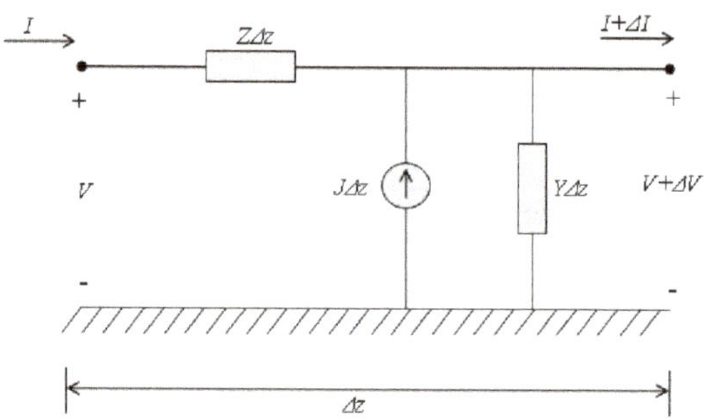

Figure 1. Per unit length equivalent transmission line circuit with corona.

By applying circuit analysis, equations representing the RI propagation can be written in a matrix form including the corona current density source, as follows:

$$\frac{d\mathbf{V}}{dz} = -\mathbf{Z}\mathbf{I},\tag{1}$$

$$\frac{d\mathbf{I}}{dz} = -\mathbf{Y}\mathbf{V} + \mathbf{J},\tag{2}$$

where $\mathbf{V}$ and $\mathbf{I}$ are the voltage and the current at any point on the line, and $\mathbf{J}$ is the column vector of corona current densities injected into the conductors. $\mathbf{Z}$ and $\mathbf{Y}$ are square matrices that represent the series impedance and the shunt admittance per unit length of the line. These parameters are determined considering the skin effect, by using the concept of complex penetration depth in the conductors and in the ground return. They can be calculated using the formulas described in [12]. (1) and (2) represent n sets of equations for the voltage and the current for a transmission line with n pole conductors. Due to the inductive and capacitive coupling between the conductors, the n sets of equations are also coupled. The modal analysis is used to simplify Equations (1) and (2) into a number of uncoupled sets of equations which can each be solved as in the case of a single conductor line.

The modal transformation matrix $\mathbf{M}$ is defined by

$$\lambda = \mathbf{M}^{-1}\mathbf{Z}\mathbf{Y}\mathbf{M},\tag{3}$$

where λ and $\mathbf{M}$ are the eigenvalue (diagonal) and eigenvector matrices of $\mathbf{Z}\mathbf{Y}$ product, respectively. The modal propagation constants $\mathbf{\Psi}$ and the modal attenuation constants α_m matrices are given as

$$\mathbf{\Psi} = \sqrt{\lambda},\tag{4}$$

$$\alpha_m = \mathrm{Re}\{\mathbf{\Psi}\}.\tag{5}$$

The corona current density vector is obtained as

$$\mathbf{J} = \frac{\mathbf{C}}{2\pi\varepsilon_0}\mathbf{\Gamma},\tag{6}$$

where $\mathbf{C}$ is the capacitance matrix of the line and $\mathbf{\Gamma}$ is the excitation function vector.

The excitation function formula obtained from [5] is

$$\Gamma = \Gamma_0 + k_1(g_{\max} - g_0) + k_2 \log_{10}\left(\frac{n_c}{n_0}\right) + 40\log_{10}\left(\frac{d}{d_0}\right),\tag{7}$$

where:

- Γ is the RI excitation function in dB above $1\,\mu A/\sqrt{m}$;
- $g_{\max}$ is the maximum bundle electric field in kV/cm;
- n_c is the number of sub-conductors in the bundle;
- d is the sub-conductor diameter in cm;
- n_0, g_0, and d_0 are reference values given by $n_0 = 6$, $d_0 = 4.064$ cm and $g_0 = 25$ kV/cm.

The formula for the maximum bundle electric field for bipolar lines obtained from [13] is

$$g_{\max} = \frac{\left[1 + (n_c - 1)\frac{r}{R}\right]V}{n_c r \ln\left(\dfrac{2H}{(n_c r R^{n-1})^{1/n} * \left[(2H/S)^2 + 1\right]^{1/2}}\right)},\tag{8}$$

where:

- V is the line to ground voltage in kV;
- r is the radius of each sub-conductor in cm;
- R is the bundle radius in cm;
- H is the height of the poles;
- S is the distance between poles.

The reference value Γ_0 and the empirical constants k_1 and k_2 in (7) are given for all seasons in different weather conditions in [5]. Since the positive conductor of the bipolar line is considered the only source of RI, the vector $\boldsymbol{\Gamma}$ may be expressed as

$$\boldsymbol{\Gamma} = \begin{bmatrix} \Gamma_+ \\ 0 \end{bmatrix}. \tag{9}$$

The modal corona current density vector is given by

$$\mathbf{J}_m = \mathbf{M}^{-1}\mathbf{J}. \tag{10}$$

Using Equations (5) and (10), the modal components of current on the conductors are obtained as

$$\mathbf{I}_m = \begin{bmatrix} \dfrac{J_{m1}}{2\sqrt{\alpha_{m1}}} \\ \dfrac{J_{m2}}{2\sqrt{\alpha_{m2}}} \end{bmatrix}, \tag{11}$$

where J_{m1} and J_{m2} are elements of the vector $\mathbf{J}_m$, while α_{m1} and α_{m2} are the modal attenuation constants, elements of the diagonal matrix $\boldsymbol{\alpha}_m$. Then, the current in each conductor is obtained as

$$\mathbf{I} = \mathbf{M}\mathbf{I_m}. \tag{12}$$

The current in each conductor is the sum of two modal components. By knowing the currents flowing in all the conductors of the line, the corresponding horizontal component of the magnetic field at any point (x, y) is calculated as

$$\mathbf{H}_x = \sum_{i=1}^{n} \frac{I_i}{2\pi}\left[\frac{h_i - y}{(h_i - y)^2 + (x_i - x)^2} + \frac{h_i + y + 2P}{(h_i + y + 2P)^2 + (x_i - x)^2} \right], \tag{13}$$

where:

- n is the number of poles;
- I_i is the current of the ith conductor;
- h_i is the ith conductor height;
- x_i is the ith conductor lateral distance from the center of the tower;
- x is the lateral distance of the measurement point from the center of the tower;
- y is the measurement point height above the ground;
- P is the complex depth of penetration for the ground return defined as

$$P = \sqrt{\frac{\rho_e}{j\omega\mu_e}}, \tag{14}$$

where ρ_e and μ_e are the resistivity and permeability of the ground, respectively. The corresponding vertical component of the electric field is calculated, assuming a quasi-TEM propagation, as

$$\mathbf{E}_y = Z_0\mathbf{H}_x, \tag{15}$$

where the wave impedance of free space is $Z_0 = 120\pi$. After determining the electric field component, the resultant field is determined by an *rms* addition:

$$E_{y,total} = \sqrt{\sum_{k=1}^{n} |E_{y,k}|^2}.$$ (16)

The electric field due to corona $E_{y,total}$ is usually expressed in dB above 1 μV/m using the following equation:

$$E_{y,total}(dB) = 20 \log_{10} \frac{E_{y,total}(\mu V/m)}{1\mu V/m}.$$ (17)

3. Comparison of Measured and Computed RI Profiles of an HVDC Test Line

In this section, the RI values produced by an HVDC test line of ±800 kV are computed using the method described in the previous section. Then, the computed values are compared with measurements reported in [14,15]. This test line is located in the National Engineering Laboratory for UHV Technology in Kunming, China, with an altitude of 2100 m. The length of the test line is 800 m and the terminals are open ends. The measurement system consists of radio receivers connected to loop antennas with a height of 1.5 m and bandwidth of 9 kHz–30 MHz. Additionally, two wave trappers were installed between generators and the test line in order to suppress RI noises. The measuring frequency was 0.5 MHz and the measurements were made under good weather in fall. The measured RI levels are represented by quasi-peak values (QP). The data of the test line considered in the calculations are presented in Table 1.

Table 1. Data of the high voltage direct current (HVDC) test line considered in the calculations.

Variable	Value
Voltage (V)	±800 kV
Number of poles (n)	2
Conductor height (H)	18 m
Distance between the poles (S)	22 m
Number of sub-conductors in the bundle (n_c)	6
Bundle spacing (a)	45 cm
Sub-conductor radius (r)	1.68 cm
Ground resistivity (ρ_e)	100 Ωm
Measuring frequency (f)	500 kHz

The RI levels are given in dB with 1 μV/m as base magnitude. The profiles are plotted 80 m from each side of the center of the tower at 1.5 m above ground level. The altitude correction factor for HVDC lines proposed in [16] was applied.

In Figure 2, both measured and computed RI profiles are shown. The position zero of the horizontal axis corresponds to the center of the tower and the positive direction is towards the positive pole. It can be seen that the computed RI values are in good agreement with the measurements, principally for the measurements made on the direction of the positive pole. The maximum measured RI value is 66.3 dB while the maximum computed RI value is 65.03 dB, corresponding to a difference of 1.27 dB. Also, it can be observed the asymmetric shape of the profiles that confirms the assumption of the positive pole as the source of RI [9]. The biggest differences between measured and computed values (3–4 dB) appeared for distances longer than 60 m in the direction of the negative pole.

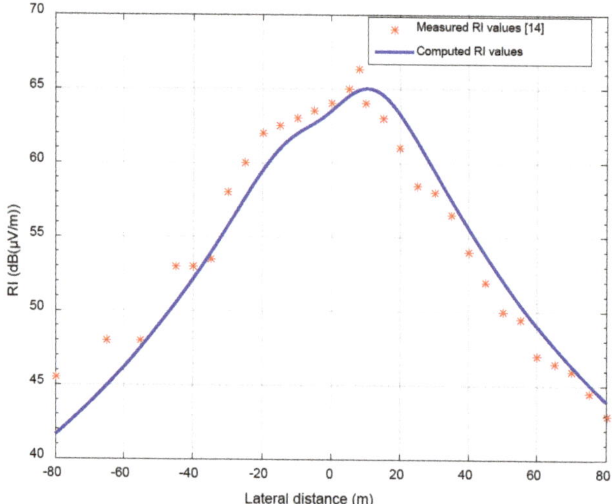

Figure 2. Comparison of measured [14] and computed radio interference (RI) profiles.

4. Computation of RI Levels in ±500 and ±600 kV HVDC Transmission Lines

4.1. Radio Interference Lateral Profiles

In this section, the RI lateral profiles of two bipolar HVDC transmission lines of ±500 and ±600 kV are presented. The method described in Section 2 was programmed in MATLAB in order to compute the RI levels expressed in dB with 1 μV/m as base magnitude. The RI profiles were plotted up to 50 m from each side of the center of the tower at 1 m above the ground. The tower geometry of a bipolar, quadrupole-bundle HVDC line is shown in Figure 3. Line and tower data are given in Table 2.

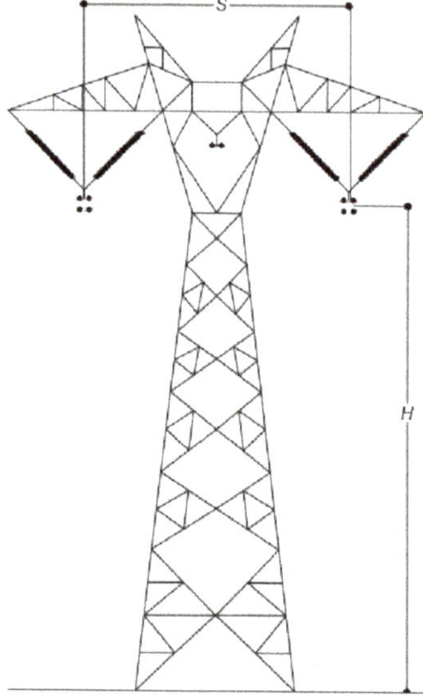

Figure 3. Tower geometry of a bipolar HVDC transmission line.

Table 2. Data of the HVDC lines considered in the simulations.

Variable	Value
Voltage (V)	±500 or ±600 kV
Number of poles (n)	2
Conductor height of ±500 kV line (H)	27 m
Conductor height of ±600 kV line (H)	34 m
Distance between the poles (S) for both voltage levels	16 m
Number of sub-conductors in the bundle (n_c)	4
Bundle spacing (a)	45 cm
Sub-conductor radius (r)	1.71 cm

The parameters used for the excitation formula (7) were those corresponding to fair weather condition in summer ($\Gamma_0 = 27$, $k_1 = 1.83$, and $k_2 = 45.8$), which is assumed to be the most critical condition for the generation of radio interference voltages in HVDC lines [5]. In the simulations, ground resistivity was considered with a value of 100 Ωm, and the RIV frequency was taken as 0.5 MHz, the frequency that is usually considered in the RI measurements [15,17].

Simulated RI lateral profiles at rated conditions are shown in Figure 4 for both HVDC lines. The zero position of the horizontal axis corresponds to the center of the tower and the positive direction is towards the positive pole. The maximum values of RI for the bipolar lines of ±500 and ±600 kV are 56.37 and 63.19 dB, respectively.

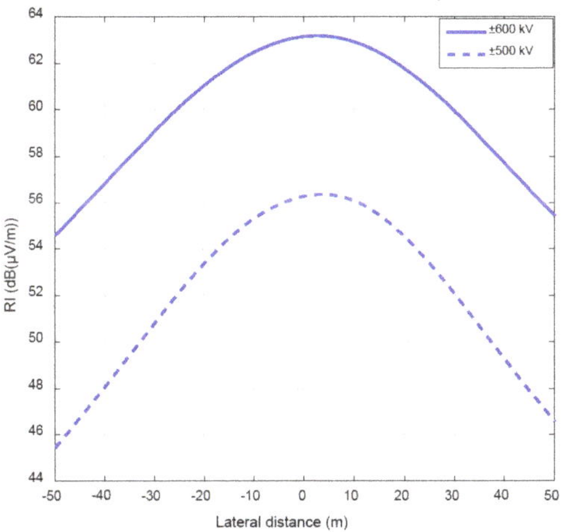

Figure 4. RI lateral profiles of the two HVDC transmission lines.

As can be seen in Figure 4, the maximum values of RI are placed closer to the positive pole because of the higher amplitude of the positive streamers in comparison with the Trichel pulses in negative polarity. In fact, the streamer pulses from positive corona are considered as the dominant source of the RI in DC transmission lines [9].

Some details of the calculations are summarized in Table 3. Note that an additional measurement at a reference point $(x, y) = (23\ \text{m}, 1.0\ \text{m})$ is included in the table, and this value will be used for comparisons in the parametric sweep computations presented in the next section.

Table 3. Results of the calculations for the bipolar HVDC lines.

Variable	Line of ±500 kV	Line of ±600 kV
Maximum bundle electric field (g_{max})	19.93 kV/cm	23.83 kV/cm
Excitation function (Γ)	6.65 dB	13.79 dB
Maximum value of RI	56.37 dB	63.19 dB
RI at point $(x, y) = (23\ m, 1.0\ m)$	53.85 dB	61.27 dB

4.2. Parametric Sweeps

In this section, parametric sweeps of five variables that affect the RI levels are presented. Both the RI and the maximum bundle electric field (g_{max}) are calculated as a function of sub-conductor radius, bundle spacing, and the number of sub-conductors in the bundle and, additionally, the RI levels are also calculated as a function of the soil resistivity and RIV frequency. In the parametric sweeps, the reported RI level was computed considering a measurement point at $(x, y) = (23\ m, 1.0\ m)$. The previous consideration is taken according to the Canadian Standards Association that specifies tolerable limits for RI in a reference point placed at 15 m from the outermost conductor of the power line [18]. For this case, the reference point is assigned as (23 m, 1.0 m), i.e., 23 m from the center of the tower and 1 m above the ground.

In Figure 5, it can be seen that both the RI and the maximum bundle electric field (g_{max}) tend to decrease as the sub-conductor radius increases. This behavior is observed for the two voltage levels, however, the decrease rate becomes smaller as the sub-conductor radius increases. Therefore, the sub-conductor radius is determined according to some other technical and economic considerations [1].

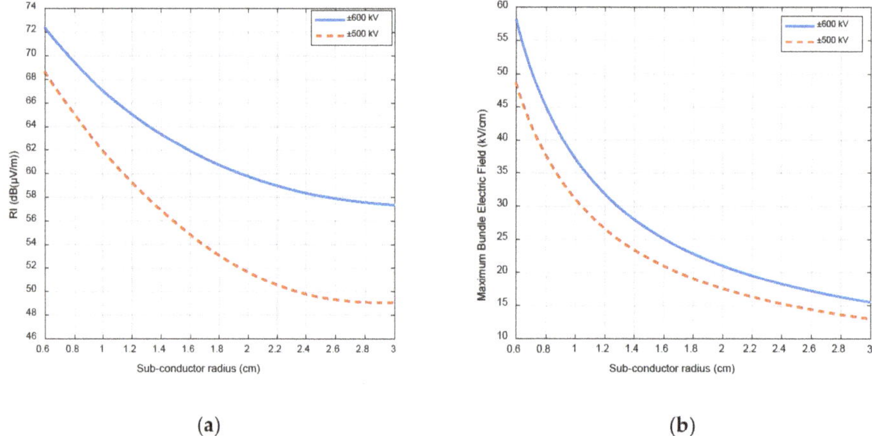

Figure 5. Parametric sweeps: (**a**) RI as a function of sub-conductor radius; (**b**) Maximum bundle electric field (g_{max}) as a function of sub-conductor radius.

Figure 6 shows the impact of bundle spacing on RI and g_{max} values. At first glance, the RI levels present a minimum for a bundle spacing around 30 cm; whereas the optimum bundle spacing for g_{max} (with a minimum value) is between 30 and 40 cm. This behavior shows that the bundle spacing is a critical variable for RI and g_{max} levels. Therefore, it should be selected carefully around the aforementioned optimal values while taking into account some other technical and constructional constraints.

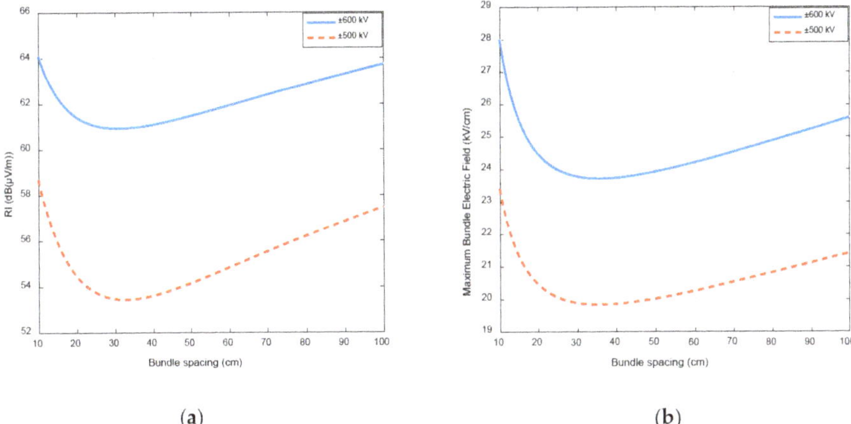

Figure 6. Parametric sweeps: (**a**) RI as a function of bundle spacing; (**b**) Maximum bundle electric field (g_{max}) as a function of bundle spacing.

Simulation results show that the RI levels go through a minimum value for a certain number of conductors in the bundle (Figure 7a), whereas g_{max} decreases continuously with the increment in the number of conductors in the bundle (Figure 7b). The lowest RI value for the ±500 and ±600 kV lines are achieved with five and six sub-conductors in a bundle, respectively. The sensitivity of RI levels to the number of conductors in the bundle is lower around the optimal point in the ±600 kV lines, which provides flexibility in the choice of the number of conductors. In fact, the increment in the RI levels—observed when the number of sub-conductors in a bundle is higher than six—can be due to the reference value of $n_0 = 6$ used in the excitation function. It is clear that the third term in Equation (7) becomes positive for values of n_c greater than 6.

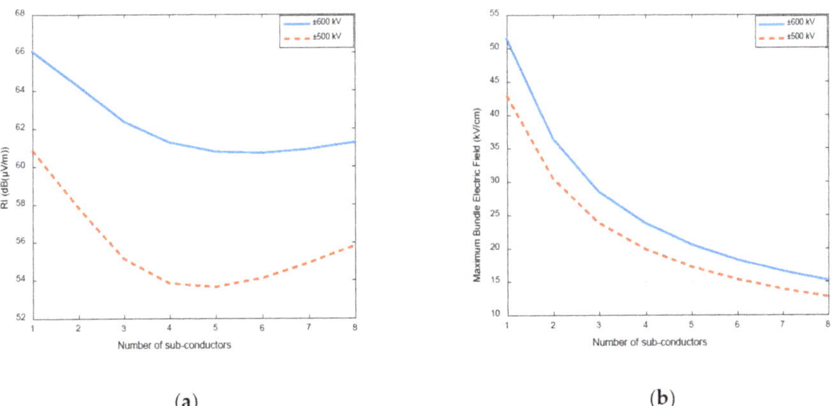

Figure 7. Parametric sweeps: (**a**) RI as a function of the number of sub-conductors in the bundle; (**b**) Maximum bundle electric field (g_{max}) as a function of the number of sub-conductors in the bundle.

The effect of soil resistivity on RI levels is illustrated in Figure 8. The RI levels show a minimum for a soil resistivity value between 2000 and 5000 Ωm for both systems. The maximum bundle electric field, g_{max}, is not dependent on the soil resistivity.

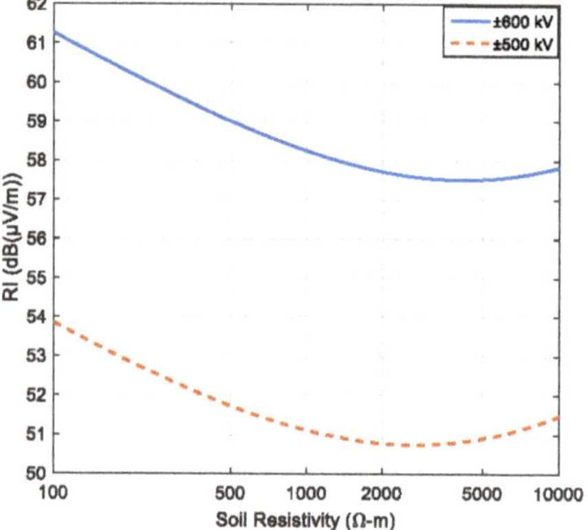

Figure 8. RI as a function of the soil resistivity.

Finally, the impact of RIV frequency on RI levels is illustrated in Figure 9. It can be seen how the RI values tend to decrease while the measuring frequency increases. This behavior is attributed to the increment in the modal attenuation constants with the frequency. g_{max} is not a function of the RIV frequency.

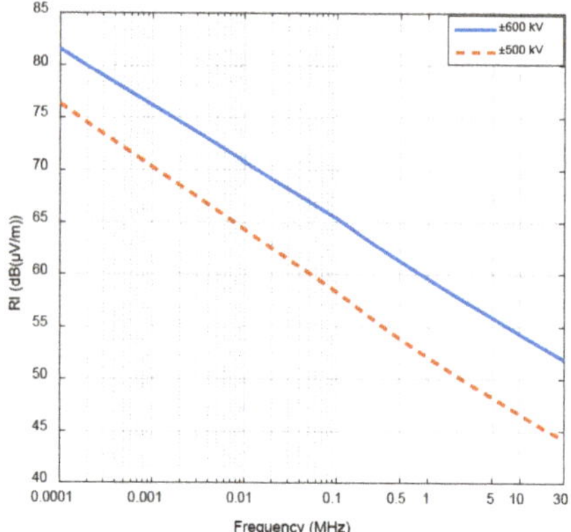

Figure 9. RI as a function of the RIV frequency.

According to the aforementioned results, it seems that an optimum HVDC line design can be obtained by varying the geometrical dimensions. Both the soil resistivity and the measurement frequency are not considered adjustable parameters during the design process; therefore, they are not included in the optimization process that is presented below.

5. Optimization of Radio Interference Levels in HVDC Lines

In this section, vector optimization is applied in order to minimize the RI levels produced by each of the two HVDC lines described in the previous section. Only the three geometric parameters—sub-conductor radius, bundle spacing, and the number of sub-conductors in a bundle—are taken into account in the optimization process. The genetic algorithm solver of the MATLAB optimization toolbox is used. The RI value at the reference point $(x, y) = (23$ m, 1.0 m$)$ is considered as the objective function. In Table 4, the parameters used in the optimization process for both ± 500 and ± 600 kV lines are shown.

Table 4. Genetic algorithm parameters for the MATLAB optimization toolbox.

Parameter	Value
Population type	Double vector
Population size	100
Scaling function	Rank
Selection function	Stochastic uniform
Elite count	1.5
Crossover fraction	0.8
Mutation function	Constraint dependent
Crossover function	Constraint dependent
Migration direction	Forward
Migration fraction	0.2
Migration interval	20
Constraint initial penalty	10
Constraint penalty factor	100

Based on the previous simulations, lower and upper limits are specified for each independent geometric variable. Optimization results are presented together with the variable limits in Tables 5 and 6 for the ±500 and ±600 kV HVDC lines, respectively.

Table 5. Optimal geometric parameters for ±500 kV HVDC line.

Variable	Lower Bound	Upper Bound	Nominal Value	Optimal Value
Sub-conductor radius	1.04 cm 477 MCM	2.21 cm 2167 MCM	1.71 cm 1272 MCM	2.21 cm 2167 MCM
Bundle spacing	20 cm	80 cm	45 cm	42 cm
Number of sub-conductors	2	8	4	3
RI at $(x, y) = (23\ \text{m}, 1.0\ \text{m})$			53.85 dB	50.23 dB
Total mass of the bundle			4 × 2135 = 8540 kg/km	3 × 3431 = 10,293 kg/km

Table 6. Optimal geometric parameters for ±600 kV HVDC line.

Variable	Lower Bounds	Upper Bounds	Nominal Values	Optimal Values
Sub-conductor radius	1.04 cm 477 MCM	2.21 cm 2167 MCM	1.71 cm 1272 MCM	2.21 cm 2167 MCM
Bundle spacing	20 cm	80 cm	45 cm	38 cm
Number of sub-conductors	2	8	4	4
RI at $(x, y) = (23\ \text{m}, 1.0\ \text{m})$			61.27 dB	58.88 dB
Total mass of the bundle			4 × 2135 = 8540 kg/km	4 × 3431 = 13,724 kg/km

According to the optimization results shown in Table 5 for the ±500 kV line, increasing the sub-conductor radius from 1.71 to 2.21 cm, reducing the bundle spacing from 45 to 42 cm, and reducing the number of sub-conductors from 4 to 3, the RI value calculated at the reference point $(x, y) = (23\ \text{m}, 1.0\ \text{m})$ is reduced from 53.85 dB with the nominal parameters to 50.23 dB with the optimal parameters. In Figure 10, the comparison of nominal and optimized RI lateral profiles for the ±500 kV line is shown. An almost constant reduction of 3.62 dB can be observed along the complete profile. Due to the increment in the sub-conductor radius, the total mass of the bundle increased from 8540 to 10,293 kg/km. The increment of 1753 kg/km should be evaluated from an economical point of view as well as from the strength of the tower to support this new load.

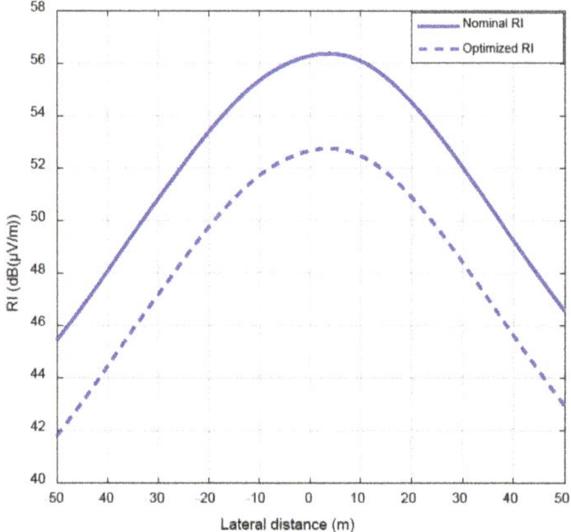

Figure 10. Comparison of nominal and optimized RI lateral profiles for ±500 kV line.

According to the optimization results for the ±600 kV line shown in Table 6, by increasing the sub-conductors radius from 1.71 to 2.21 cm, reducing the bundle spacing from 45 to 38 cm, and keeping the number of sub-conductors as 4, the RI value computed at the measuring point was reduced from 61.27 dB for the nominal parameters to 58.88 dB for the optimal parameters. In Figure 11, the comparison of nominal and optimized RI lateral profiles for the ±600 kV line is shown. The increment in the sub-conductor radius produces an increment in the total mass of the conductor bundle from 8540 to 13,724 kg/km. Similar to the ±500 kV line case, the increment of 5184 kg/km should be evaluated from the economical and mechanical point of view.

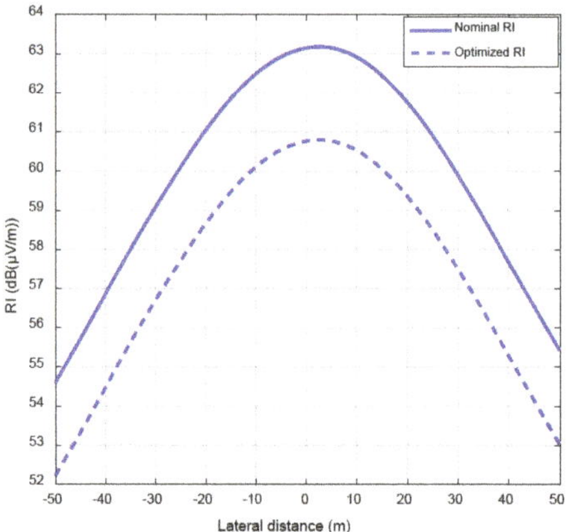

Figure 11. Comparison of nominal and optimized RI lateral profiles for ±600 kV line.

6. Discussion

With regard to the ±500 kV line, it is observed that this line presents an RI level (53.85 dB) below the maximum tolerable limit recommended by the relevant standards, which is usually 60 dB in the reference measuring point [18]. Nevertheless, optimization studies, such as the one applied in this case, allow for maintaining an acceptable RI performance of the line at even higher altitude regions. Considering an RI altitude correction factor of 1 dB/300 [19], the initial HVDC line design would exceed 60 dB at an altitude of 1900 m above the sea level; however, with the optimization applied, this line would present an acceptable RI performance (59.23 dB) until 2700 m above the sea level. On the other hand, even though the number of sub-conductors was reduced from 4 to 3, because of the increment in the sub-conductor radius, the total mass of the bundle increased from 8540 kg/km to 10,293 kg/km. This increment of 1753 kg/km (20.5%) in the total mass of the bundle should be evaluated from an economic point of view as well as from the strength of the tower to support this new load [20,21].

Regarding the ±600 kV HVDC transmission line, the RI value computed at the reference point (23 m, 1.0 m) is 61.27 dB, exceeding the limit of 60 dB for this voltage level. With the adjustments in the sub-conductor radius and bundle spacing derived from the optimization study, the RI value was reduced to 58.88 dB. This optimal case involves the use of larger conductors, increasing the total mass of the conductors by 5184 kg/km (60.7%), therefore, for this option, the adjustments derived from the optimization study seem to be feasible only in those cases that consist in the conversion of an AC line to DC operation, where one of the phases is removed.

In future work, it would be interesting to consider other variables, such as the height and the distance between the poles, a comparison of different configurations of the line (horizontal and vertical), and the inclusion of shielding wires in the model. Also, to determine the feasibility of the increment of the sub-conductor radius, it is necessary to consider the maximum mass that the towers can support as a constraint in the optimization study.

7. Conclusions

Radio interference lateral profiles in HVDC lines were computed using a method that considers a transmission line model which includes the skin effect in the conductors and in the ground plane. The attenuation constants were determined from the line parameters, and the bipolar system is decoupled using modal decomposition theory. As application cases, two HVDC bipolar transmission lines were analyzed, one of ±500 kV and the other of ±600 kV. The parametric sweeps of five prospective parameters revealed that, as expected, the RI levels decrease when increasing the sub-conductor radius. However, in the case of bundle spacing and the number of sub-conductors, the results show that these parameters should be carefully selected because these present a more complicated relationship with the magnitude of the RI level. Finally, vector optimization was applied to minimize RI levels and the feasibility of the obtained optimal designs discussed. This kind of electromagnetic compatibility study (EMC) can be used to estimate the magnitude of the impact on the environment in the vicinity of a transmission line during the design stage.

Author Contributions: Investigation, C.T.-M., F.P.E.-C., S.I. and A.O.; Methodology, C.T.-M. and F.P.E.-C.; Project administration, F.P.E.-C. and A.O.; Software, C.T.-M.; Validation, S.I. and A.O.

Funding: This research was funded by "The Scientific and Technological Research Council of Turkey TUBITAK" and "The National Council of Science and Technology-CONACYT of Mexico".

Acknowledgments: This research is funded as a part of "215E262/263956 Corona discharge characterization and electromagnetic effects of HVDC transmission lines" project under the framework of the "Bilateral Research and Technology Cooperation Turkey-Mexico" organized by "The Scientific and Technological Research Council of Turkey TUBITAK" and "The National Council of Science and Technology-CONACYT of Mexico".

Conflicts of Interest: The authors declare no conflict of interest.

References

1. Font, A.; Ilhan, S.; Ismailoglu, H.; Espino-Cortes, F.P.; Ozdemir, A. Design and Technical Analysis of 500–600 kV HVDC Transmission System for Turkey. In Proceedings of the 10th International Conference on Electrical and Electronics Engineering ELECO 2017, Bursa, Turkey, 30 November–2 December 2017.
2. Rodrigues, E.; Pontes, R.S.T.; Bandeira, J.; Aguiar, V.P.B. Analysis of the Incidence of Direct Lightning over a HVDC Transmission Line through EFD Model. *Energies* **2019**, *12*, 555. [CrossRef]
3. Pei, X.; Tang, G.; Zhang, S. A Novel Pilot Protection Principle Based on Modulus Traveling-Wave Currents for Voltage-Sourced Converter Based High Voltage Direct Current (VSC-HVDC) Transmission Lines. *Energies* **2018**, *11*, 2395. [CrossRef]

4. Naeem, R.; Cheema, M.S.; Ahmad, M.; Haider, S.A.; Shami, U.T. Impact of HVDC grid segmentation topology on transient stability of HVDC-segmented electric grid. In Proceedings of the 2015 International Conference on Open Source Systems & Technologies (ICOSST), Lahore, Pakistan, 17–19 December 2015; pp. 132–136.

5. Maruvada, P.S. *Corona Performance of High-Voltage Transmission Lines*; Research Studies Press Ltd.: Baldock, UK, 2000.

6. Electric Power Transmission and the Environment: Fields, Noise and Interference. In *CIGRE Technical Brochure*; No. 74; National Grid Research and Development Centre: Paris, France, 2000.

7. Tejada-Martinez, C.; Espino-Cortes, F.P.; Ilhan, S.; Ozdemir, A. Computation of Corona Radio Interference Levels in HVDC Transmission Lines. In Proceedings of the 10th International Conference on Electrical and Electronics Engineering ELECO 2017, Bursa, Turkey, 30 November–2 December 2017.

8. Tejada-Martinez, C.; Espino-Cortes, F.P.; Ilhan, S.; Ozdemir, A. Optimization of Corona Radio Interference Levels in HVDC Transmission Lines. In Proceedings of the 2018 International Conference on High Voltage Engineering and Application, ICHVE 2018, Athens, Greece, 10–13 September 2018.

9. Interferences Produced by Corona Effect of Electric Systems. In *CIGRE Technical Brochure*; No. 61; CIGRE Committee Report: Paris, France, 1996.

10. Impacts of HVDC Lines on the Economics of HVDC Projects. In *CIGRE Technical Brochure*; No. 388; Task Force: Paris, France, 2009.

11. Tejada-Martinez, C.; Gómez, P.; Escamilla, J.C. Computation of Radio Interference Levels in High Voltage Transmission Lines with Corona. *IEEE Lat. Am. Trans.* **2009**, *7*, 54–61. [CrossRef]

12. Paul, C.R. *Analysis of Multiconductor Transmission Lines*, 2nd ed.; Wiley-Interscience: New York, NY, USA, 2008.

13. Padiyar, K.R. *HVDC Power Transmission Systems: Technology and System Interactions*; Wiley: New York, NY, USA, 1990.

14. Tian, F.; Yu, Z.; Zeng, R. Radio Interference and Audible Noise of the UHVDC test line under high altitude condition. In Proceedings of the 2012 Asia-Pacific Symposium on Electromagnetic Compatibility, Singapore, 21–24 May 2012; pp. 449–452.

15. Yu, Z.; Zeng, R.; Li, M.; Li, R.; Liu, L.; Yang, D.; Zhang, Z.; Zhang, B.; Tian, F. Radio Interference of Ultra HVDC Transmission Lines in High Altitude Region. In Proceedings of the 2010 Asia-Pacific International Symposium on Electromagnetic Compatibility, Beijing, China, 12–16 April 2010; pp. 1672–1675.

16. Zhao, L.; Cui, X.; Xie, L.; Lu, J.; He, K.; Ju, Y. Altitude Correction of Radio Interference of HVdc Transmission Lines Part II: Measured Data Analysis and Altitude Correction. *IEEE Trans. Electromagn. Compat.* **2017**, *59*, 284–292. [CrossRef]

17. Baoquan, W.; Dichen, L.; Xiong, W.; Yao, L. The study on the radio interference from ±800kV Yun Guang UHVDC transmission line. In Proceedings of the 2006 International Conference on Power System Technology, Chongqing, China, 22–26 October 2006; pp. 1–5.

18. *Limits and Measurement Methods of EM Noise from AC Power Systems, 0.15–30 MHz*; Standard CAN3-C108.3.1-M84; Canadian Standard Association: Toronto, ON, Canada, 2014.

19. *IEC TR, CISPR 18-3, Radio Interference Characteristics of Overhead Power Lines and High Voltage Equipment—Part 3: Code of Practice for Minimizing the Generation of Radio Noise*; IEC: Geneva, Switzerland, 2010; pp. 30–31.

20. Yang, J.; Wu, J.; Li, M.; Han, J. Structural Optimization for the China UHV Transmission Steel Tower. In Proceedings of the 2010 Asia-Pacific Power and Energy Engineering Conference, Chengdu, China, 28–31 March 2010; pp. 1–4.

21. Vivek, K.S.; Nagendra, V. Optimized Design of Steel Transmission Line Tower by Limit State Methodology. *Int. J. Mod. Eng. Res. IJMER* **2015**, *5*, 81–100.

Article

Electrical Detection of Creeping Discharges over Insulator Surfaces in Atmospheric Gases Under AC Voltage Application

Michail Michelarakis [1,*]**, Phillip Widger** [1]**, Abderrahmane Beroual** [2] **and Abderrahmane (Manu) Haddad** [1]

[1]   Advanced High Voltage Engineering Research Centre, School of Engineering, Cardiff University, The Parade, Cardiff CF24 3AA, UK

[2]   École Centrale de Lyon, University of Lyon, Ampère CNRS UMR 5005, 36 Avenue Guy Collongue, 69134 Écully, France

*   Correspondence: MichelarakisM@cardiff.ac.uk

Received: 30 June 2019; Accepted: 30 July 2019; Published: 1 August 2019

Abstract: Creeping discharges over insulator surfaces have been related to the presence of triple junctions in compressed gas insulated systems. The performance of dielectric materials frequently utilised in gaseous insulating high voltage applications, stressed under triple junction conditions, has been an interesting topic approached through many different physical perspectives. Presented research outcomes have contributed to the understanding of the mechanisms behind the related phenomena, macroscopically and microscopically. This paper deals with the electrical detection of creeping discharges over disc-shaped insulator samples of different dielectric materials (polytetrafluoroethylene (PTFE), epoxy resin and silicone rubber) using atmospheric gases (dry air, N_2 and CO_2) as insulation medium in a point-plane electrode arrangement and under AC voltage application. The entire approach implementation is described in detail, from the initial numerical field simulations of the electrode configuration to the sensing and recording devices specifications and applications. The obtained results demonstrate the dependence of the generated discharge activity on the geometrical and material properties of the dielectric and the solid/atmospheric gas interface. The current work will be further extended as part of a future extensive research programme.

Keywords: creeping discharge; AC voltage; point-plane; atmospheric gases; flashover voltage; polytetrafluoroethylene (PTFE); epoxy resin; silicone rubber

---

1. Introduction

In gaseous insulation applications, triple junctions are defined as points where the gaseous medium, dielectric material and electrode meet leading to local electric field enhancements. When a certain electric field level is reached, the initiation of discharge activity is possible. In the case of their appearance, such phenomena become present at voltages below the optimised rated operating and withstand levels of the affected apparatus, introducing additional concerns in the overall effort of designing effective and reliable insulating systems. Some significant examples of triple junctions are incorporated into the design of Gas Insulated Lines (GIL), Gas Insulating Switchgear (GIS), transformer bushings and Gas Circuit Breakers (GCB).

The most common discharge phenomenon linked with enhanced electric fields in the vicinity of solid dielectric-gas interface is the streamer creeping discharge. Over the years, major contributions have been published, approaching the development and propagation of streamers over dielectric surfaces from several different experimental and computational perspectives. Characteristics of single streamer events in homogenous electric field arrangements have been reported [1,2] where the dependence of the propagation velocity and associated electric field over the dielectric material is described. In [3], the effect of the dielectric permittivity on streamer propagation along insulating surfaces using electrical and optical techniques is reported. The morphology and propagation length of creeping discharges and their dependence on voltage waveform, voltage levels, dielectric material and gaseous medium have been extensively studied in more recent works [4–7]. Reported works using dust figure [8] and Pockels effect methods [9] always constitute very interesting optical approaches. Another very important aspect, related to the phenomena described above, is the surface charge accumulation and its impact on the flashover voltage [10,11] and degradation of solid insulators under tests [12].

This paper examines the development of creeping discharge over disc-shaped insulator samples of different dielectric materials, namely polytetrafluoroethylene (PTFE), epoxy resin and silicone rubber, which are frequently utilised in high voltage technology applications. These samples are also insulated by a gaseous medium of either dry air, nitrogen (N_2) or carbon dioxide (CO_2). A needle-plane electrode configuration is employed with a strongly non-uniform electric field at 50 Hz AC voltage applied to the needle electrode, in order to replicate triple junction conditions. The implementation of the test procedure is described from the early stages of the numerical field simulation process of the electrodes arrangement using a Finite Element Solver (FEM) software package. The aim of the accurate modeling of the electrode geometry is the optimisation of the electric field distribution along the insulator surface. Detection of surface discharges is performed by means of sensing associated currents using a high-sensitivity, high frequency current transformer (HFCT) ranging from a few kHz up to several MHz bandwidth. The current transformer sensing technique provides several performance advantages over other techniques implementing different physical principles [13], such as low power loss, high accuracy and no need for further amplification of the output. Additionally, the overall convenience of a HFCT installation, together with the provided electrical isolation between the system under test and the high-cost recording devices, make the technique a very practical choice for laboratory-based high voltage testing.

2. Design and Implementation of the Test Procedure

2.1. Electrodes Configuration Simulation and Electric Field Computation

As mentioned in the previous section, the aim of the test procedure is to replicate triple junction conditions in the vicinity of a dielectric insulator surface surrounded by a gaseous insulating medium. For that reason, a needle-plane electrode configuration is employed. A two-dimensional illustration along with an actual picture of the electrode configuration, showing the disc-shaped insulator sample under test conditions are shown in Figure 1.

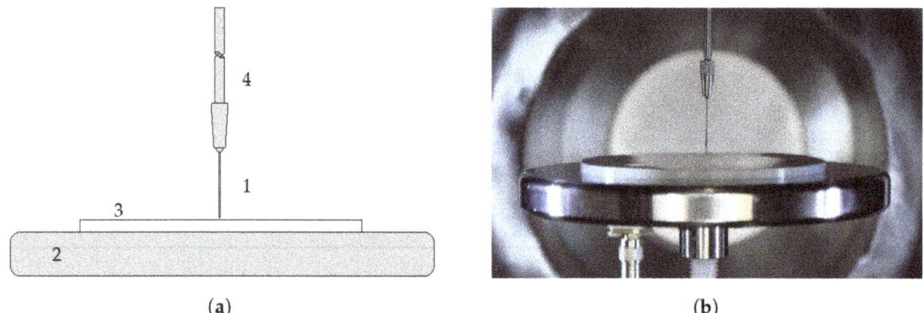

(a)   (b)

Figure 1. Needle-plane electrode configuration with a disc-shaped insulator sample: (**a**) Two-dimensional illustration of the configuration with the following numbered features: 1-needle electrode, 2-plane electrode, 3-disc-shaped insulator sample and 4-needle holder; (**b**) Actual picture of the practical test configuration as set-up inside the pressure vessel.

The needle is made of tungsten with a high-precision tip diameter of 20 ± 0.5 μm, 0.51 mm shaft diameter and a length of 32 mm. It is attached to a stainless-steel needle holder placed perpendicularly and in close proximity to the insulator sample disc center. The plane electrode incorporates a dull polished stainless-steel planar surface of 150 mm diameter and 15 mm thickness and is electrically separated from the rest of the mounting system with a nylon threaded rod. The entire configuration is placed vertically and centered inside a cylindrical 90 l dull polished stainless-steel pressure vessel rated up to 10 bars operational gauge pressure, which has a diameter of 480 mm and a height of 500 mm.

The described design is further examined from the numerical electric field perspective using a Finite Element Method (FEM) simulation model. The detailed geometry is transferred to COMSOL Multiphysics® through its integrated three-dimensional geometry builder, with some geometry simplifications that can improve significantly the computational time without affecting the quality and reliability of the generated result. As the design is largely symmetrical around the central axis of the entire configuration, only half of the geometry needs to be computationally solved without affecting the accuracy of the computed results. It is expected that the maximum electric field stress will appear at the needle tip because of its small diameter and short distance from the zero potential plane electrode. Figure 2 shows the simulated full geometry of the test system, as designed within COMSOL Multiphysics®, together with a generated illustration of the three-dimensional equipotential surfaces (isosurfaces) after solving the simulation model. For the example shown in Figure 2b, the considered insulator sample is made of PTFE with a 4 mm thickness; the insulating gaseous medium being air at standard atmospheric conditions. The applied voltage is set to 1 V.

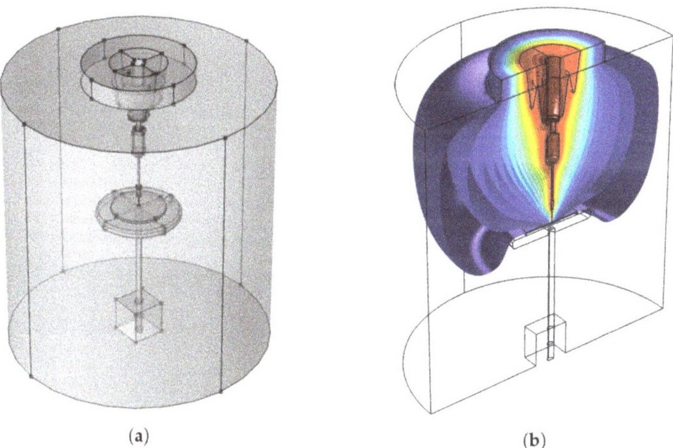

(a)   (b)

Figure 2. Finite Element Method (FEM) simulation model: (**a**) Full geometry including electrodes arrangement and surrounding volume; (**b**) Three-dimensional equipotential surfaces (isosurfaces) within the half symmetrical volume of the initial geometry.

The distribution of the electric field on the surface of the insulator sample is of great significance as the possibility of large irregularities may lead to inaccurate observations. In a broad sense, in case of discharge activity appearance, it will initiate from the sharp needle tip and will have equal chance of propagating in any direction towards the edges of the disc insulator. Figure 3 shows the computed electric field distribution along the diameter of the insulator sample top surface. The presented electric field values are normalised to the maximum surface electric field that appears at the point with the shortest distance from the needle tip.

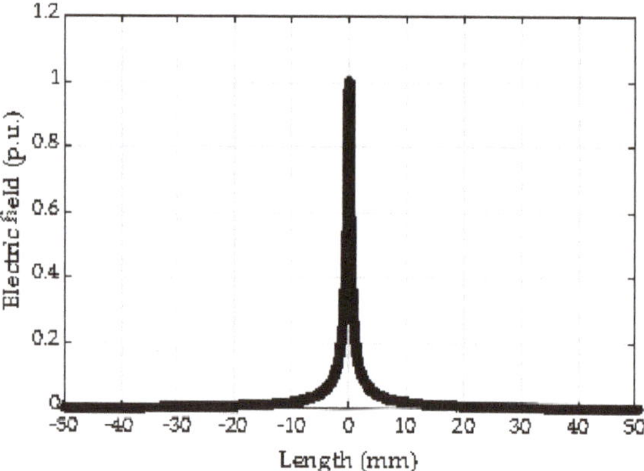

Figure 3. Electric field distribution along insulator surface. Here, 0 mm represents the center of the top surface of the insulator sample while ±50 mm the edges of it. The presented values are normalised based on the maximum computed value.

2.2. Experimental Set-Up

Figure 4 summarises the main parts of the experimental configuration used for the purposes of this work. For the AC voltage application, a 50 kV/3.75 kVA transformer is used, which is controlled through an isolating transformer and a voltage regulator with adjustment of the voltage level on the low-voltage side. An RC voltage divider of ratio 3750:1 is used to scale down the generated voltage to safe measurable levels. Voltage is applied to the pressure vessel through a bushing rated up to 39 kV AC rms. As described previously, the electrodes configuration, together with the insulator samples under test, are vertically and horizontally centered inside the cylindrical stainless-steel dull polished chamber. The side apertures are covered with stainless-steel blanking plates, preventing ambient light entering into the pressure vessel.

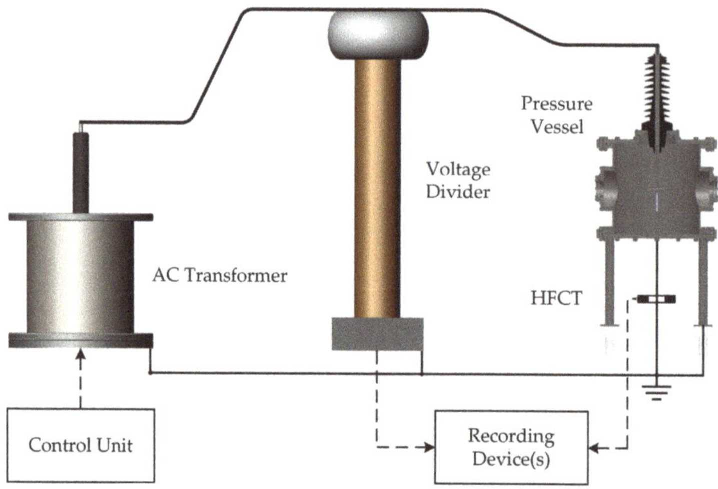

Figure 4. Experimental set-up for AC voltage application to the needle-plane configuration and insulator samples.

The insulator samples tested are disc-shaped with a diameter of 100 mm and thicknesses of 4 and 6 mm. Three different materials are considered namely polytetrafluoroethylene (PTFE), silicone rubber and epoxy resin with relative permittivities (ε_r) of 2.1, 2.9 and 3.5, respectively. The PTFE and epoxy resin samples surfaces were polished with sandpaper of 800 grit gradually increasing up to 7000 grit size. The silicone rubber samples were vacuum casted using a highly polished stainless-steel mold, followed by at least 24 h curing at 50 °C. For all cases, arithmetical mean height (R_a) is maintained between 0.6–0.8 µm. The insulator samples are cleaned with high-purity isopropyl alcohol and dried for at least 4 h at 50 °C. Between drying and placing the samples inside the final test set-up, a brief time interval of 30 min is kept allowing the samples to return to room temperature levels. After each test series, each insulator is examined for traces, indicating degradation/damage, and if any is observed on the surface, it is replaced with a new unused one. As described previously, the needles are made of pure tungsten, a material well-known for its high melting temperature, allowing for a sufficiently large number of tests without degradation of the needle tip quality. During the creeping discharge detection tests, a maximum of twenty voltage applications are made before the needle is replaced. For the case of flashover tests, the needle is used for a maximum of ten attempts prior to being replaced, assuming that no abrupt deviations in the readings of successive attempts occur which would indicate a damaged needle. For the purposes of the presented work, the test vessel housing the test electrode configuration is filled and tested using three different atmospheric gases, respectively: dry air, nitrogen (N_2) and carbon dioxide (CO_2) at 1 bar absolute pressure. Preceding gas injection, the pressure vessel is vacuumed for 30 min after a vacuum level of -1000 mbar gauge pressure is reached. This additional vacuum is held in an attempt to maintain low humidity levels inside the test chamber.

Two different, although similar, patterns for the AC voltage application are followed. For flashover tests, the applied voltage is manually increased with a rate of 1 kV/s until flashover occurs. A lower sensitivity, 0.1 V/A, current transformer is used for triggering the recording device. The last full AC-cycle recorded, preceding the flashover event, corresponds to the measured rms flashover voltage. Ten flashover tests are performed for each insulator sample/gaseous medium combination. The arithmetic mean value and standard deviation for the flashover voltage for each case are specified as implied from the relevant standards [14]. For creeping discharge detection tests, test voltages are specified as a percentage of the flashover voltage for each case study. Following that, the test voltage is gradually reached with an increment of 2% per second resulting in a ramp duration of 50 s for each case. After that, the voltage level is maintained for a maximum of 10 s in order to avoid overstress of the insulator sample and the utilised needle. Between two successive voltage applications, a time interval of at least 2 min is maintained.

2.3. Current Sensing & Recording

Several methods have been reported for the detection of electrical phenomena within the high-frequency (HF), very-high-frequency (VHF) and ultra-high-frequency (UHF) regions, including acoustic, optical, chemical and electrical methods. These methods find wide application in the detection of partial discharges [15–19]. In this work, direct measurement of the currents associated with creeping discharge is performed using a high-sensitivity 5 V/A, high frequency current transformer (HFCT) ranging from 4.8 kHz up to 400 MHz. Considering that the test configuration is installed inside a grounded metal fabricated pressure vessel, it is important that the generating signals will exit the test chamber and, consequently, reach the sensing device undistorted and unattenuated. For that purpose, a flange mount SMA connector is installed at the bottom side of the plane electrode of Figure 1. A 50 Ω RG-405 coaxial cable connects the plane electrode with a pre-installed high-pressure rated coaxial feedthrough. A coaxial cable connected to the low-pressure side of the feedthrough is then connected to the configuration shown in Figure 5, similar implementations of which were presented in [20,21].

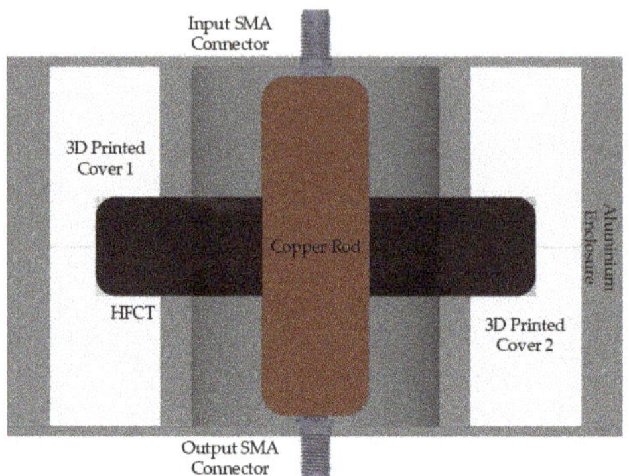

Figure 5. HFCT enclosure configuration.

Two SMA connectors are mounted on the top and bottom of a C101 alloy copper rod. The diameter of this central conductor was calculated based on the 50 Ω coaxial characteristic impedance principle using Equation (1) [15],

$$Z_0 = \frac{138}{\sqrt{\varepsilon_{air}}} \log \left(\frac{d_1}{d_2} \right) \tag{1}$$

where, Z_0 is the characteristic impedance, d_1 the diameter of the HFCT aperture, d_2 the diameter of the copper conductor and ε_{air} the relative permittivity of atmospheric air. The entire configuration is enclosed in an EMI/RFI shielded aluminium enclosure. The scattering parameters of the configuration were measured using an R&S® ZVL Vector Network Analyzer (9 kHz − 6 GHz), following a 50 Ω two-port calibration, and the results are presented in Figure 6. As shown, the upper cut-off frequency is measured at 359.46 MHz, which is quite close to the bandwidth rating of the HFCT. The output side is terminated through a short-circuited 20 dB/50 W, DC-8.5 GHz, high-power fixed attenuator, preventing possible reflections reaching the sensing configuration.

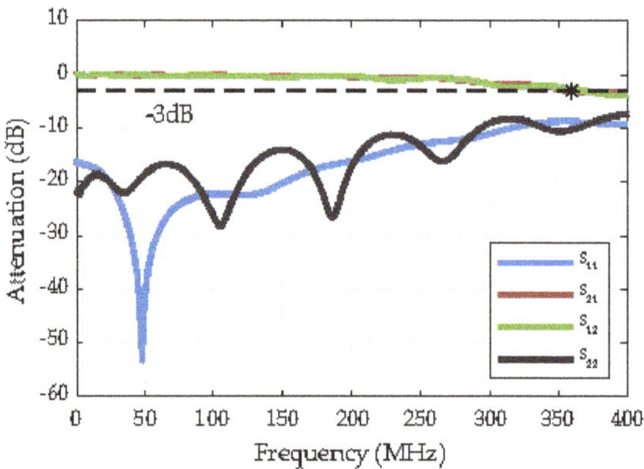

Figure 6. S-parameters of the configuration shown in Figure 5, for the frequency range corresponding to the rating of the used HFCT.

The recording of the voltage waveform and associated currents was performed in two different stages: high-resolution single trigger recordings were made using a Teledyne LeCroy HDO6104, 1 GHz, 2.5 GS/s, 12-bit oscilloscope while, for multiple successive trigger events, a large buffer PicoScope® 5000 series, 200 MHz, 250 MS/s, 12-bit was used.

3. Experimental Results and Discussion

3.1. Flashover Tests

A series of flashover tests were performed using all the solid dielectric-insulating gas combinations considered in this work. Ten flashover events were recorded for each case. These tests were implemented in order to quantify the equivalent threshold voltage level for creeping discharges as a percentage of the corresponding mean flashover voltages of each case. The obtained results are summarised in Figure 7. For the case of the 6 mm thick PTFE sample in dry air, no flashover events were recorded within the allowed application voltage levels up to 39 kV rms, hence, it is not included in the graph.

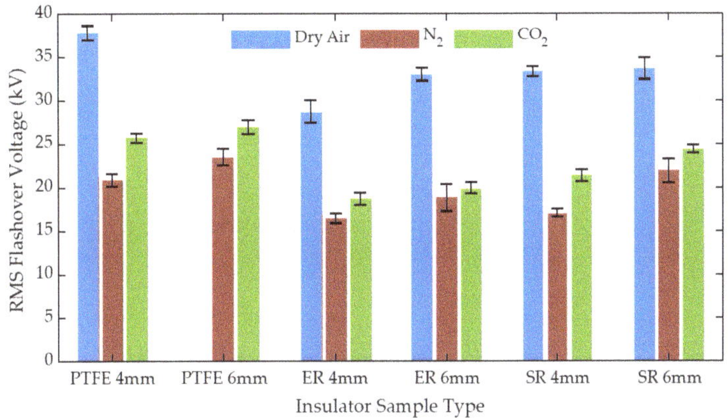

Figure 7. Flashover voltage (FOV) test results for six different insulator samples in dry air, nitrogen (N$_2$) and carbon dioxide (CO$_2$), all at 1 bar absolute pressure. Here, arithmetic mean values are depicted together with the corresponding standard deviations after 10 voltage applications. ER stands for epoxy resin while SR for silicone rubber.

For all the different sample types, dry air showed the best insulating performance as compared to the two other gaseous mediums. Additionally, pure CO$_2$ showed a better performance compared to pure nitrogen, with the difference in FOVs being larger for 4 mm samples compared to those of 6 mm thickness. Focusing on the calculated standard deviations, N$_2$ showed a more stable insulating behaviour for thinner samples, for all the materials compared to thicker ones insulated with the same gas, while CO$_2$ seemed to have on average the lowest standard deviations. The electrode system with PTFE ($\varepsilon_{PTFE} = 2.1$) had the highest resistance to flashover for all gaseous mediums and sample thicknesses considered. Despite the results of the other two materials being close, the configuration with silicone rubber ($\varepsilon_{SR} = 2.9$) seemed to have a slightly higher withstand voltage compared to that with epoxy resin ($\varepsilon_{ER} = 3.5$). Figure 8 shows the dependence of the computed electric field and measured flashover voltages on the different values of dielectric permittivity. The values of each curve are normalised to the, computed or measured, corresponding value for PTFE. It is obvious that, as the dielectric permittivity increases, the electric field on the center of the disc insulator sample also increases, while, the flashover voltage levels decrease for all the cases considered. Such an observation can be correlated with results reported in research works [5] where the stopping length of creeping discharges is examined, showing that insulator discs made of higher dielectric permittivity materials are responsible for the propagation of longer streamer channels on their surface for both N$_2$ and CO$_2$, but also for SF$_6$. AC breakdown test results of point-plane arrangements for different gap distances, presented in [7], also showed that the insulating performance of CO$_2$ is better when compared to N$_2$, without a dielectric material between the electrodes.

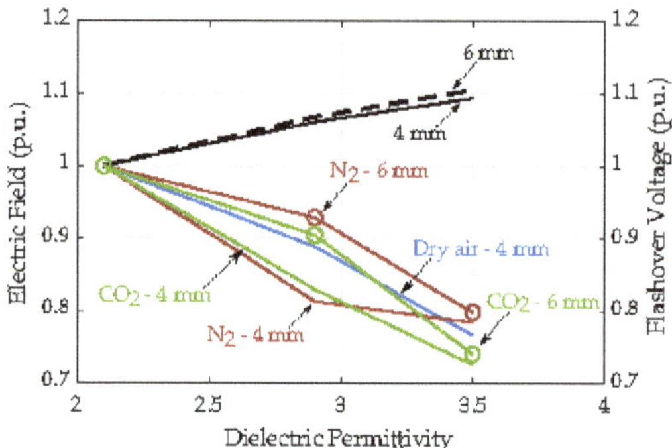

Figure 8. Computed electric field (black curves) and measured flashover voltages over dielectric permittivity variation. All the curves are normalised over the corresponding value of $\varepsilon_r = 2.1$ for PTFE.

3.2. High-Resolution Recordings

High-resolution recordings were obtained through single trigger events of the recording device at dual-channel, sampling at 1.25 GS/s rate. Two signals of 40 ms duration each were captured, corresponding to two full AC-cycles. The applied voltage levels considered for each case correspond to 85% of the flashover levels presented in previous section. That way, it is possible to compare between the different cases, where the variations in the FOVs do not allow the application of the same test voltage levels. Additionally, a 15% safety margin from flashover events was maintained, avoiding unwanted stress of the test configuration in case of flashover. Figure 9 shows the obtained recordings for PTFE of 4 mm thickness in dry air, nitrogen (N_2) and carbon dioxide (CO_2) at 1 bar absolute pressure. Because similar patterns were observed for the remaining insulator samples types, only the above selected results were shown here.

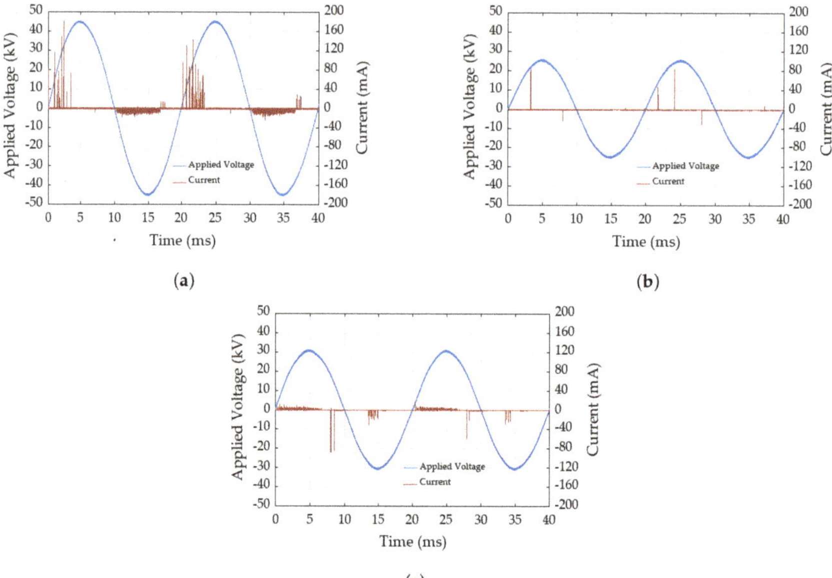

Figure 9. PTFE of 4 mm thickness in (**a**) dry air at 1 bar absolute pressure and 31.74 kV rms applied voltage, (**b**) N_2 at 1 bar absolute pressure and 17.74 kV rms applied voltage and (**c**) CO_2 at 1 bar absolute pressure and 21.61 kV rms applied voltage.

For the case of dry air in Figure 9a, a high density of positive and negative polarity current pulses is observed. It is obvious that positive polarity current peaks amplitudes dominate over the negative spikes. However, significant differences in the current pulse characteristics were observed. For nitrogen (N_2), as shown in Figure 9b, a very different behaviour is observed when compared to dry air. During the positive AC half cycle of the applied voltage, a small number of positive current pulses, although quite high in amplitude, were detected while absence of activity during the negative half cycle is obvious. Finally in Figure 9c, the behaviour of carbon dioxide (CO_2) differs significantly from both dry air and N_2. Low amplitude positive current pulses were detected during the positive half cycle of the applied voltage while, during the negative half cycle, higher amplitude but less dense negative polarity current pulses were observed.

Wide bandwidth sensing and recording devices combined with high sampling rates allow the capturing of current pulses corresponding to creeping discharge events. For the presented test results, applied voltages were very close to the FOVs and most of the recordings consist of superimposed pulses, indicating multiple discharge events occurring in very fast time frames on the surface of the insulator sample. Single streamer current pulses are also present however, less frequent. Current pulses are classified into two different categories: those having the same polarity as the applied voltage and those with opposite polarity. Positive polarity pulses during the positive half cycle of the AC voltage application are present for all the cases with varying amplitudes and characteristics. Examples of positive polarity pulses exported from the recordings in Figure 9 are shown in Figure 10 with their time resolved characteristics being summarised in Table 1. Negative polarity current pulses are also detected, the recordings of which are shown in Figure 11, with their corresponding characteristics summarised in Table 2. Similar pulse shapes were observed for all the insulator sample materials. The apparent charge, which is also depicted, was calculated by integrating the recorded pulse over the corresponding time domain.

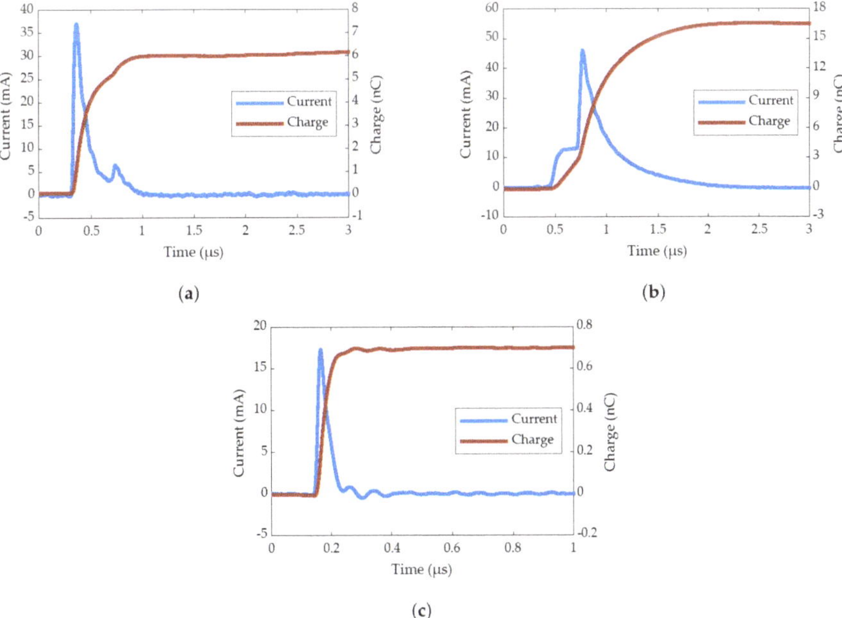

Figure 10. Positive polarity current pulses and calculated apparent charge for PTFE of 4 mm thickness in: (**a**) dry air, (**b**) nitrogen (N_2) and (**c**) carbon dioxide (CO_2).

Table 1. Positive polarity current pulses characteristics.

Gaseous Medium	Current Peak (mA)	Rise Time (ns)	Fall Time (ns)	Width (ns)	Duration (ns)
Dry air	37.07	28.49	263.42	110.26	326.21
N_2	46.14	274.25	662.93	192.94	975.02
CO_2	17.35	13.77	49.06	35.54	73.08

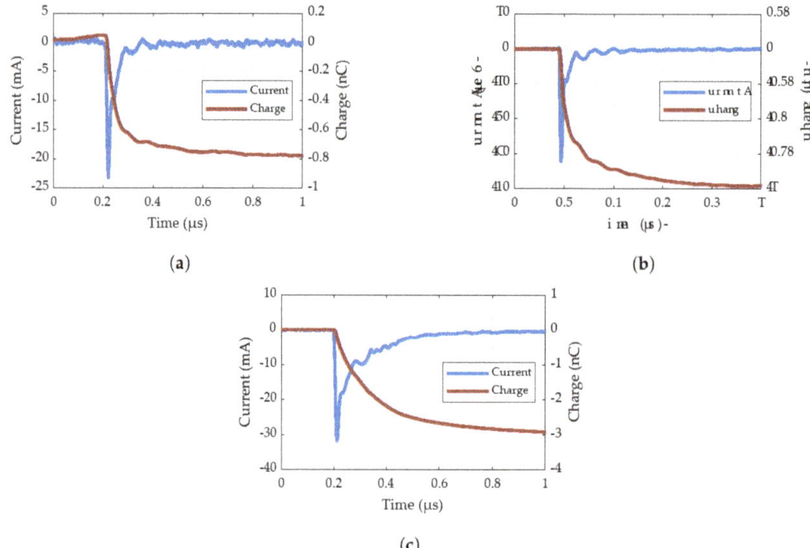

(a)

(b)

(c)

Figure 11. Negative polarity current pulses and calculated apparent charge for PTFE of 4 mm thickness in: (a) dry air, (b) nitrogen (N_2) and (c) carbon dioxide (CO_2).

Table 2. Negative polarity current pulses characteristics.

Gaseous Medium	Current Peak (mA)	Rise Time (ns)	Fall Time (ns)	Width (ns)	Duration (ns)
Dry air	−23.19	51.81	7.95	16.14	65.00
N_2	−32.23	41.63	5.76	12.94	52.04
CO_2	−31.63	232.38	8.53	38.58	247.65

Significant differences in the measured rise and fall times are observed. For all the tests performed for the purposes of the presented work, the fastest rise and fall times, for these kinds of pulses, were observed in carbon dioxide (CO_2) while the slowest were seen in nitrogen (N_2). The calculated apparent charge appears to be dependent and proportional to these values, with the highest charge being observed for the case of N_2 in Figure 10b. In [22], the authors describe the correlation between rise and fall times with the electron avalanche process during partial discharge (PD) and electron drift after the full extension of the discharge activity. For the single double-exponential shaped pulses, such an approach can provide valuable information. Figure 11 includes captures of negative polarity current pulses during the negative half cycle of the AC applied voltage for dry air and CO_2, while for nitrogen the negative polarity current pulse appears during the positive AC half cycle. Pulses of opposite polarity to that of the applied voltage were observed in all measurements. Similar findings, using the Pockels effect optical method, were reported in [9] and identified as back-discharges. Following the provided description for the negative discharges during positive polarity of the applied voltage, the negative charge expands uniformly on the positively charged surface neutralizing the positive charges. The opposite process occurs for positive discharges during the negative half cycles.

3.3. Current Peaks Density Recordings

High resolution recordings, within the GS/s sampling rate levels, usually do not allow capturing of multiple trigger events which are able to demonstrate the repeatability of the generated phenomena, during the same applied voltage test. For that purpose, a lower sampling rate, larger buffer size, fast-triggering device was employed. In this way, the quick recording of multiple full AC-cycle trigger events for the same applied voltage attempt was possible without overstressing the test object. Post-processing of these recordings involves isolating the peak values of the current pulses within the time domain and incorporate them in a concatenating plot where colour grading was applied based on the density of these peaks. The colour identification was normalised over the total number of detected peaks based on their polarity. Figures 12–14 illustrate part of the results using the described technique for the cases of 4 mm thickness insulator samples while Figure 15 includes selected datasets for 6 mm thick samples. Here, 25 trigger events of 20 ms duration each are considered, resulting in a total duration of 500 ms for each case. For all the presented measurements, the applied voltage is approximately equal to the 85% of the corresponding FOV.

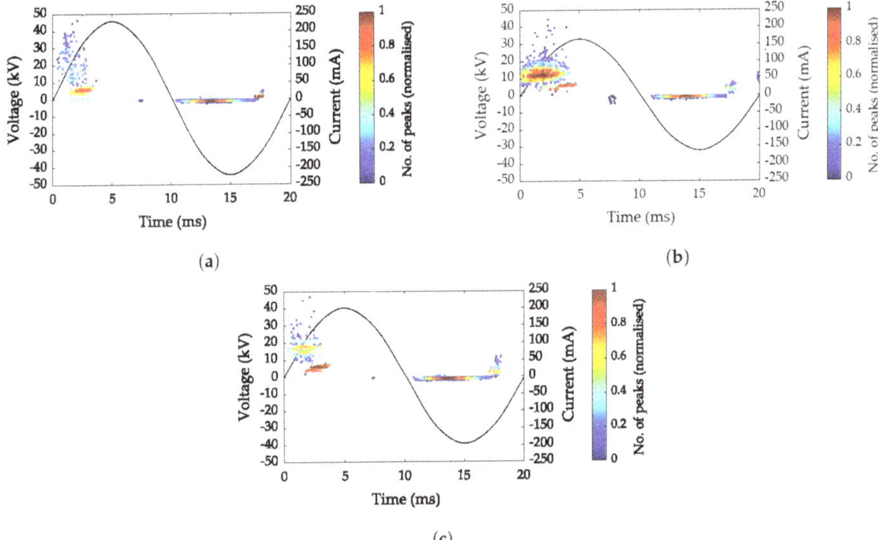

Figure 12. Current peaks densities for 4 mm thickness insulator samples in dry air at 1 bar absolute pressure: (**a**) PTFE, (**b**) epoxy resin and (**c**) silicone rubber.

In Figure 12, for samples in dry air, 4 mm thickness samples of PTFE and silicone rubber have fairly similar behaviour while epoxy resin shows a higher density of discharge activity during the positive half cycle of the applied voltage. For samples of the same thickness in N_2 in Figure 13, negative polarity pulses during the negative half cycle were not detected. For silicone rubber, a high density of low amplitude positive polarity discharges was found during the positive half cycles while, for epoxy resin samples, very high positive peaks were captured, possibly indicating that any increase in the applied voltage could have resulted in a flashover. PTFE shows a well distributed activity within the first quarter of the applied voltage waveform. It could be said that CO_2 in Figure 14 shows the most stable behaviour. PTFE and silicone rubber once again are very similar while PTFE shows less dense activity during the negative polarity of the applied voltage. For tests with the epoxy resin sample, no detectable pulses were recorded during the second half of the AC waveform for that specific applied voltage level. Tests with thicker samples in Figure 15 show that PTFE in CO_2 shows decreased peak amplitudes during the negative half cycle, while epoxy resin in N_2 shows considerably lower peaks and high repeatability for the 6 mm samples compared to the 4 mm. Silicone rubber of 6 mm thickness in dry air shows increased resistance to creeping discharge activity in comparison with the 4 mm sample of the same material, especially when the polarity of the applied voltage is positive. Overall, the dependence of the insulator sample thickness is visible for the presented cases. For tests with epoxy resin of 4 mm thickness in CO_2, the applied voltage margin for occurrence of the discharge activity in the negative half cycle until flashover is small. Results of N_2 combined with 4 mm thickness samples need further investigation, especially for the case of epoxy resin.

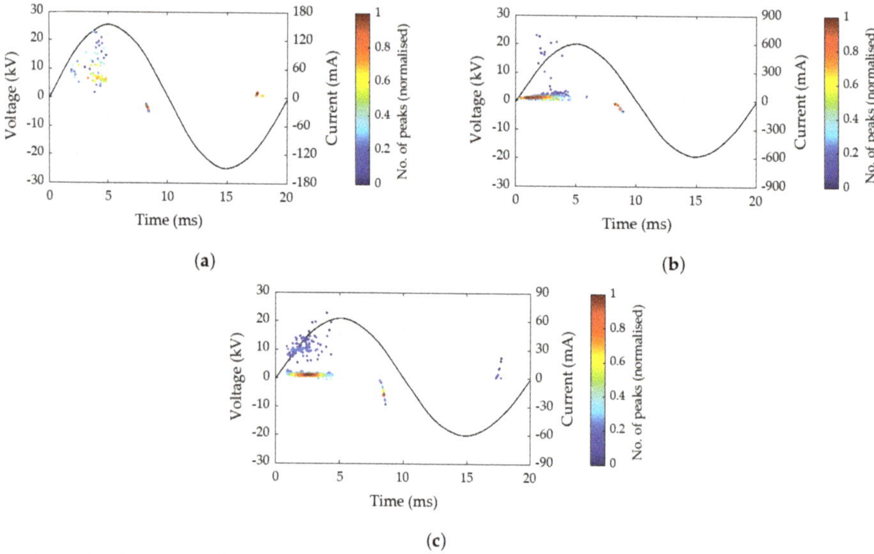

Figure 13. Current peaks densities for 4 mm thickness insulator samples in nitrogen (N_2) at 1 bar absolute pressure: (**a**) PTFE, (**b**) epoxy resin and (**c**) silicone rubber.

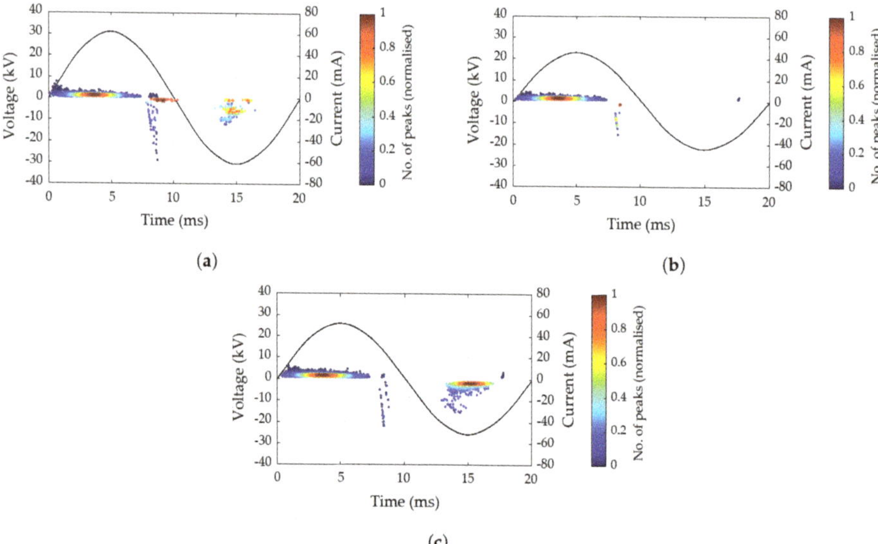

Figure 14. Current peaks densities for 4 mm thickness insulator samples in carbon dioxide (CO_2) at 1 bar absolute pressure: (**a**) PTFE, (**b**) epoxy resin and (**c**) silicone rubber.

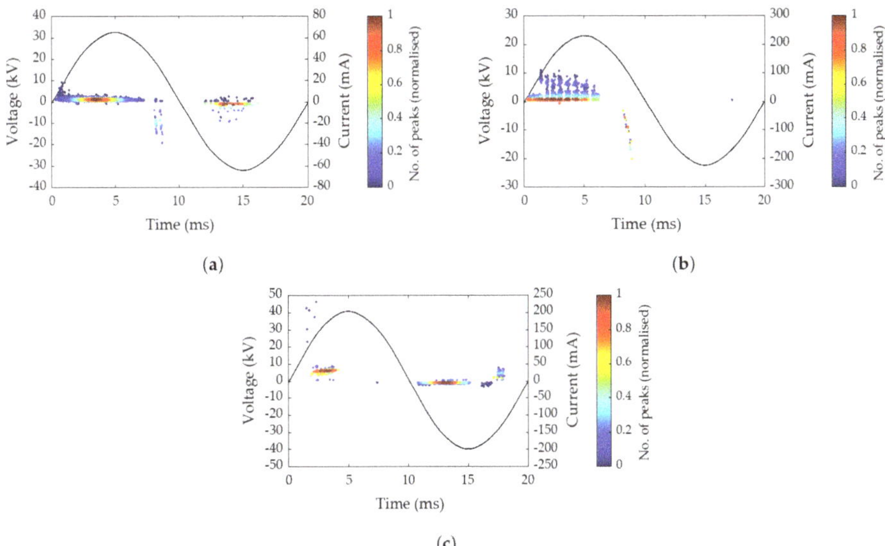

Figure 15. Current peaks densities for 6 mm thickness insulator samples: (**a**) PTFE in CO_2 at 1 bar absolute pressure, (**b**) epoxy resin in N_2 at 1 bar absolute pressure and (**c**) silicone rubber in dry air at 1 bar absolute pressure.

4. Conclusions

In this paper, electrical detection of creeping discharge under AC voltage application was presented. The complete procedure, starting from the numerical simulation of the electrode configuration until the final test implementation was described in detail. Emphasis was given to the selection of current sensing and recording devices so that their cut-off frequency is kept as high as possible.

Flashover tests results showed that, for the examined electric field distribution and for all the examined insulating gaseous media (dry air, CO_2 and N_2), lower flashover voltages were measured for electrode configurations with insulating materials having higher relative permittivity. Furthermore, high-frequency recordings of the current pulses associated with creeping discharges clearly indicated that there is more significant electrical discharge activity when the relative permittivity of the insulating material is higher. For lower permittivity insulator sample materials, fewer current pulses were detected however, the activity was spread greatly within a range corresponding to relatively higher amplitudes. It could be said that, the higher the difference between the relative permittivities of the insulator sample material and the insulating gaseous medium, the stronger the detected creeping discharge events are, especially for the cases of dry air and N_2. For the case of CO_2, the higher the relative permittivity of the insulator material the narrower the margin between creeping discharge and flashover is. Back-discharges were detected for all the cases tested in this investigation.

Future research will extend the current work and further explore the presented scenario using different physical perspectives and detection techniques.

Author Contributions: Conceptualization, M.M., P.W., A.B. and A.H.; methodology, M.M.; software, M.M.; validation, P.W., A.B. and A.H.; formal analysis, M.M.; investigation, M.M.; data curation, M.M.; writing—original draft preparation, M.M.; writing—review and editing, M.M., P.W., A.B. and A.H.; visualization, M.M.; supervision, P.W., A.B. and A.H.

Funding: The initial research leading to this work received funding from the Royal Academy of Engineering through a DVF to Professor A. Beroual to visit Professor Manu Haddad at Cardiff University, "Environmentally-friendly insulating gases for high voltage equipment", Grant number (DVF1617\5 \28): https://www.raeng.org.uk/grants-and-prizes/grants/international-research-and-collaborations/ distinguished-visiting-fellowships. In support to the DVF award, the PhD studentship awarded to Michail Michelarakis was funded by the School of Engineering at Cardiff University.

Conflicts of Interest: The authors declare no conflict of interest. The funders had no role in the design of the study; in the collection, analyses, or interpretation of data; in the writing of the manuscript, or in the decision to publish the results.

References

1. Allen, N.L.; Mikropoulos, P.N. Streamer propagation along insulating surfaces. *IEEE Trans. Dielectr. Electr. Insul.* **1999**, *6*, 357–362, doi:10.1109/94.775623. [CrossRef]
2. Allen, N.L.; Mikropoulos, P.N. Dynamics of streamer propagation in air. *J. Phys. D Appl. Phys.* **1999**, *32*, 913–919, doi:10.1088/0022-3727/32/8/012. [CrossRef]
3. Akyuz, M.; Gao, L.; Cooray, V.; Gustavsson, T.G.; Gubanski, S.M.; Larsson, A. Positive streamer discharges along insulating surfaces. *IEEE Trans. Dielectr. Electr. Insul.* **2001**, *8*, 902–910, doi:10.1109/94.971444. [CrossRef]
4. Sadaoui, F.; Beroual, A. DC creeping discharges over insulating surfaces in different gases and mixtures. *IEEE Trans. Dielectr. Electr. Insul.* **2014**, *21*, 2088–2094, doi:10.1109/TDEI.2014.004486. [CrossRef]
5. Sadaoui, F.; Beroual, A. AC creeping discharges propagating over solid—gas interfaces. *IET Sci. Meas. Technol.* **2014**, *8*, 595–600, doi:10.1049/iet-smt.2014.0050. [CrossRef]
6. Beroual, A.; Coulibaly, M.L.; Aitken, O.; Girodet, A. Investigation on creeping discharges propagating over epoxy resin and glass insulators in the presence of different gases and mixtures. *Eur. Phys. J. Appl. Phys.* **2011**, *56*, 30802, doi:10.1051/epjap/2011110122. [CrossRef]
7. Beroual, A.; Khaled, U.; Coulibaly, M.L. Experimental Investigation of the Breakdown Voltage of CO_2, N_2, and SF_6 Gases, and CO_2–SF_6 and N_2–SF_6 Mixtures under Different Voltage Waveforms. *Energies* **2018**, *11*, doi:10.3390/en11040902. [CrossRef]
8. Murooka, Y.; Takada, T.; Hiddaka, K. Nanosecond surface discharge and charge density evaluation Part I: review and experiments. *IEEE Electr. Insul. Mag.* **2001**, *17*, 6–16, doi:10.1109/57.917527. [CrossRef]
9. Zhu, Y.; Takada, T.; Inoue, Y.; Tu, D. Dynamic observation of needle-plane surface-discharge using the electro-optical Pockels effect. *IEEE Trans. Dielectr. Electr. Insul.* **1996**, *3*, 460–468, doi:10.1109/94.506221. [CrossRef]
10. Kumara, S.; Alam, S.; Hoque, I.R.; Serdyuk, Y.V.; Gubanski, S.M. DC flashover characteristics of a polymeric insulator in presence of surface charges. *IEEE Trans. Dielectr. Electr. Insul.* **2012**, *19*, 1084–1090, doi:10.1109/TDEI.2012.6215116. [CrossRef]
11. Winter, A.; Kindersberger, J. Surface charge accumulation on insulating plates in SF_6 and the effect on DC and AC breakdown voltage of electrode arrangements. In Proceedings of the Annual Report Conference on Electrical Insulation and Dielectric Phenomena, Cancun, Mexico, 20–24 October 2002; pp. 757–761.
12. Nakanishi, K.; Yoshioka, A.; Arahata, Y.; Shibuya, Y. Surface Charging On Epoxy Spacer At DC Stress In Compressed SF_6 Gas. *IEEE Trans. Power Appar. Syst.* **1983**, *PAS-102*, 3919–3927, doi:10.1109/TPAS.1983.317931. [CrossRef]
13. Ziegler, S.; Woodward, R.C.; Iu, H.H.; Borle, L.J. Current Sensing Techniques: A Review. *IEEE Sens. J.* **2009**, *9*, 354–376, doi:10.1109/JSEN.2009.2013914. [CrossRef]
14. British Standard, B.S. *High-Voltage Test Techniques—Part 1: General Definitions and Test Requirements*; BSI; BS EN 60060-1:2010; 28-02-2011.
15. Reid, A.J.; Judd, M.D.; Stewart, B.G.; Fouracre, R.A. Partial discharge current pulses in SF_6 and the effect of superposition of their radiometric measurement. *J. Phys. D Appl. Phys.* **2006**, *39*, 4167–4177, doi:10.1088/0022-3727/39/19/008. [CrossRef]
16. Saitoh, H.; Morita, K.; Kikkawa, T.; Hayakawa, N.; Okubo, H. Impulse partial discharge and breakdown characteristics of rod-plane gaps in N_2/SF_6 gas mixtures. *IEEE Trans. Dielectr. Electr. Insul.* **2002**, *9*, 544–550, doi:10.1109/TDEI.2002.1024431. [CrossRef]
17. Judd, M.D.; Farish, O. High bandwidth measurement of partial discharge current pulses. In Proceedings of the Conference Record of the 1998 IEEE International Symposium on Electrical Insulation (Cat. No.98CH36239), Washington, DC, USA, 7–10 June 1998.
18. Mansour, D.A.; Kojima, H.; Hayakawa, N.; Hanai, M.; Okubo, H. Physical mechanisms of partial discharges at nitrogen filled delamination in epoxy cast resin power apparatus. *IEEE Trans. Dielectr. Electr. Insul.* **2013**, *20*, 454–461, doi:10.1109/TDEI.2013.6508747. [CrossRef]

19. Rodrigo Mor, A.; Castro Heredia, L.C.; Muñoz, F.A. A Novel Approach for Partial Discharge Measurements on GIS Using HFCT Sensors. *Sensors* **2018**, *18*, 4482, doi:10.3390/s18124482. [CrossRef] [PubMed]

20. Zachariades, C.; Shuttleworth, R.; Giussani, R.; MacKinlay, R. Optimization of a High-Frequency Current Transformer Sensor for Partial Discharge Detection Using Finite-Element Analysis. *IEEE Sens. J.* **2016**, *16*, 7526–7533, doi:10.1109/JSEN.2016.2600272. [CrossRef]

21. Hu, X.; Siew, W.H.; Judd, M.D.; Peng, X. Transfer function characterization for HFCTs used in partial discharge detection. *IEEE Trans. Dielectr. Electr. Insul.* **2017**, *24*, 1088–1096, doi:10.1109/TDEI.2017.006115. [CrossRef]

22. Okubo, H.; Hayakawa, N.; Matsushita, A. The relationship between partial discharge current pulse waveforms and physical mechanisms. *IEEE Electr. Insul. Mag.* **2002**, *18*, 38–45, doi:10.1109/MEI.2002.1014966. [CrossRef]

Article

Effect of Polycyclic Compounds Fillers on Electrical Treeing Characteristics in XLPE with DC-Impulse Voltage

Lewei Zhu [1,2], Boxue Du [2,*], Hongna Li [1] and Kai Hou [2]

[1]   Maritime College, Tianjin University of Technology, Tianjin 300384, China
[2]   Key Laboratory of Smart Grid of Education Ministry, School of Electrical and Information Engineering, Tianjin University, Tianjin 300072, China
*   Correspondence: duboxue@tju.edu.cn; Tel.: +86-22-2740-5477

Received: 27 June 2019; Accepted: 16 July 2019; Published: 18 July 2019

Abstract: Electrical tree is an important factor in the threat of the safety of cross-linked polyethylene (XLPE) insulation, eventually leading to the electrical failure of cables. Polycyclic compounds have the potential to suppress electrical treeing growth. In this paper, three types of polycyclic compounds, 2-hydroxy-2-phenylacetophenone, 4-phenylbenzophenone, and 4,4′-difluorobenzophenone are added into XLPE, denoted by A, B, and C. Electrical treeing characteristics are researched with DC-impulse voltage at 30, 60, and 90 °C, and the trap distribution and carrier mobility are characterized. It has been found that although three types of polycyclic compounds can all suppress the electrical tree propagation at different voltages and temperatures, the suppression effect of these polycyclic compounds with the same DC-impulse polarity is worse than with the opposite polarity. As the temperature increases, the suppression effect becomes weak. The energy level and deep trap density are the largest in XLPE-A composite, leading to a decrease in the charge transportation and resulting in the suppression of electrical treeing growth. Experimental results reveal that the polycyclic compound A has great application prospects in high voltage direct current (HVDC) cables.

Keywords: electrical tree; XLPE; polycyclic compound; DC-impulse voltage; temperature; trap distribution

---

1. Introduction

Cross-linked polyethylene (XLPE) is wildly used in HVDC cables as the insulation material. Electrical tree degradation is an important issue to deteriorate overall insulation level, eventually leading to the electrical failure of cables [1–3]. Researchers have tried many ways to improve the electrical tree breakdown resistance of high voltage (HV) cables, including material blending modification, nanoparticle modification, and polycyclic compound modification [4–6]. Results showed that polycyclic compound had an excellent function of suppressing electrical treeing in polymeric insulating material [7,8]. The America Dow Chemical Company studied the effects of a series of siloxane polycyclic compounds containing pendant aromatic groups on the growth characteristics of electrical tree. It was found that the aromatic ring side groups are indispensable groups for suppressing electrical tree [9]. It was also found that aromatic ketones and diketones had great effects on inhibiting electrical treeing growth in XLPE and polyethylene (PE) [5,10]. Although much research has been carried out on the suppression effect of polycyclic compounds on electrical tree growth, the relationship between polycyclic compounds, charges, trapping levels, and electrical treeing is still unclear.

Research presented a lot of methods to analyze the electrical treeing process. The acoustic emission method and artificial neural networks were used in the detection of the treeing process [11,12]. In addition, imaging techniques such as transmission electron microscope (TEM), scanning electron microscope (SEM), or X-ray computed tomography (XCT) were used to reveal the more complete representation of the electrical treeing phenomenon [13]. Electrical treeing is a kind of cumulative breakdown which is also associated with the charge movement and trap distribution [14,15]. In the process of electron trapping and recombination, the hot electrons gain energy, impact, and destroy the molecular chains of polymer, forming the new electrical tree channel [16,17]. The deep trap level can capture mobile hot electrons, resulting in the decrease of the internal free charge [18]. However, little research has been done on the relationship between electrical tree and trap distribution in the XLPE/polycyclic compound composite.

In the DC transmission system, due to the on and off operations of power electronic devices, impulse overvoltage occurs. The lightning events and operating conditions may also cause lightning and operating overvoltage [19–21]. These impulse voltages are superimposed on the rated DC voltage to produce the DC-impulse voltage that affects the insulation electrical tree degradation process [22,23]. The electrical initiation and treeing characteristics in epoxy resin and polypropylene (PP) with DC-impulse voltage were studied, respectively [24,25]. Results revealed that the electrical treeing characteristics with DC-impulse voltage was different from those with DC voltage, and the polarity of impulse voltage had a significant effect on the electrical initiation and growth characteristics. However, at present, research on the effect of polycyclic compound on electrical tree suppression mostly focuses on AC voltage, and the electrical tree dependence on polycyclic compounds with DC-impulse voltage still needs to be studied. In addition, due to the heat generated by the large current, during the operation of HVDC cables, cables operate under high temperature conditions for a long time [26]. The maximum long-term design temperature of XLPE cables is generally 90 °C, and the operating temperature is usually around 50–60 ° C [27]. It was found that high temperature affected the charge movement and partial discharge characteristics in the insulation, thus affecting the electrical treeing process [28,29]. However, most studies on polycyclic compounds are carried out at room temperature, and the effect of high temperature on polycyclic compounds still needs to be further studied.

In this paper, XLPE is employed as the polymer matrix, and three types of polycyclic compounds with the content of 0.5% are added into the XLPE. The electrical treeing properties of XLPE/ polycyclic compound composites are researched with DC-impulse voltage at 30, 60 and 90 °C. In addition, the trap distribution and carrier mobility behaviors are also studied to further reveal the mechanism of polycyclic compounds fillers on electrical treeing growth.

2. Experiment

2.1. Test Samples and Electrode Arrangement

This paper selects three types of polycyclic compounds. The polycyclic compound 2-hydroxy-2-phenylacetophenone is denoted by A, which is produced by J&K Scientific Ltd. The polycyclic compound 4-phenylbenzophenone is denoted by B, which is produced by Shanghai Macklin Biochemical Co., Ltd. The polycyclic compound 4,4′-difluorobenzophenone is denoted by C, which is produced by J&K Scientific Ltd. The molecular structures of these polycyclic compounds are shown in Figure 1. The neat XLPE was supplied by Borealis Company. At 110 °C, we put the appropriate amount of XLPE into the mixer and mixed for 3 min. Then we weighed an amount of 0.5 wt% polycyclic compound and added it to the internal mixer. We then mixed for 10 min to make it evenly dispersed in XLPE. Then we used the flat vulcanizing machine and the special mold to make the needle-plate electrode samples [30]. The distance between the tip and the ground electrode was 2 ± 0.1 mm. The needle electrode diameter and curvature radius were 300 μm and 3 μm, respectively.

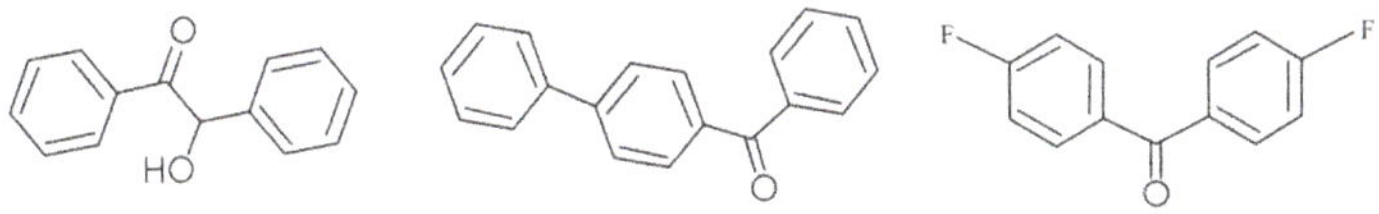

(a) 2-hydroxy-2-phenylacetophenone　　(b) 4-phenylbenzophenone　　(c) 4,4'-difluorobenzophenone

Figure 1. Molecular structures of polycyclic compounds.

2.2. Experimental Apparatus and Procedure

Figure 2 shows the schematic diagram of the experimental setup. The electrical tree experiment was carried out in the high temperature environment experimental platform. The sample was placed in a heat-resistant glass cylinder with dimethyl silicone oil, which could effectively prevent surface flashover of the sample during the experiment. In addition, silicone oil could also fill the surface of the sample to increase the transparency of the sample, improving the clarity of the electrical tree observation. The high temperature environment was provided by the resistance wires on both sides of the heat-resistant glass cylinder, and the maximum operating temperature was 200 °C. The experimental temperature was set to 30, 60, and 90 °C to make a temperature gradient. After the temperature reached the set temperature during the experiment, the sample was allowed to stand in the incubator for 10 min to ensure that the temperature of the material was consistent with the environment. The DC-impulse voltage was generated by the DC-impulse power source, which consisted of a DC power source and impulse power source [31]. The DC voltage was applied at a rate of 1 kV/s. After 1 min, the impulse voltage was applied at a rate of 1 kV/s. In order to reduce the error, each set of experiments was repeated 20 times. Because the electrical treeing initiation and breakdown phenomenon had a great relationship with the relative polarity of the impulse voltage with DC-impulse voltage [25], the electrical treeing characteristics with −25 kV DC and ±35 kV impulse voltage were measured to analyze the suppression effect of polycyclic compounds with DC-impulse voltage. The impulse voltage frequency was 400 Hz. The equivalent circuit of the experimental configuration is shown in Figure A1 in our Appendix A. The electrical tree imaging system included a computer, microscope unit, and a cold light source. The microscope unit consisted of an objective lens, an eyepiece, and a charge-coupled device (CCD). The CCD was a high-resolution imaging device. The camera multiplier was 1× and the highest resolution was 1024 × 768 pixels. The three objective magnifications were 4×, 10×, 40×, and the eyepiece magnification was 10×. The accumulated damage was used to analyze the electrical treeing characteristics. The accumulated damage refers to the number of pixels in the area covered by the electrical tree, which can be used to characterize the damage area of electrical tree to insulating materials, and to describe the development trend of electrical tree in space [29,32]. The accumulated damage was calculated using Matlab language. The specific calculation method was divided into three steps. The first step was to take a photo of the entire electrical branch with a pixel value of 500 × 500 pixels. In the second step, the image was subjected to filtering binarization to obtain a black and white image. The third step was to count the total number of pixels in the black area of the image. The value obtained is the accumulated damage value of the electrical tree.

The surface potential decay (SPD) technique is an effective method for measuring trap distribution behaviors [33]. The surface charge test system consisted of an HVDC power supply, a pin-gate-plate electrode system, a TREK type surface potentiometer (including a Kelvin type vibrating probe), and a constant temperature and humidity chamber. The vertical distance from the tip of the needle electrode to the gate electrode, and the vertical distance from the gate electrode to the surface of the sample were both 5 mm. The Kelvin vibrating probe was fixed with an epoxy holder with a vertical distance of 3 mm from the surface of the specimen. Charge was injected onto the surface of the sample for a charge time of 10 min. After the corona was over, we moved the center of the sample quickly below the surface potential measurement probe. The experimental temperature was set to 30, 60, and 90 °C with a relative humidity of ~20%. In this paper, the distribution of trap energy level (E_t)

and the trap density (Nt) are calculated to analyze the electrical property behaviors [34]. The carrier mobility was also calculated [35], which can obtain the characteristics of charge transportation.

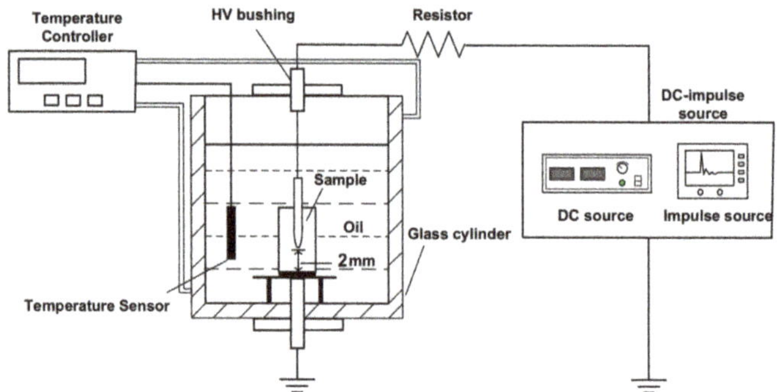

Figure 2. Schematic diagram of experimental setup.

3. Results

3.1. Electrical Tree Degradation

3.1.1. Tree Structure

The structure of electrical tree in polymer is related to certain combined factors: voltage waveform, temperature, and polymer fillers [36–38]. Table 1 shows the electrical tree structure distribution at 1 min. Figure 3 shows the electrical tree structures of the neat XLPE and polycyclic compound modified samples. Figure 3a,b show the compared electrical tree structures with −25 kV DC and +35 kV impulse voltages (the opposite polarity). Figure 3c,d show the compared electrical tree structures with −25 kV DC and −35 kV impulse voltages (the same polarity). The treeing times for both are 1 min for the convenience of comparison. In order to save space, only modified samples with a type A polycyclic compound were selected for comparison with neat XLPE samples. The electrical tree structures of XLPE-B and XLPE-C are shown in Figure A2 in our Appendix A. It can be seen that the electrical tree structures of neat XLPE samples are not affected by the temperature and the DC-impulse voltage waveform, which are branch trees. These results are consistent with the results in the literature [22], of which the electrical tree structures of PP samples are not affected by the DC-impulse voltage waveform at room temperature. After the addition of the polycyclic compound, the electrical tree structure changes, which are related to the polycyclic compound type. For XLPE-A samples, with −25 kV DC and +35 kV impulse voltage (the opposite polarity), the tree structures change to bush trees at 30 and 60 °C and it is double tree at 90 °C. With −25 kV DC and −35 kV impulse voltage (the same polarity), the tree structures change to bush trees at 30 °C. However, they are still branch trees at 60 and 90 °C. For XLPE-B and XLPE-C, the tree structures are double tree at 30 °C. However, they are still branch trees at 60 and 90 °C.

Table 1. Tree structure distribution at 1 min.

DC (kV)	Pulse (kV)	Temperature (°C)	XLPE	XLPE-A	XLPE-B	XLPE-C
−25	+35	30	Branch	Bush	Bush-branch	Bush-branch
		60	Branch	Bush	Branch	Branch
		90	Branch	Bush-branch	Branch	Branch
	−35	30	Branch	Bush	Bush-branch	Bush-branch
		60	Branch	Branch	Branch	Branch
		90	Branch	Branch	Branch	Branch

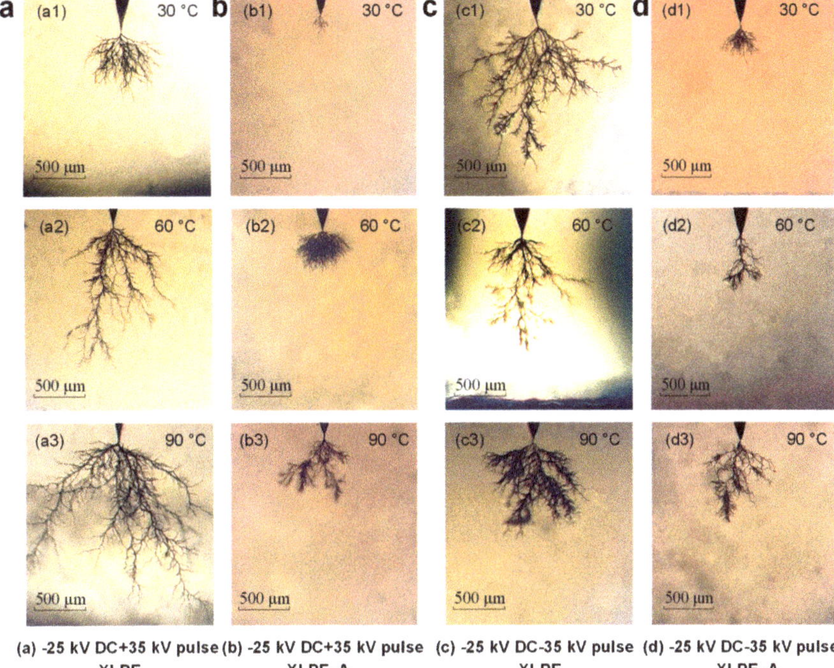

Figure 3. The electrical tree structures of the neat cross-linked polyethylene (XLPE) and XLPE–A composite.

3.1.2. Electrical Treeing Characteristics with Opposite Polarity DC-Impulse Voltage

Figure 4 shows the electrical treeing characteristics of a representative electrical tree with opposite polarity DC-impulse voltage; Figure 4a is the relationship of the electrical tree length and the treeing time; Figure 4b is the relationship of the accumulated damage and the treeing time. The treeing voltage is +35 kV impulse superimposed −25 kV DC voltage, of which the impulse voltage polarity is opposite to the DC voltage polarity. The treeing time is 60 s, and the experimental temperatures are respectively 30, 60, and 90 °C. The electrical tree length of XLPE-A composite is 167 μm at 30 °C in 60 s and is reduced by 74.8% compared to the neat XLPE; the accumulated damage of XLPE-A is 0.5×10^3 pixels at 30 °C in 60 s and is reduced by 97.6% compared to the neat XLPE. The electrical tree length of XLPE-A is reduced by 69.9% at 60 °C in 60 s compared to the neat XLPE and the accumulated damage is reduced by 64%. At 90 °C, the electrical tree length of XLPE-A is reduced by 68.5% and the accumulated damage is reduced by 59.6%. For XLPE-B composites, the electrical tree length and accumulated damage are respectively reduced by 51.7% and 87.9% at 30 °C, 51.4% and 36.4% at 60 °C, and 50.3% and 31.6% at 90 °C in 60 s. For XLPE-C composites, the electrical tree length and accumulated damage is respectively reduced by 34.3% and 41.7% at 30 °C, 32.2% and 19.1% at 60 °C, and 30.8% and 26.3% at 90 °C. The experimental results reveal that the three type of polycyclic compounds all inhibit the length and accumulated damage of the electrical tree. Among them, the polycyclic compound A has the best effect, the type B is second, and the type C has the worst effect. It can be concluded that as the temperature increases, although the three polycyclic compounds still inhibit the growth of electrical trees, the effect of inhibition becomes weak.

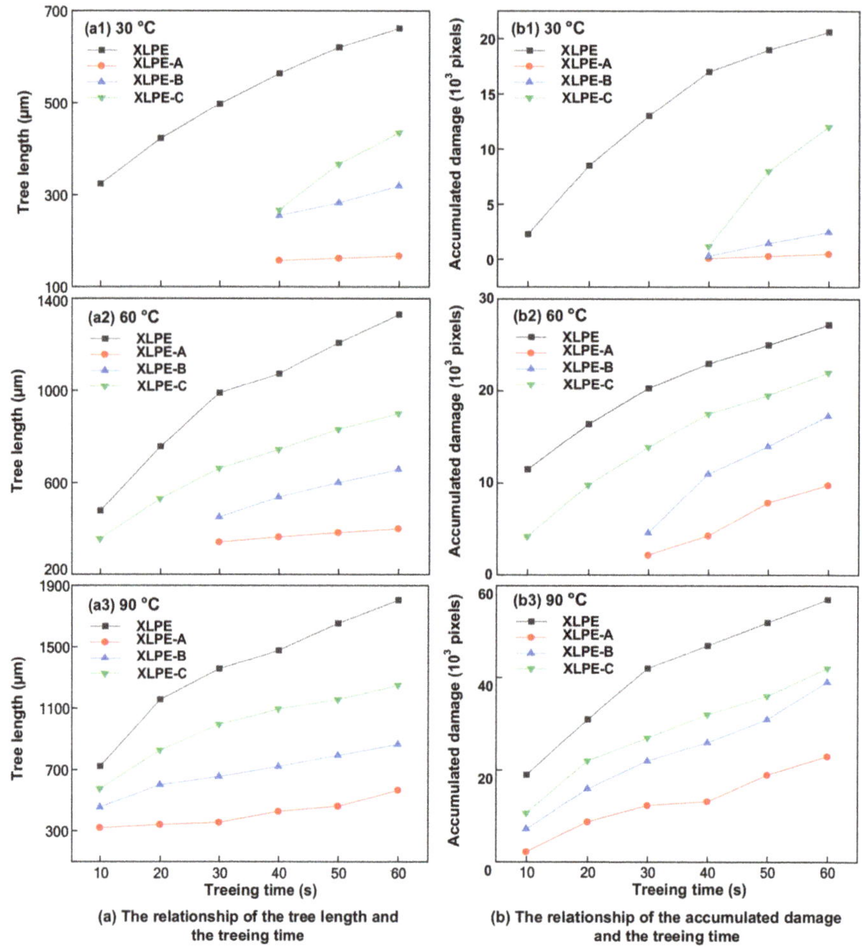

Figure 4. Electrical treeing characteristics with opposite polarity DC-impulse voltage.

3.1.3. Electrical Treeing Characteristics with the Same Polarity DC-Impulse Voltage

Figure 5 shows the electrical treeing characteristics of a representative electrical tree with the same polarity DC-impulse voltage; Figure 5a is the relationship of the electrical tree length and the treeing time; Figure 5b is the relationship of the accumulated damage and the treeing time. The treeing voltage is −35 kV impulse superimposed −25 kV DC voltage, of which the impulse voltage polarity is the same as the DC voltage polarity. The electrical tree length of XLPE-A is 379 μm at 30 °C in 60 s and is reduced by 71.9% compared to the neat XLPE; The accumulated damage of XLPE-A is 9.1×10^3 pixels at 30 °C in 60 s and is reduced by 72.1% compared to the neat XLPE. The electrical tree length of XLPE-A is reduced by 57.3% at 60 °C in 60 s compared to the neat XLPE and the accumulated damage is reduced by 63.2%. At 90 °C, the electrical tree length of XLPE-A is reduced by 22.9% and the accumulated damage is reduced by 46.8%. For XLPE-B composites, the electrical tree length and accumulated damage are respectively reduced by 46.7% and 44.8% at 30 °C, 20.5% and 32.5% at 60 °C, and 14% and 29.1% at 90 °C. For XLPE-C composites, the electrical tree length and accumulated damage are respectively reduced by 31.9% and 26.4% at 30 °C, 23.8% and 16.9% at 60 °C, and 3% and 10.2% at 90 °C. The three polycyclic compounds all inhibit the growth of the electrical tree, and the polycyclic compound A has the best effect, which is consistent with the results that have an opposite polarity DC-impulse voltage. The effect of temperature on the polycyclic compound is still the same as that with the opposite polarity. As the temperature increases, the suppression effect of the three polycyclic compounds becomes weak. It can be concluded that the suppression effect of the three types of polycyclic compounds with the same polarity is worse than with the opposite polarity.

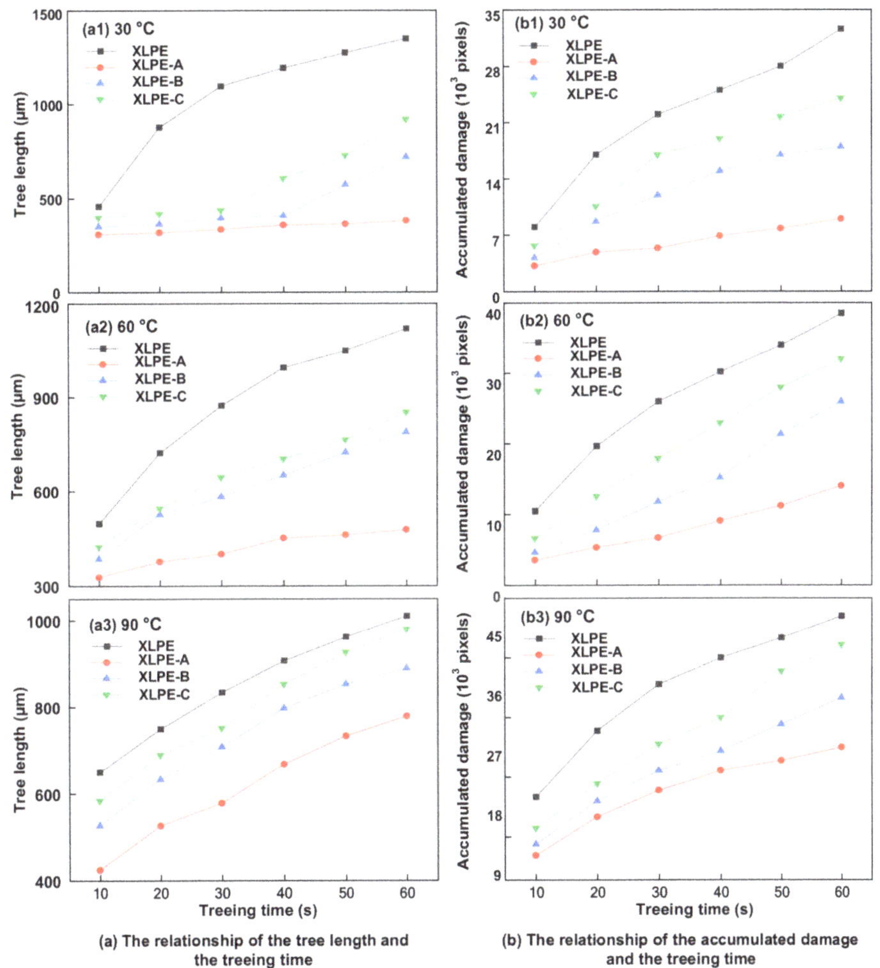

Figure 5. Electrical treeing characteristics with the same polarity DC-impulse voltage.

3.2. Trap Distribution and Carrier Mobility Behaviors

There are many trap levels in the forbidden band of polymer materials. The formation of trap levels is complicated, and many factors affect the trap level, including molecular chain end groups, branches, amorphous regions and crystallization regions, polarizing groups, impurities, etc., as well as various physical chemistry effects that cause structural defects in materials [39]. After the polycyclic compound is added to the XLPE, the trap distribution and the carrier mobility behaviors change accordingly. In this section, the variation of the trap distribution and the carrier mobility behaviors of representative XLPE/polycyclic compounds at different temperatures are obtained, as shown in Figure 6. Figure 6(a1) compares the trap distribution behaviors of different samples at 30 °C. It can be seen that the trap distribution of the neat XLPE sample exhibits a *double peak* shape, for which a peak appears at a shallower trap level, and a peak appears at a deeper trap level. However, there is only one deep trap peak of the XLPE/polycyclic compound composites, indicating that the number of shallow traps is small. It can be seen from Table 2 that the deep trap depth of the neat XLPE sample is 0.88 eV at 30 °C. After adding the polycyclic compound, the deep trap depth is among 0.93 eV to 0.97 eV at 30 °C. It can be concluded that the deep trap level of the XLPE/polycyclic compounds composites increases, and the corresponding trap density increases, indicating that the addition of the polycyclic compounds introduces deep traps inside the XLPE sample. Among them, trap depth of the XLPE-A composite is the largest, XLPE-B is second largest, and trap depth of XLPE-C is the smallest. Figure 6(b1) compares the carrier mobility of different samples at 30 °C. The neat XLPE sample has a carrier mobility of 12.5×10^{-14} $m^2V^{-1}s^{-1}$, the XLPE-A has a carrier mobility of 0.5×10^{-14} $m^2V^{-1}s^{-1}$, the XLPE-B has a carrier mobility of 2×10^{-14} $m^2V^{-1}s^{-1}$, and the XLPE-C has a carrier mobility of 5.6×10^{-14} $m^2V^{-1}s^{-1}$. It is more difficult for charges to escape from deep traps than shallow traps [35]. After the polycyclic compound is added, the depth of the deep trap becomes larger, and the charge trapped by the deep trap is more difficult to transfer from the deep trap to the ground electrode, so that the carrier mobility of samples becomes smaller [11]. Figure 6(a2,a3) compare the trap distribution behaviors of different samples at 60 and 90 °C. The neat XLPE has a shallow trap depth of 0.85 eV and a deep trap depth of 0.91 eV at 60 °C. After adding A, B, and C polycyclic compounds, only deep traps are measured and the depths are 0.97, 0.94, and 0.93 eV, respectively. The neat XLPE has a shallow trap depth of 0.91 eV and a deep trap depth of 1 eV at 90 °C. After adding three types of polycyclic compounds, A, B, and C, the shallower trap depths are 0.98, 0.95, and 0.92 eV, respectively. The deeper trap depths are 1.02, 1.02 and 1.01 eV, respectively. Figure 6(b2,b3) compare the carrier mobility behaviors of different samples at 60 and 90 °C. The carrier mobility of XLPE is the largest, 24×10^{-14} $m^2V^{-1}s^{-1}$ and $100 \times 10^{-14} m^2V^{-1}s^{-1}$, respectively. The carrier mobility of XLPE-A is the smallest, 0.9375×10^{-14} $m^2V^{-1}s^{-1}$ and 13.65×10^{-14} $m^2V^{-1}s^{-1}$, respectively. It can be concluded that the addition of the polycyclic compounds increases the trap depth and the corresponding trap density of the XLPE and reduces the carrier mobility. Among them, the XLPE-A composite has the deepest trap depth, trap density, and the smallest carrier mobility.

Table 2. Cross-linked polyethylene (XLPE) and its polycyclic compounds composites trap depth.

Type	Temperature (°C)	Shallow Trap Depth / (eV)	Deep Trap Depth / (eV)
	30	0.81	0.88
XLPE	60	0.85	0.91
	90	0.91	1
	30	/	0.98
XLPE-A	60	/	0.97
	90	0.98	1.02
	30	/	0.93
XLPE-B	60	/	0.94
	90	0.95	1.02
	30	/	0.9
XLPE-C	60	/	0.93
	90	0.92	1.01

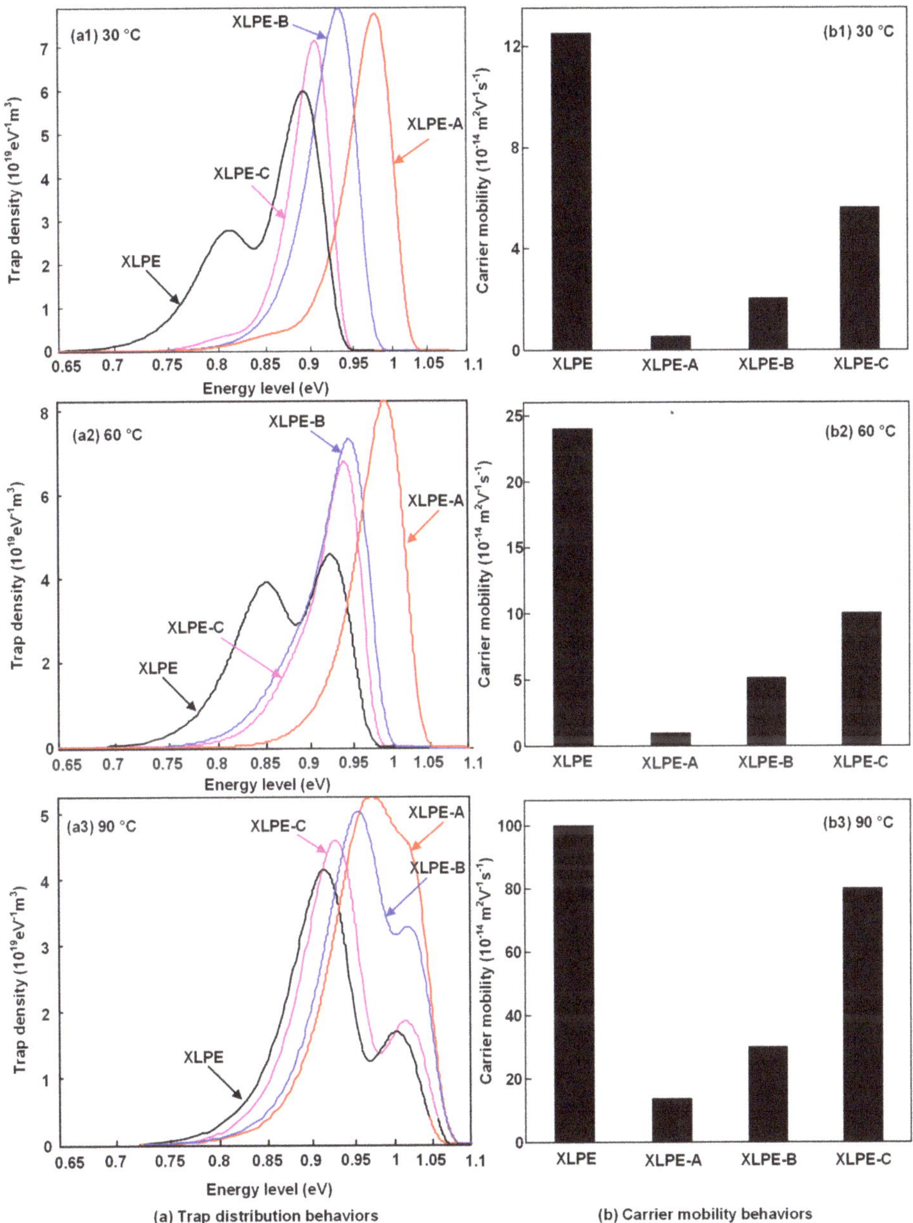

Figure 6. Trap distribution and carrier mobility behaviors of XLPE and its polycyclic compounds composites.

4. Discussion

4.1. Mechanism of Polycyclic Compounds Fillers Inhibiting Electrical Treeing Growth

The growth process of the electrical tree with the DC-impulse voltage is closely related to the hot electrons motion and trap distribution [40]. After the addition of the polycyclic compound, the trap depth and density of samples increase significantly. The traps enhance the charge trapping ability of the needle tip and easily forms the same polarity charge accumulation to weaken the external electric field, resulting in a decrease in the injected charge amount [41]. At the same time, with the deep trap level and the trap density increasing, the internal charge trapping of the sample enhances, resulting in a decrease in the internal free charge of the sample. For the growth of the electrical tree, the charge injection process is suppressed and the free charge inside the sample is reduced. Therefore, the molecular chain breaking process is suppressed, and the electrical tree deterioration resistance of the XLPE/ polycyclic compounds composites is improved. In addition, in the process of electrical tree growth, some of the electrical tree channels are carbonized, and the carbonization channel delivers charges as a conductive current to the end of the channel, enhancing the electric field strength [42]. The deeper trap captures the transport charges in the carbonized electrical tree channels, reducing the electric field strength at the end of the channel, thereby decreasing the growth of the electrical tree [29]. The carrier migration process is closely related to the trap distribution of the composite material. The free charge may be captured by the local state (trap) during the extended state migration process. The charge transfer and exchange process between the extended state and the local state affects the migration characteristics of carriers [41]. After the addition of the polycyclic compound, the carrier mobility is reduced, and the carrier mobility of XLPE-A is the smallest. The deep trap energy level increases, leading to the increase of trap barrier to overcome for charge trapping. The deep trap density increases, leading to a decrease of the charge average free travel and energy obtained from migration, increasing the probability of charge trapping, and leading to a decrease in mobility. Therefore, the internal collision ionization probability of the composite material is reduced, leading to the improvement in the resistance to electrical tree.

4.2. Electrical Tree Dependence on Polycyclic Compounds with DC-Impulse Voltage

It can be seen from the results in Section 3.1 that the effect of the polycyclic compound is related to the polarity of the DC-impulse voltage, and the suppression effect of the polycyclic compound with the opposite polarity is better than with the same polarity. When −35 kV impulse is superimposed on −25 kV DC voltage, after the DC voltage is applied, electrons are accumulated at the needle tip. After the impulse voltage is applied, a large number of electrons are injected into polymer over a short period of time to push accumulated charges to move. During the process of charge trapping, as the state of charge changes from high energy to low energy state, excess energy is transferred to other charges, making them hot electrons, destroying the XLPE molecular chain, and causing growth of the electrical tree. When +35 kV impulse is superimposed on −25 kV DC voltage, and after the polarity of the voltage changes, part of the positive charge is injected and the positive and negative charges neutralize to generate energy, which accelerates the growth of the electrical tree [43]. In addition, as the polarity of the voltage changes from negative to positive, the electrons change from the same-polar space charge to the hetero-polar space charge, causing electric field distortion, aggravating partial discharge, causing molecular chain breakage, and accelerating electrical tree growth [44]. The mechanism of electrical tree growth with different polarities is different, which results in different effects of polycyclic compounds with different polarities. With the opposite DC-impulse polarity voltage, and after the polycyclic compound is added, on the one hand, the electric field distortion caused by the polarity change is reduced due to the decrease of the injected charges; on the other hand, since free charges inside the sample are reduced, the energy generated by the positive and negative charges is reduced. Therefore, the polycyclic compound effect with the opposite polarity is better.

4.3. Electrical Tree Dependence on Polycyclic Compounds at Different Temperatures

Electrical treeing is an electro-thermal aging phenomenon, which is a comprehensive process including charge motion, partial discharge, local high pressure, and local high temperature. The hot electrons accelerate in the free volume, which impacts the molecular chain of the polymer and accelerates the formation of low-density regions. Charges collide with ionization in low-density regions, releasing energy to destroy more molecular chains and forming micropores. Subsequent discharges then occur in the micropores. Partial discharge produces local temperature rise in a short time. When the ambient temperature plus local temperature rise is greater than the local softening temperature of the insulating material ($\Delta T_{ambient\ temperature} + \Delta T_{discharge\ temperature} > T_{softening\ temperature}$), local thermal breakdown will occur, and the air gaps form cracks along the applied field direction [45]. Under the action of charged particles, the XLPE molecular chains are broken rapidly and proceed to decompose and gasify. With these gases generated in a short time, the gas pressure in the micropores and cracks increases rapidly, resulting in the material around the micropores or cracks to undergo an expansion stress. Under the action of this stress, the micropores and cracks rapidly expand toward the amorphous region where the mechanical strength is weak, resulting in the growth of the electrical tree. As the temperature increases, on the one hand, the carrier mobility increases, the ionization probability increases, and the polycyclic compound's ability to capture hot electrons is relatively reduced; on the other hand, the local thermal breakdown increases, and the influence of hot electrons on the growth characteristics of the electrical tree is relatively reduced. Therefore, the effect of the polycyclic compound decreases with temperature increase.

5. Conclusions

In this paper, three different types of polycyclic compounds were chosen to prepare the XLPE/polycyclic compounds composites. The effects of these polycyclic compounds on the tree structure and electrical treeing characteristics with DC-impulse voltage were studied at 30, 60, and 90 °C. Their trap distribution and carrier mobility behaviors were also investigated. The experimental results reveal that polycyclic compound A has great application prospects in HVDC cables. The following are our main conclusions:

(1) The addition of the polycyclic compound changes the electrical tree structure, which is related to the type of the polycyclic compound, the temperature and the relative polarity of DC and impulse voltage.

(2) The electrical treeing characteristics are related to the type of polycyclic compounds. The three type of polycyclic compounds all inhibit the length and accumulated damage of the electrical tree with DC-impulse voltages at 30, 60, and 90 °C. The energy level and trap density are the largest in XLPE-A composite, decreasing the charge transport, and leading to the suppression of the electrical treeing growth, improving the lifetime of XLPE. The polycyclic compound A has great application prospects in HVDC cables.

(3) The effect of the polycyclic compound is related to the relative polarity of the applied DC-impulse voltage. Although three types of polycyclic compounds can suppress the electrical tree propagation at different DC-impulse voltages, the suppression effect with the same polarity is worse than with the opposite polarity, which is related to the difference in the mechanism of electrical tree growth with different impulse polarities.

(4) The effect of the polycyclic compound is related to the temperature. With the temperature increasing, the suppression effect to electrical treeing growth of polycyclic compounds decreases. However, the three types of polycyclic compounds can still suppress the electrical treeing growth at 30, 60, and 90 °C, which can improve the lifetime of XLPE cables.

Author Contributions: L.Z. and B.D. came up with the idea and designed the structure of the paper. L.Z. finished the experiments and wrote the manuscript. H.L. and K.H. analyzed the data and provided analytical theory for electrical breakdown.

Acknowledgments: This work is supported by the National Key Research and Development Program of China (Grant 2016YFB0900701); the Chinese National Natural Science Foundation under the Grant 51537008.

Conflicts of Interest: The authors declare no conflict of interest.

Appendix A

Figure A1 shows the equivalent circuit of the experimental configuration.
Figure A2 shows the electrical tree structures of XLPE-B and XLPE-C.

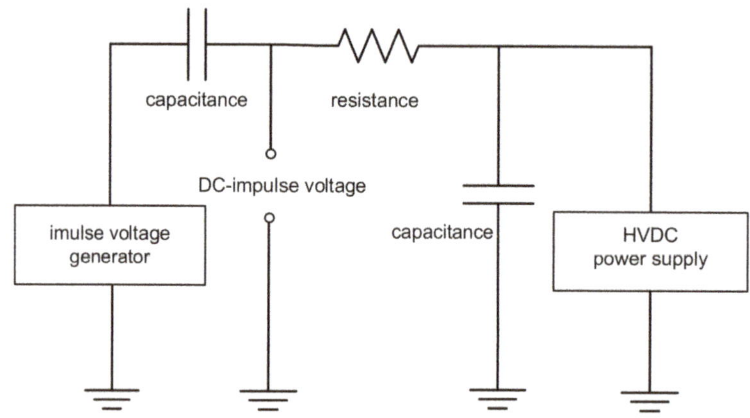

Figure A1. The equivalent circuit of the experimental configuration.

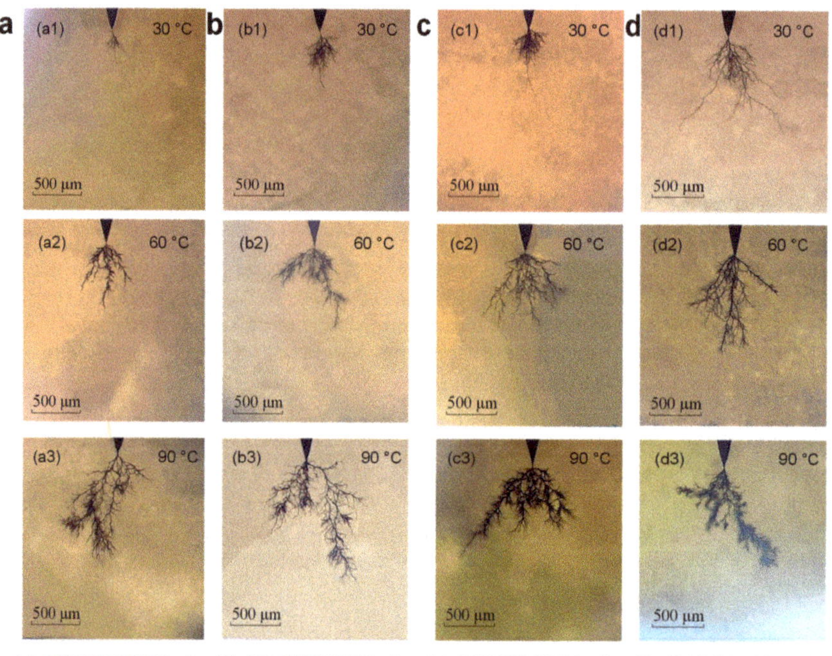

Figure A2. The electrical tree structures of XLPE–B and XLPE–C composite.

References

1. Danikas, M.G.; Tanaka, T. Nanocomposites-a review of electrical treeing and breakdown. *IEEE Trans. Dielectr. Electr. Insul.* **2009**, *25*, 19–24. [CrossRef]
2. Saito, Y.; Fukuzawa, M.; Nakamura, H. On the mechanism of tree initiation. *IEEE Trans. Electr. Insul.* **1977**, *1*, 31–34. [CrossRef]
3. Han, T.; Du, B.X.; Ma, T.T.; Wang, F.Y.; Gao, Y.; Lei, Z.P.; Li, C.Y. Electrical tree in HTV silicone rubber with temperature gradient under repetitive pulse voltage. *IEEE Access* **2019**, *7*, 41250–41260. [CrossRef]
4. Jiang, C.; Hou, Y.X.; Yu, J.Z.; Wei, Z.J.; Chen, X.R.; Zhou, H. Electrical treeing of polyethylene blends with/without voltage stabilizer. In Proceedings of the IEEE 2018 12th International Conference on the Properties and Applications of Dielectric Materials, Xi'an, China, 20–24 May 2018; pp. 266–269.
5. Kisin, S.; Doelder, J.D.; Eaton, R.F. Quantum mechanical criteria for choosing appropriate voltage stabilization additives for polyethylene. *Polym. Degrad. Stab.* **2009**, *94*, 171–175. [CrossRef]
6. Han, T.; Du, B.X.; Su, J.G.; Gao, Y.; Xing, Y.Q.; Fang, S.C.; Li, C.Y.; Lei, Z.P. Inhibition effect of graphene nanoplatelets on electrical degradation in silicone rubber. *Polymers* **2019**, *11*, 968. [CrossRef] [PubMed]
7. Yamano, Y. Roles of polycyclic compounds in increasing breakdown strength of LDPE film. *IEEE Trans. Electr. Insul.* **2006**, *13*, 773–781. [CrossRef]
8. Jarvid, M.; Johansson, A.; Kroon, R.; Bjuggren, J.M.; Wutzel, H.; Englund, V.; Gubanski, S.; Andersson, M.R.; Muller, C. A new application area for fullerenes: Voltage stabilizers for power cable insulation. *Adv. Mater.* **2015**, *27*, 897–902. [CrossRef]
9. Vincent, G.A. Anti-Treeing Additives. U.S. Patent 4,840,983, 20 June 1989.
10. Jarvid, M.; Johansson, A.; Bjuggren, J.M.; Wutzel, H.; Englund, V.; Gubanski, S.; Muller, C.; Andersson, M.R. Tailored side-chain architecture of benzil voltage stabilizers for enhanced dielectric strength of cross-linked polyethylene. *J. Polym. Sci. Part B Polym. Phys.* **2014**, *52*, 1047–1054. [CrossRef]
11. Opydo, W.; Dobrzycki, A. Detection of electric tree of solid dielectrics with the method of acoustic emission. *Electr. Eng.* **2012**, *94*, 37–48. [CrossRef]
12. Dobrzycki, A.; Mikulski, S.; Opydo, W. Using ANN and SVM for the detection of acoustic emission signals accompanying epoxy resin electrical treeing. *Appl. Sci.* **2019**, *9*, 1523. [CrossRef]
13. Schurch, R.; Rowland, S.M.; Bradley, R.S.; Withers, P.J. Imaging and analysis techniques for electrical trees using X-ray computed tomography. *IEEE Trans. Dielectr. Electr. Insul.* **2014**, *21*, 53–63. [CrossRef]
14. Li, S.T.; Min, D.M.; Wang, W.W.; Chen, G. Linking traps to dielectric breakdown through charge dynamics for polymer nanocomposites. *IEEE Trans. Dielectr. Electr. Insul.* **2016**, *23*, 2777–2785. [CrossRef]
15. Tanaka, Y.; Ohnuma, N.; Katsunami, K. Effects of crystallinity and electron mean-free-path on dielectric strength of low-density polyethylene. *IEEE Trans. Dielectr. Electr. Insul.* **1991**, *26*, 258–265. [CrossRef]
16. Shimizu, N.; Laurent, C. Electrical tree initiation. *IEEE Trans. Dielectr. Electr. Insul.* **1998**, *5*, 651–659. [CrossRef]
17. Kao, K.C. New theory of electrical discharge and breakdown in low-mobility condensed insulators. *J. Appl. Phys.* **1984**, *55*, 752–755. [CrossRef]
18. Wang, W.W.; Min, D.M.; Li, S.T. Understanding the conduction and breakdown properties of polyethylene nanodielectrics: Effect of deep traps. *IEEE Trans. Dielectr. Electr. Insul.* **2016**, *23*, 564–572. [CrossRef]
19. Hingorani, N.G. Transient overvoltage on a bipolar HVDC overhead line caused by DC line faults. *IEEE Trans. Power Appar. Syst.* **1970**, *PAS-89*, 592–610. [CrossRef]
20. Lu, W.X.; Ooi, B.T. DC overvoltage control during loss of converter in multiterminal voltage-source converter-based HVDC (M-VSC-HVDC). *IEEE Trans. Power Deliv.* **2003**, *18*, 915–920.
21. Zhou, C.H.; Wang, P. A study of temporary overvoltage at HVDC rectifier stations. In Proceedings of the 2011 IEEE Electrical Power and Energy Conference, Winnipeg, MB, Canada, 3–5 October 2011; pp. 211–215.
22. Hagiwara, M.; Akagi, H. Control and experiment of pulsewidth-modulated modular multilevel converters. *IEEE Trans. Power Electron.* **2009**, *24*, 1737–1746. [CrossRef]
23. Murata, Y.; Katakai, S.; Kanaoka, M. Impulse breakdown superposed on ac voltage in XLPE cable insulation. *IEEE Trans. Dielectr. Electr. Insul.* **1996**, *3*, 361–365. [CrossRef]

24. Du, B.X.; Su, J.G.; Xue, J.S. Tree growth characteristics of epoxy resin in LN2 under DC superimposed pulse voltage. *IEEE Trans. Appl. Superconduct.* **2018**, *28*, 1–5. [CrossRef]
25. Zhu, L.W.; Du, B.X.; Su, J.G.; Han, T.; Danikas, M.G. Electrical treeing initiation and breakdown phenomenon in polypropylene under DC and pulse combined voltages. *IEEE Trans. Dielectr. Electr. Insul.* **2019**, *26*, 202–210. [CrossRef]
26. Bozzo, R.; Gemme, C.; Guastavino, F. The effects of temperature on the tree growth phenomena and relevant PD. In Proceedings of the IEEE Conference on Electrical Insulation and Dielectric Phenomena, Virginia Beach, VA, USA, 22–25 October 1995; pp. 69–72.
27. Shimizu, N.; Shibata, Y.; Ito, K.; Imai, K.; Nawata, M. Electrical tree at high temperature in xlpe and effect of oxygen. In Proceedings of the IEEE Conference on Electrical Insulation and Dielectric Phenomena, Victoria, BC, Canada, 15–18 October 2000; pp. 329–332.
28. Wang, Y.; Li, G.; Wu, J. Effect of temperature on space charge detrapping and periodic grounded DC tree in cross-linked polyethylene. *IEEE Trans. Dielectr. Electr. Insul.* **2017**, *23*, 3704–3711. [CrossRef]
29. Du, B.X.; Zhu, L.W.; Han, T. Effect of ambient temperature on electrical treeing and breakdown phenomenon of polypropylene with repetitive pulse voltage. *IEEE Trans. Dielectr. Electr. Insul.* **2017**, *24*, 2216–2224. [CrossRef]
30. Du, B.X.; Zhu, L.W.; Han, T. Effect of low temperature on electrical treeing of polypropylene with repetitive pulse voltage. *IEEE Trans. Dielectr. Electr. Insul.* **2016**, *23*, 1915–1923. [CrossRef]
31. Han, T.; Du, B.X.; Su, J.G. Electrical tree initiation and growth in silicone rubber under combined dc-pulse voltage. *Energies* **2018**, *11*, 764. [CrossRef]
32. Chen, G.; Tham, C. Electrical treeing characteristics in XLPE power cable insulation in frequency range between 20 and 500 Hz. *IEEE Trans. Dielectr. Electr. Insul.* **2009**, *16*, 179–188. [CrossRef]
33. Zhou, F.S.; Li, J.Y.; Liu, M.X.; Min, D.M.; Li, S.T.; Xia, R. Characterizing traps distribution in LDPE and HDPE through isothermal surface potential decay method. *IEEE Trans. Dielectr. Electr. Insul.* **2016**, *23*, 1174–1182. [CrossRef]
34. Li, J.Y.; Zhou, F.S.; Min, D.M.; Li, S.T.; Xia, R. The energy distribution of trapped charges in polymers based on isothermal surface potential decay model. *IEEE Trans. Dielectr. Electr. Insul.* **2015**, *22*, 1723–1732. [CrossRef]
35. Zhou, F.S.; Li, J.Y.; Yan, Z.M.; Zhang, X.; Yang, Y.Q.; Liu, M.X.; Min, D.M.; Li, S.T. Investigation of charge trapping and detrapping dynamics in LDPE, HDPE and XLPE. *IEEE Trans. Dielectr. Electr. Insul.* **2016**, *23*, 3742–3751. [CrossRef]
36. Chen, X.R.; Hu, L.B.; Xu, Y.; Cao, X.L.; Gubanski, S.M. Investigation of temperature effect on electrical trees in XLPE cable insulation. In Proceedings of the Electrical Insulation and Dielectric Phenomena, Montreal, QC, Canada, 14–17 October 2012; pp. 612–615.
37. Liu, Y.; Cao, X.L. Electrical tree initiation in XLPE cable insulation by application of DC and impulse voltage. *IEEE Trans. Dielectr. Electr. Insul.* **2013**, *20*, 1691–1698.
38. Ashcraft, A.C.; Eichhorn, R.M.; Shaw, R.G. Laboratory studies of treeing in solid dielectrics and voltage stabilization of polyethylene. In Proceedings of the IEEE International Conference on Electrical Insulation, Montreal, QC, Canada, 14–16 June 2016; pp. 213–218.
39. Du, B.X.; Han, C.L.; Li, J. Effect of voltage stabilizers on the space charge behavior of XLPE for HVDC cable application. *IEEE Trans. Dielectr. Electr. Insul.* **2019**, *26*, 34–42. [CrossRef]
40. Serra, S.; Tosatti, E.; Iarlori, S.; Scandolo, S.; Santoro, G. Interchain electron states in polyethylene. *Phys. Rev. B* **2000**, *62*, 4389–4393. [CrossRef]
41. Du, B.X.; Su, J.G.; Tian, M.; Han, T.; Li, J. Understanding trap effects on electrical treeing phenomena in EPDM/POSS composites. *Sci. Rep.* **2018**, *8*, 8481–8495. [CrossRef]
42. Zheng, X.Q.; Chen, G. Propagation Mechanism of Electrical Tree in XLPE Cable Insulation by Investigating a Double Electrical Tree Structure. *IEEE Trans. Dielectr. Electr. Insul.* **2008**, *15*, 800–807. [CrossRef]
43. Bamji, S.S.; Bulinski, A.T.; Densley, R.J. Degradation mechanism at XLPE/semicon interface subjected to high electrical stress. *IEEE Trans. Dielectr. Electr. Insul.* **1991**, *26*, 278–284. [CrossRef]

44. Zhang, S.; Zhang, H.; Liu, P.; Peng, Z.R. Space charge characteristics of epoxy-based nanocomposites filled with graphene oxide. In Proceedings of the IEEE Conference on Electrical Insulation and Dielectric Phenomena, Toronto, ON, Canada, 16–19 October 2016; pp. 691–694.
45. Zhao, L.; Su, J.C.; Pan, Y.F.; Li, R.; Zheng, L.; Zhang, Y.; Wu, X.L.; Zhao, P.C. Calculation on heating effect due to void discharge in polymers in cumulative breakdown process. *IEEE Trans. Dielectr. Electr. Insul.* **2017**, *24*, 3113–3121. [CrossRef]

Article

Assessment of Surface Degradation of Silicone Rubber Caused by Partial Discharge

Kazuki Komatsu [1],*, Hao Liu [1], Mitsuki Shimada [2] and Yukio Mizuno [1]

[1] Department of Electrical and Mechanical Engineering, Nagoya Institute of Technology, Nagoya 466-8555, Japan
[2] Information and Analysis Technologies Division, Nagoya Institute of Technology, Nagoya 466-8555, Japan
* Correspondence: 30413080@stn.nitech.ac.jp

Received: 20 June 2019; Accepted: 15 July 2019; Published: 18 July 2019

Abstract: This paper reports experimental and analytical results of partial discharge degradation of silicone rubber sheets in accordance with proposed procedures. Considering the actual usage condition of silicone rubber as an insulating material of polymer insulators, an experimental procedure is established to evaluate long-term surface erosion caused only by partial discharge. Silicone rubber is subjected to partial discharge for 8 h using an electrode system with air gap. Voltage application is stopped for subsequent 16 h for recovery of hydrophobicity. The 24 h cycle is repeated 50 or 100 times. Deterioration of sample surface is evaluated in terms of contact angle and surface roughness. It is confirmed the proposed experimental procedure has advantage of no arc discharge occurrence, good repeatability of results, and possible acceleration of erosion. Surface erosion of silicone rubber progresses gradually and finally breakdown of silicone rubber occurs. Alumina trihydrate (ATH), an additive to avoid tracking and erosion by discharge, is not necessarily effective to prevent breakdown caused by partial discharge when localized electric field in air is enhanced by adding ATH. In such a situation, lower permittivity and higher resistance of silicone rubber seem dominant factors to prevent partial discharge breakdown and a careful insulation design should be required.

Keywords: silicone rubber; partial discharge; degradation; breakdown; contact angle; surface roughness; FTIR; ATH

---

1. Introduction

Silicone rubber has been widely used as an electrical insulating material of polymer insulators of power transmission and distribution lines, because of advantages like light weight, easy handling, and water repellency. Electrical and mechanical degradation of silicone rubber is caused by discharges on its surface, exposure to ultraviolet ray, surface contamination, and heavy rain.

With respect to degradation of silicone rubber caused by discharge, arc discharge gives serious and irreversible damage to silicone rubber when compared with partial discharge. To improve resistance to tracking and erosion of silicone rubber caused by arc discharge, alumina trihydrate (ATH) is usually added to silicone rubber [1]. Water of crystallization contained in ATH evaporates by absorbing energy of arc discharge, resulting in good performance. Arc discharge usually occurs on silicone rubber surface under sever condition such as high electric field, heavy contamination and wetting [2]. However, occurrence probability of such conditions may be low.

On the contrary, partial discharge occurs more easily under less sever condition. Partial discharge is a localized breakdown observed in a small portion of an insulation system subjected to high electric field. It does not cause short-circuit of the insulation system. In the case of polymer insulators, partial discharge may be observed at the triple junction point-like interface of silicone rubber/electrode/air and silicone rubber/water droplet/air [3,4]. Partial discharge generated in air attacks the surface of silicone rubber. It is difficult for partial discharge to breakdown silicone rubber by one attack because its energy is not sufficient enough. However, silicone rubber subjected to partial discharge over a long period of time will be eroded gradually from the surface to the bulk and finally breakdown. Knowledge of partial discharge degradation of silicone rubber is considered essential to electrical insulation design of composite insulator. However, researches focused on partial discharge degradation of silicone rubber have not been performed much.

Analysis of contact angle and SEM images of surface after 48 and 96 h partial discharge treatment show that the resistance to partial discharge is improved by adding nano-sized silica to silicone rubber [5]. Effect of partial discharge on rheological and chemical properties of silicone rubber has been discussed in terms of oxidization and localized temperature rise, moisture, and depolymerization [6]. Another paper reports that deposited charge on silicone rubber surface by partial discharge may have an impact on its flashover voltage, but the material degradation caused by partial discharge is limited [7]. To understand partial discharge degradation phenomena of silicone rubber, further fundamental long-term experiments under various conditions in accordance with appropriate procedures seem necessary.

The authors have carried out experiments in rather actual conditions, where silicone rubber was subjected to partial discharge in salt or clean fog [8,9]. To simplify the experimental condition by avoiding the effect of water droplet on silicone rubber, we started experiments in the atmosphere without fog [10]. The present paper reports a series of results obtained in accordance with proposed experimental procedures, which are suitable to study degradation of hydrophobic silicone rubber caused only by partial discharge. The repeatability of results is good and the acceleration rate can be changed in a certain range by adjusting applied voltage and spacing of the gap where partial discharge occurs. The effect of alumina trihydrate (ATH)—a typical additive for mitigating tracking and erosion of silicone rubber—on partial discharge degradation and breakdown is studied by using five kinds of silicone rubber samples including different amount of additives. Results of chemical analysis show that ATH is decomposed by partial discharge as reported in the case of arc discharge. However, it is suggested addition of ATH lowers permittivity and resistivity of silicone rubber, which results in active partial discharge in air gap and shorter lifetime to breakdown.

2. Experimental Procedures

2.1. Sample and Electrode

Five kinds of silicone rubber samples of the same size (45 × 45 × 2 mm) were used as samples. The matrix of each sample was identical. Respective samples had different amount and surface treatment of additives: ATH and silica. The physical properties of samples [11] are summarized in Table 1.

Table 1. Physical properties of samples.

Sample	A	B	C	D	E
ATH [1,2]	0	50 Treated	100 Treated	100 Untreated	100 Untreated
Silica [2]	Treated	Treated	Treated	Untreated	Treated
Density (g/cm^3)	1.14	1.38	1.53	1.53	1.52
Resistivity (Ω-cm)	6.73×10^{17}	5.29×10^{15}	3.38×10^{14}	8.68×10^{12}	1.75×10^{14}
Relative permittivity	2.9	3.7	4.2	4.4	4.4
Breakdown strength (kV/mm)	14.8	15.4	15.1	14.0	14.2
Hardness (Hs)	62	72	83	77	80

[1] Part by weight. [2] Treated stands for rounded surface of particle.

A bundle of four steel round bars of 6 mm diameter was used as high voltage electrode; each bar had a hemispherical tip of 3 mm radius of curvature. The ground electrode was a stainless steel disc of 30 mm in diameter. A silicone rubber sample was placed on the ground electrode. An air gap of 1 or 2 mm spacing was formed between the tip of the round bars (high voltage electrode) and a sample surface. The electrode system is shown in Figure 1. Five electrode systems in total, each sample was placed in one electrode system, were set in an acrylic chamber of 1 m^3.

The level of applied ac voltage was determined so that no arc discharge occurred throughout the experiment and surface degradation progressed as fast as possible only by partial discharge. In the present study, 6.6 and 8.5 kVrms (60 Hz) were selected for air gap spacing of 1 and 2 mm, respectively. Voltage was applied simultaneously to five electrode systems for 8 h and then interrupted for 16 h for recovery of hydrophobicity of sample surface. The 24-h cycle was repeated 50 or 100 times, during these cycles degradation performance of five samples is categorized into 3 groups. Applied voltage and current of each sample were monitored and recorded. A schematic diagram of the experimental setup is shown in Figure 2.

In the present study, samples were not tested under the same partial discharge conditions (for example, amount of charge, and number of pulses). The magnitude of applied voltage was fixed instead, considering the actual usage condition of polymer insulators at the site. It result in different partial discharge activities among samples because localized electric field in the air gap depends on permittivity and conductivity of the sample.

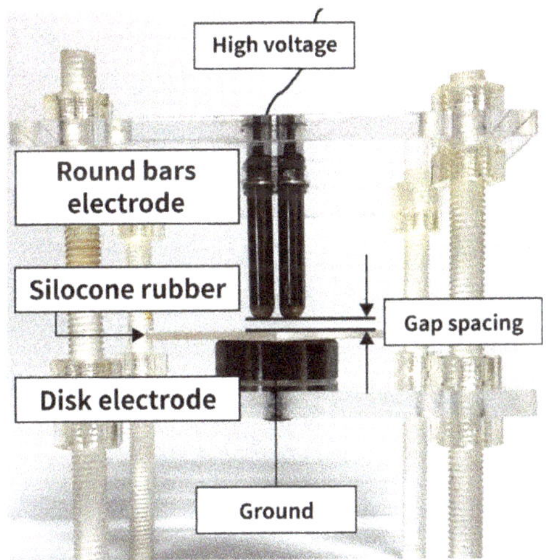

Figure 1. Electrode system.

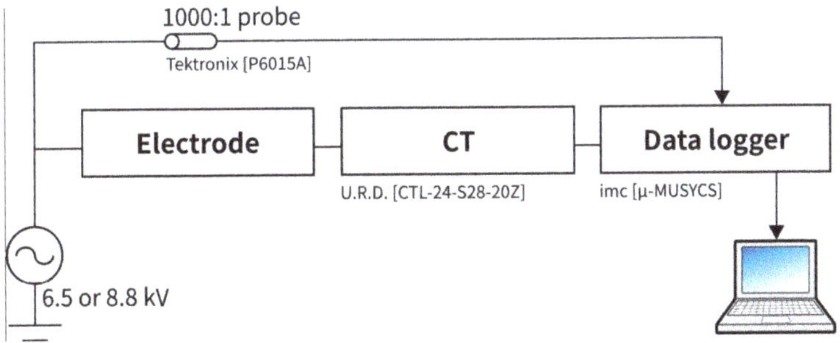

Figure 2. Schematic diagram of experimental setup.

2.2. Analyses of Surface Erosion

Contact angle was measured with a Drop Master DM500 (Kyowa Interface Science Co., LTD.) just before voltage application and just after voltage interruption of every cycle. A water droplet of 1 µl (conductivity: ~50 µS/cm) was dropped on a horizontally placed sample. Measurement was performed at five spots on the eroded area of a sample by partial discharge and the average contact angle was calculated.

Surface erosion of a sample was evaluated after 8-h voltage application with a surface roughness meter (TOKYO SEIMITSU CO., LTD., Surfcom 1400D) every three cycles. Measurement was carried out by 3 mm-long linear motion of a sensor on a sample surface.

Confocal microscope (Lasertec Corporation, OPTELICS HYBRID C3) enabled close observation of a sample surface of the area 1.5 mm × 1.5 mm. Images of samples were obtained every ten cycles. The arithmetic average roughness Ra of the surface in the observed area was also available. Ra gives the average of the absolute values of the roughness profile ordinates [12].

Infrared absorption spectrum was obtained every three cycles with a Fourier transformation infrared spectrophotometer (JASCO Corporation, FT/IR 6300).

3. Results

3.1. Sample Breakdown

No arc discharge was observed during the experiments. Samples C, D, and E were broken-down under some conditions as shown in Table 2. Higher relative permittivity of these samples compared with samples A and B as shown in Table 1 generates higher localized electric field in the air gap, resulting in active partial discharge and shorter lifetime to breakdown. Also, lower resistivity of these samples may have some relation to breakdown.

Table 2. Number of cycles of sample breakdown.

Sample	50-Cycle Test 1 mm Air Gap	50-Cycle Test 2 mm Air Gap	100-Cycle Test 1 mm Air Gap	100-Cycle Test 2 mm Air Gap
C	No breakdown	No breakdown	No breakdown	65
D	No breakdown	30	82	47
E	47	33	80	39

3.2. Contact Angle

Figure 3a,b shows the change in contact angle with number of cycle just before voltage application and just after voltage interruption in the 50-cycle test with 2 mm air gap spacing, respectively. Characteristics obtained in the 100-cycle test with 2 mm gap spacing are shown in Figure 4a,b. The initial value of contact angle is about 100 degrees for any sample. After 8-h voltage application, it decreased at most 20 degrees due to reduction of hydrophobicity of sample surface by partial discharge. Reduction in contact angle of Sample A is much lower than those of other samples. This corresponds to lower surface erosion of Sample A as described below. Recovery of contact angle to almost the initial value is confirmed after 16-h voltage interruption. It is considered that 16 h in the present experimental procedures are appropriate for low-molecular weight silicone oil to migrate from the bulk to the surface of a sample and to recover hydrophobicity. No remarkable reduction is observed with the number of cycle. Samples A and B are superior to the other samples when repeating cycles.

Large fluctuation of contact angle may be attributed to microscopic uneven surface caused by partial discharge, not only erosion like pitting but also accumulation of discharge by-product on the surface as suggested by the results of surface roughness measurement shown later in Figure 7. Contact angle does not necessarily reflect the surface profile in a small area, because the size of a water droplet for evaluation of contact angle is large enough to mask microscopic uneven surface and consequently an average information on surface in a larger area will be given.

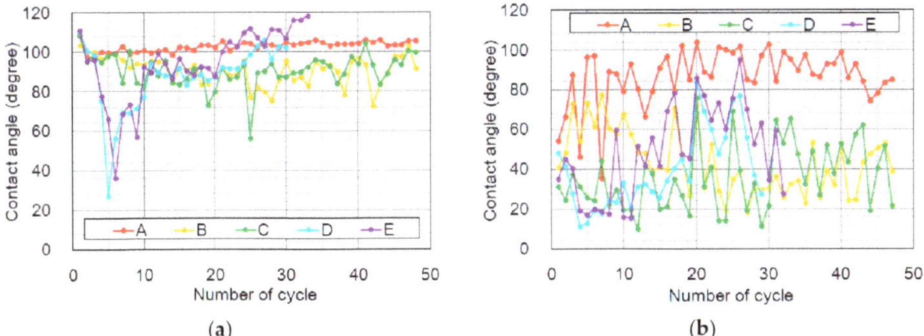

Figure 3. Change in contact angle with number of cycles in 50-cycle test with gap spacing 2 mm. (**a**) Just before commencement of voltage application. (**b**) Just after voltage interruption.

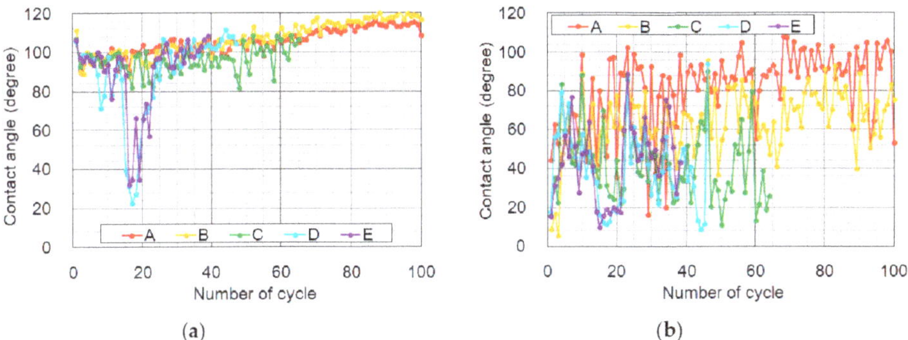

Figure 4. Change in contact angle with number of cycles in 100-cycle test with gap spacing 2 mm. (**a**) Just before commencement of voltage application. (**b**) Just after voltage interruption.

3.3. Observation of Sample Surface

Optical and confocal microscope images of sample surface are shown in Figure 5, which indicate change in surface erosion in the area of 1.5 mm × 1.5 mm of samples A and E in the 100-cycle test with 1 mm air gap. It is clearly shown especially by confocal microscope images that surface erosion progresses gradually with the number of cycle and degree of erosion is much different between samples A and E; surface erosion of sample E is serious and sample A is eroded slightly even after completing 80th cycle.

Figure 6 shows confocal microscopic images of all samples taken after completing 80 cycles in the 100-cycle test with 1 mm air gap. Appearance of surface erosion of Sample B is similar to that of Sample C. Sample D has eroded surface similar to Sample E. Samples D and E are severely damaged, which relates to the shorter lifetime to breakdown as shown in Table 2.

A quantitative discussion of surface erosion will be given in the next section.

Number of cycle	Sample A		Sample E	
	Optical	Confocal	Optical	Confocal
0				
10				
40				
80				

Figure 5. Optical and confocal microscope images of surface of samples A and E obtained in the 100-cycle test with 1 mm air gap.

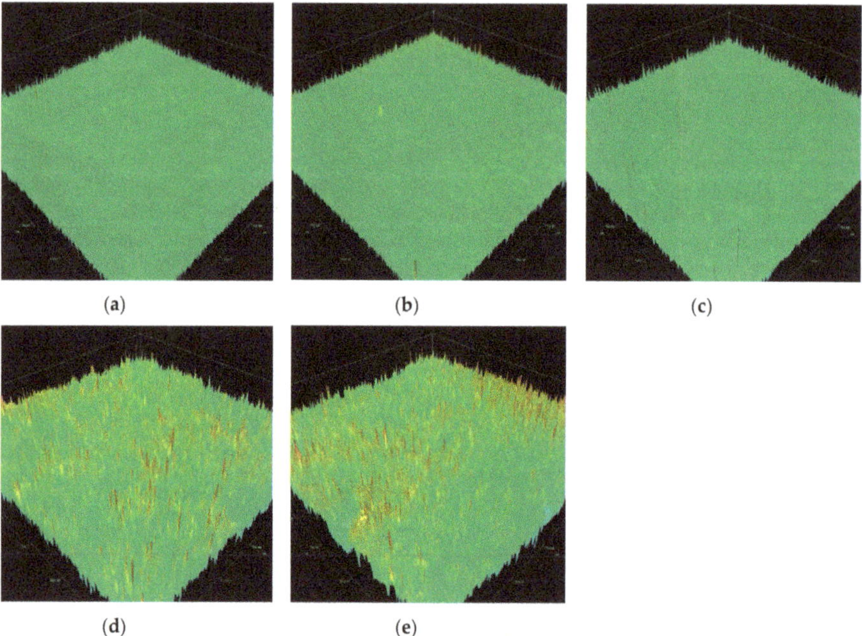

Figure 6. Confocal microscope images of sample surface obtained after completing 80 cycles in the 100-cycle test with 1 mm air gap. (**a**) Sample A; (**b**) Sample B; (**c**) Sample C; (**d**) Sample D; (**e**) Sample E.

3.4. Surface Erosion

3.4.1. Analysis Based on Surface Roughness Meter

Figure 7 shows surface roughness profiles of samples A and E, which are obtained after completing 6 and 60 cycles in the 100-cycle test with 1 mm air gap. Negative surface roughness means pitting on a sample surface generated by partial discharge. Discharge by-product accumulated on the sample surface may be a possible reason for the positive surface roughness. No remarkable change in surface roughness of sample A is observed between after 6 and 60 cycles. On the contrary, the surface of sample E is more eroded after completing 60 cycles; surface roughness is ~10 times larger compared with that after completing six cycles. These results correspond well to lower contact angle and shorter lifetime to breakdown of sample E compared with those of sample A, which is described above.

To evaluate quantitatively surface roughness shown in Figure 7, the arithmetic mean roughness Ra is used. Change in Ra in the 50-cycle tests with 1 and 2 mm air gap are shown in Figure 8a,b, respectively. When applied voltage is low, it is clear from Figure 8a that erosion of samples B to E progresses gradually with the number of cycle. Surface erosion is progressed in the order of samples D and E, samples B and C, and A, but difference among samples is not significant. In the case of 2 mm air gap, higher applied voltage intensifies partial discharge activity in the air gap and samples are clearly divided into three groups from standpoint of surface erosion. Samples D and E are eroded seriously by partial discharge and finally break down before completing 100 cycles. Erosion of samples B and C progresses steadily with the number of cycle. Sample A is eroded little. It is understood from Figure 8a,b that acceleration of partial discharge erosion is achieved by increasing applied voltage and gap spacing, especially for samples D and E.

Ra obtained in the 100-cycle test is shown in Figure 9 for 1 mm air gap. Ra increases gradually with the number of cycle except sample A. The characteristics of samples B and C are similar to each other. The same can be said of samples D and E. The result for 2 mm air gap are not shown because only limited data were available due to failure of the measuring instrument. Comparing Figure 9 with Figure 8a, obtained under the same applied voltage and the same gap spacing, surface roughness after completing 40 cycles, for example, is almost the same for any sample. This shows a good repeatability of results, suggesting the proposed experimental procedures are suitable to investigate partial discharge degradation of silicone rubber.

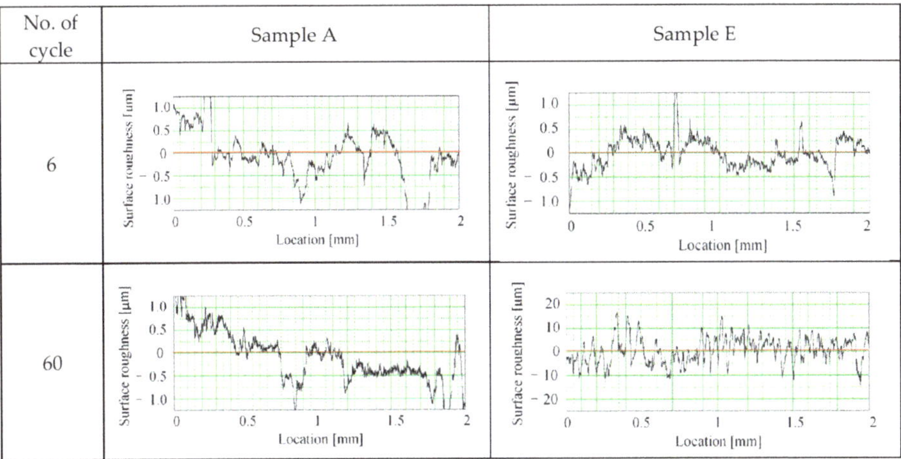

Figure 7. Examples of surface roughness profiles obtained with surface roughness meter in the 100-cycle test with 1 mm air gap.

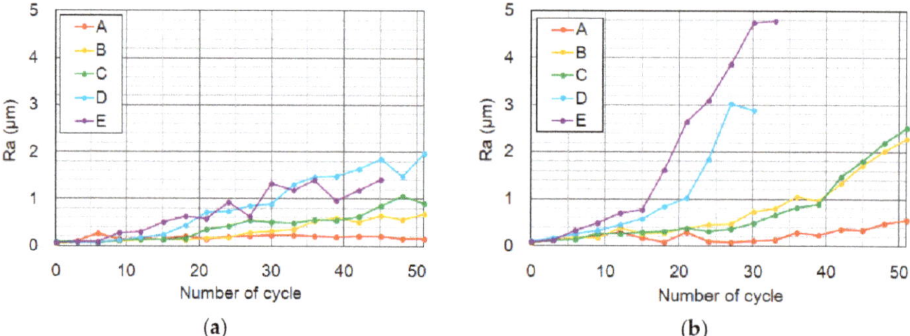

Figure 8. Change in Ra with number of cycle in the 50-cycle tests. (**a**) 1 mm air ga; (**b**) 2 mm air gap.

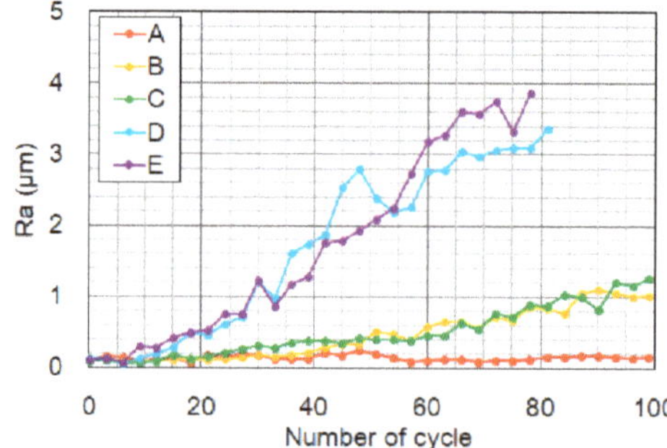

Figure 9. Change in Ra with number of cycle in the 100-cycle test with 1 mm air gap.

3.4.2. Analysis Based on Confocal Microscopic Images

In the previous section, surface roughness along the 3 mm straight line is discussed. Analysis of surface roughness in the area (1.5 mm × 1.5 mm) is available with confocal microscopic images shown in Section 3.2. Three-dimensional coordinate can be obtained at the time of surface scanning, where X- and Y-coordinates indicate the location on the sample surface and Z-coordinate shows height with reference to the sample surface without erosion.

Figure 10a,b shows changes in the arithmetic average roughness Ra over the 1.5 mm × 1.5 mm area with the number of cycle in the 100-cycle tests with 1 mm and 2 mm air gap, respectively. Samples D and E give larger Ra when compared with the other samples, which is the same tendency with results obtained with the surface roughness meter. Ra in Figure 10 is much larger than that in Figures 8 and 9. A possible reason is the difference in resolution of measuring instruments used. As the surface roughness meter is a stylus type apparatus, a fine probe needle is moved on a sample surface. Surface roughness is obtained based on up-and-down movement of the probe. Thus, it is considered difficult to measure precisely the depth of a narrow pitting smaller than the probe diameter, resulting in lower value of Ra. Meanwhile, LASER beam is used to evaluate surface roughness in the confocal microscope. Since the resolution is higher and a long and narrow pitting can be evaluated precisely, larger Ra is obtained.

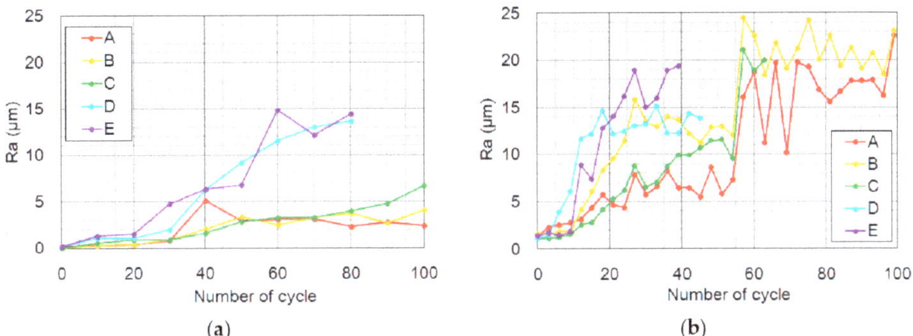

Figure 10. Change in Ra with number of cycle in the 100-cycle tests. (**a**) 1 mm air gap; (**b**) 2 mm air gap.

3.5. FTIR Spectrum

Change in FTIR spectrum with the number of cycle is shown in Figure 11 for all samples, which are obtained in the 100-cycle test with 2 mm air gap. Characteristics of Samples B and C are similar to each other. The same can be said of Samples D and E. These results correspond to progress of surface roughness described above.

In the case of sample A without ATH, absorbance around 786 and 1257 cm^{-1} attributed to hydrophobic Si-CH$_3$ [13] decreases with the number of cycle. Almost no change is observed in absorbance around 1008 cm^{-1} attributed to Si-O-Si. Increase of silica-related absorbance [13] is observed around 1060 cm^{-1} with increase in the number of cycle. A decrease in absorbance of hydrophobic Si-CH$_3$ and almost no change in absorbance of Si-O-Si, shown in Figure 12a, suggest that a side chain of silicone rubber is cleaved by partial discharge but main chain is damaged little, leading to limited erosion.

In sample E, eroded more by partial discharge compared with sample A, absorbance around 3620–3370 cm^{-1} vanishes when the number of cycle is large, which is related to ATH [13]. At the same time, absorbance around 940 cm^{-1} increases, which is attributed to Si-H [13]. It is suggested that ATH changes into aluminum hydroxide by losing hydroxyl groups. Not only absorbance of hydrophobic group Si-CH$_3$ but also that of Si-O-Si of main chain of silicone rubber decrease rather rapidly with the number of cycle as shown in Figure 12b, suggesting progress of silicone rubber decomposition and short lifetime to breakdown.

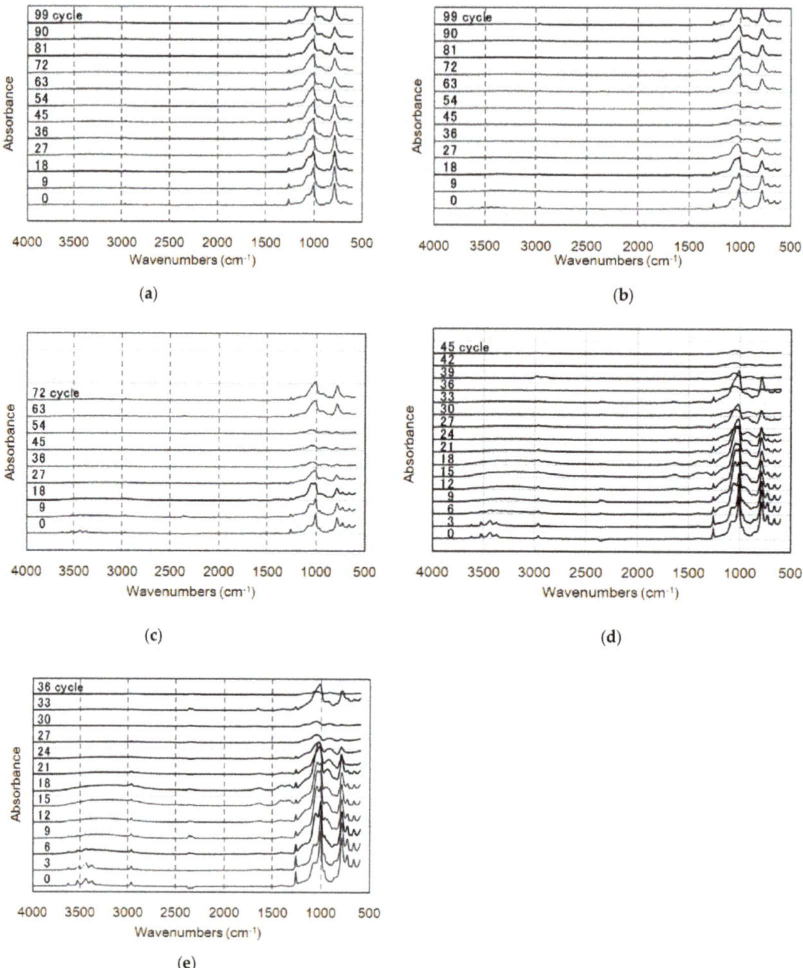

Figure 11. Change in FTIR spectra with the number of cycle obtained in the 100-cycle test with 2 mm air gap. (**a**) Sample A; (**b**) Sample B; (**c**) Sample C; (**d**) Sample D; (**e**) Sample E.

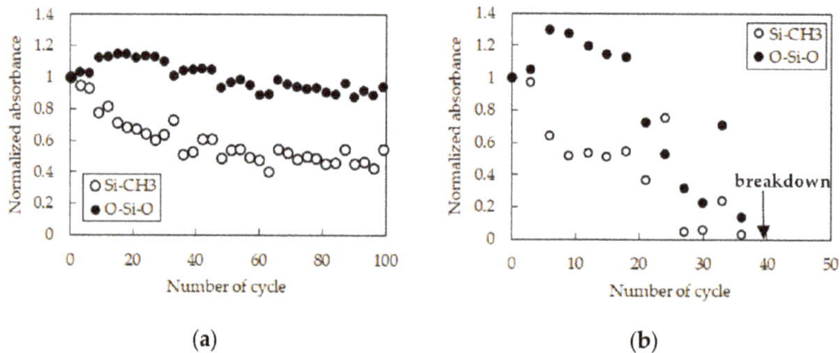

Figure 12. Change in normalized absorption of Si-CH$_3$ and O-Si-O with the number of cycle obtained in the 100-cycle test with 2 mm air gap. (**a**) Sample A; (**b**) Sample E.

4. Discussion

The proposed experimental procedures containing a cycle of 8-h voltage application and 16-h interruption seem suitable to evaluate partial discharge degradation of hydrophobic material like silicone rubber because hydrophobicity recovers almost to the initial value during 16 h. It reflects a practical usage condition of polymer insulators that hydrophobicity recovers if factors affecting hydrophobicity are removed.

Consistent results are obtained in 50- and 100-cycle tests. In both tests, surface erosion is larger in the order of samples D and E, samples B and C, and sample A. Progress of partial discharge degradation of silicone rubber, especially samples D and E, is accelerated in the case of 2 mm gap spacing compared with that of 1 mm in both tests.

Figure 13 shows change in Ra of five kinds of samples up to 50 cycles measured with the surface roughness meter, which are obtained by carrying out experiments 3 times under the condition of 1 mm air gap. Each sample shows almost the same performance in any experiment. Also Ra is larger in the order of samples D and E, samples B and C, and sample A in any experiment. It is considered the proposed experimental procedures give acceptable repeatability of results.

Surface erosion of a sample increases gradually by partial discharge with the number of cycle. In samples D and E, decrease in hydrophobic Si-CH$_3$ and Si-O-Si of main chain is confirmed by FTIR analysis. This is a possible reason for low contact angle, sever surface erosion, and consequently short cycles to breakdown. Samples D and E contain ATH, which is considered effective to enhance resistance to tracking and erosion by discharge because heat is absorbed by releasing the water of hydration from ATH molecule when the temperature of an ATH filled polymer reaches ~200 degrees [1,14]. This role of ATH seems effective for partial discharge because decrease in ATH is suggested through FTIR analysis in the present study. Nevertheless, surface erosion of samples D and E containing ATH is serious compared with sample A without ATH. This is attributed to intensified partial discharge activities in the air gap for samples D and E because their higher permittivity generates higher localized electric field in the air gap of the electrode system. Lower resistivity of samples D and E may also relate to shorter lifetime to breakdown.

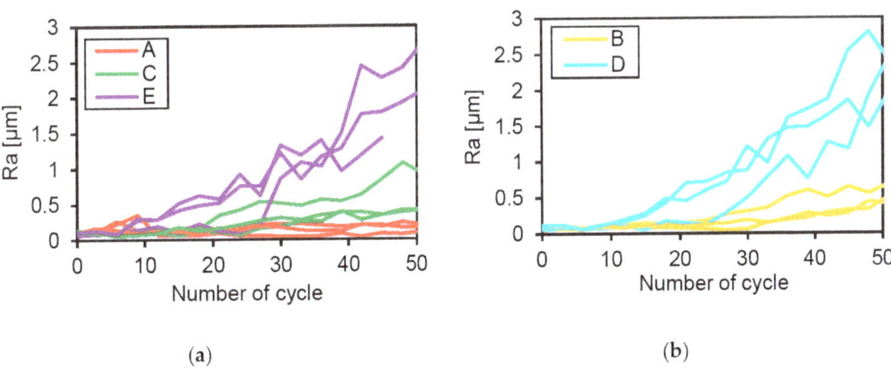

(a)   (b)

Figure 13. Repeatability of change in Ra with number of cycle obtained for 1 mm gap spacing. (a) samples A, C, and E; (b) samples B and D.

5. Conclusions

The proposed experimental procedures of partial discharge degradation of silicone rubber are evaluated by experimental results using five kinds of silicone rubber samples. Also the difference in resistance to partial discharge among samples are discussed. The results are summarized as follows.

1. The procedures proposed in this study is considered suitable for investigation of partial discharge degradation of hydrophobic material like silicone rubber. It has advantages of no arc discharge

2.   It is confirmed that surface erosion of silicone rubber progresses gradually and finally breakdown occurs only by partial discharge, though instantaneous damage is smaller compared with arc discharge. Partial discharge is one of the important factors in the long-term electrical insulation design of polymer insulators.

3.   It is suggested ATH is not necessarily effective to prevent breakdown of silicone rubber by partial discharge when localized electric field in air gap is enhanced by adding ATH. In such a situation, lower permittivity and higher resistance of silicone rubber seem more dominant factors to prevent damage by partial discharge and a careful insulation design should be required.

4.   Considering the actual usage condition of polymer insulators, applied voltage is fixed in the present study. Quantification of partial discharge activities and discussion of erosion from this standpoint are also important. Furthermore, degradation phenomena when discharge is kept constant across different sample is another important topic. The authors would like to investigate them in the near future.

Author Contributions: Conceptualization, K.K. and Y.M.; Methodology, K.K., H.L. and M.S.; Validation, K.K.; Formal Analysis, K.K., H.L. and M.S.; Investigation, K.K. and H.L.; Resources, Y.M.; Data Curation, K.K.; Writing-Original Draft Preparation, K.K.; Writing-Review & Editing, Y.M.; Visualization, K.K. and M.S.; Supervision, Y.M.; Project Administration, K.K. and Y.M.; Funding Acquisition, Y.M.

Funding: This research received no external funding.

Acknowledgments: The silicone rubber samples were supplied in connection with activities of an Investigating R & D Committee of the Institute of Electrical Engineers of Japan. The authors would like to thank the members of the committee.

Conflicts of Interest: The authors declare no conflict of interest.

References

1.   Kumagai, S.; Yoshimura, N. Tracking and Erosion of HTV Silicone Rubber and Suppression Mechanism of ATH. *IEEE Trans. Dilectr. Electr. Insul.* **2001**, *8*, 203–211. [CrossRef]

2.   Aeshad; Nekahi, A.; McMeekin, S.G.; Farzaneh, M. Flashover Characteristics of Silicone Rubber Sheets under Various Environmental Conditions. *Energies* **2016**, *9*, 683. [CrossRef]

3.   Nazemi, M.H.; Hinrichsen, V. Experimental Investigations on Water Droplet Oscillation and Partial Discharge Inception Voltage on Polymeric Insulating Surface under the Influence of AC Electric Field Stress. *IEEE Trans. Dielectr. Electr. Insul.* **2013**, *20*, 443–453. [CrossRef]

4.   Sarathi, R.; Mishra, P.; Gautam, R.; Vinu, R. Understanding the Influence of Water Droplet Initiated Discharges in Damage Caused to Corona-Aged Silicone Rubber. *IEEE Trans. Dielectr. Electr. Insul.* **2017**, *24*, 2421–2431. [CrossRef]

5.   Nazir, M.T.; Phung, B.T.; Hoffman, M. Performance of Silicone Rubber Composites with SiO_2 Micro/Nano-filler under AC Corona Discharge. *IEEE Trans. Dielectr. Electr. Insul.* **2016**, *23*, 2084–2815. [CrossRef]

6.   Habas, J.P.; Arrouy, J.M.; Perrot, F. Effect of Electric Partial Discharges on the Rheological and Chemical Properties of Polymers Used in HV Composite Insulators after Railway Service. *IEEE Trans. Dielectr. Electr. Insul.* **2009**, *16*, 1444–1454. [CrossRef]

7.   Gubanski, S.M. Outdoor Polymeric Insulators: Role of Corona in Performance of Silicone Rubber Housings. In Proceedings of the IEEE International Conference on Electrical Insulation and Dielectric Phenomena (CEIDP), Ann Arbor, MI, USA, 18–21 October 2015; pp. 1–9.

8.   Ueda, F.; Oue, K.; Mizuno, Y. Degradation of Silicone Rubber Caused by Partial Discharge in Clean Fog. In Proceedings of the 11th International Conference on the Properties and Applications of Dielectric Materials, Sydney, Australia, 11–22 July 2015. Paper No. AL0-5.

9.   Ueda, F.; Tamamura, M.; Mizuno, Y. Degradation of Silicone Rubber Caused by Partial Discharge in Clean Fog and in Atmosphere. In Proceedings of the Conference on Electrical Insulation and Dielectric Phenomena, Toronto, ON, Canada, 16–19 October 2016. Paper No. 8A-12.

10. Komatsu, K.; Shimada, M.; Mizuno, Y. Interaction of Partial Discharge in Air with Silicone Rubber. In Proceedings of the International Conference on High Voltage Engineering, Athens, Greece, 10–13 September 2018. Paper No. P-AM-4.

11. Hikita, M.; Hishikawa, S.; Kondo, T.; Yaji, K. Surface Modification and Performance of Polymer Composite Materials for Outdoor Insulation. In Proceedings of the Annual Meeting of IEEJ, Nagoya, Japan, 20–22 March 2013. Paper No. 2-S1-4.

12. Surface Roughness Terminology and Parameters. Available online: https://www.predev.com/pdffiles/surface_roughness_terminology_and_parameters.pdf (accessed on 18 June 2019).

13. Gao, Y.; Wang, J.; Liang, X.; Yan, Z.; Liu, Y.; Cai, Y. Investigation on Permeation Properties of Liquids into HTV Silicone Rubber Materials. *IEEE Trans. Dielectr. Electr. Insul.* **2014**, *21*, 2428–2437. [CrossRef]

14. Meyer, L.; Grishko, V.; Jayaram, D.; Cherney, E.; Duley, W.W. Thermal Characteristics of Silicone Rubber Filled with ATH and Silica under Laser Heating. In Proceedings of the Conference on Electrical Insulation and Dielectric Phenomena, Cancun, Mexico, 20–24 October 2002; pp. 848–852.

Article

Empirical Conductivity Equation for the Simulation of the Stationary Space Charge Distribution in Polymeric HVDC Cable Insulations [†]

Christoph Jörgens * and Markus Clemens

Chair of Electromagnetic Theory, School of Electrical, Information and Media Engineering,
University of Wuppertal, 42119 Wuppertal, Germany
* Correspondence: joergens@uni-wuppertal.de; Tel.: +49-202-439-1666
† This paper is an extended version of our paper published in 2018 IEEE International Conference on High Voltage Engineering and Application (ICHVE), Athene, Greece, 10–13 September 2018.

Received: 4 July 2019; Accepted: 1 August 2019; Published: 5 August 2019

Abstract: Many processes are involved in the accumulation of space charges within the insulation materials of high voltage direct current (HVDC) cables, e.g., the local electric field, a conductivity gradient inside the insulation, and the injection of charges at both electrodes. An accurate description of the time dependent charge distribution needs to include these effects. Furthermore, using an explicit Euler method for the time integration of a suitably formulated transient model, low time steps are used to resolve fast charge dynamics and to satisfy the Courant–Friedrichs–Lewy (CFL) stability condition. The long lifetime of power cables makes the use of a final stationary charge distribution necessary to assess the reliability of the cable insulations. For an accurate description of the stationary space charge and electric field distribution, an empirical conductivity equation is developed. The bulk conductivity, found in literature, is extended with two sigmoid functions to represent a conductivity gradient near the electrodes. With this extended conductivity equation, accumulated bulk space charges and hetero charges are simulated. New introduced constants to specify the sigmoid functions are determined by space charge measurements, taken from the literature. The measurements indicate accumulated hetero charges in about one quarter of the insulation thickness in the vicinity of both electrodes. The simulation results conform well to published measurements and show an improvement to previously published models, i.e., the developed model shows a good approximation to simulate the stationary bulk and hetero charge distribution.

Keywords: high voltage direct current; polymeric insulation; space charges; nonlinear electric conductivity

---

1. Introduction

The charge transportation behavior and accumulation in high voltage direct current (HVDC) cable insulations result in reliability problems of these components, due to an increased local electric field strength. In comparison to measurements, a cheap alternative is the simulation of such a charge distribution, using a conductivity model for the insulation. Typical conductivity models show a dependency on the electric field and the temperature [1]. For an accurate description of the space charge density, conductivity models need to include different effects. Short term effects are injection and extraction processes at both electrodes and charge packets. Relatively long term effects are the accumulation of bulk charges within the insulation and the presence of homo or hetero charges in the vicinity of both electrodes. Due to the long operation time of direct current (DC) power cables that are in service from several years up to decades, only long term effects are considered and the stationary charge distribution is simulated to determine the reliability of the insulation material.

Commonly used cable insulation materials are cross-linked polyethylene (XLPE) and low-density polyethylene (LDPE). These materials are in use due to their good electrical characteristics, ease of processing, and acceptable cost [1,2].

Depending on the charge type, high electric fields occur inside the insulation or in the vicinity of the electrodes. At low electric fields, the charge movement is higher than the charge injection. The injected charges move across the insulation, resulting in accumulated hetero charges and an increased electric field at both electrodes. At high electric fields, the charge movement is less than the charge injection and homo charges accumulate, resulting in decreased electric fields at the electrodes and increased electric fields within the insulation bulk [3,4].

Typically, the insulation material in power cables contains a semiconducting layer at the anode and another one at the cathode. Moving injected charges are blocked at these layers and form hetero charges. Furthermore, with an applied electric field, ionized impurities move towards the electrodes and form hetero charges as well. The electric field of the applied voltage is superimposed by the electric field of the hetero charges. As a consequence, the resulting electric field stress can exceed the breakdown strength of the insulation [5].

Without an applied temperature gradient, most of the charges accumulate in the vicinity of the electrodes. With an applied temperature gradient, bulk space charges accumulate, where the magnitude increases with increasing temperature gradient and, thus, with increasing conductivity gradient.

Accumulated space charges, either homo or hetero charges, result from different sources, e.g., the electrode material, the applied temperature gradient, or the local electric field. Due to common applied electric fields of less than 20 kV/mm within polymeric cable insulations (see [4]) and measured hetero charges in XLPE and LDPE insulations at an average electric field below 20 kV/mm, only the accumulation of hetero charges is considered in the simulations [6,7].

Typical conductivity-based cable models show a dependency on the electric field and the temperature. Thus, only bulk space charges are simulated and charges in the vicinity of the electrodes are neglected. Due to many processes (e.g., injection and extraction of charges at the electrodes), it is difficult to predict accumulated charges in the vicinity of the electrodes. In some cases, charges at the electrodes are higher compared to bulk space charges (see Figures 4a and 5). Thus, the resulting electric field is also increased, considering charges at the electrodes and common conductivity models that neglect homo or hetero charges are less reliable [4].

In [8], a conductivity model is used to predict the time dependent charge behavior. Inside XLPE insulation materials, a varying conductivity and permittivity in the vicinity of electrodes is discussed in [9]. In this paper, the stationary charge distribution in polymeric cable insulations is simulated, using an improved conductivity model. To validate the developed model and to find a general expression for the bulk and hetero space density in polymeric cable insulations, space charge measurements, taken from literature, are used.

The next section is about the development of an empirical conductivity model equation and the basic equations to simulate the stationary charge distribution. The simulation results are discussed and are compared against measurements in the third section. The obtained results are concluded in a short summary in section IV.

2. Empirical Conductivity Model Equation for the Simulation of the Stationary Charge Distribution

The stationary space charge and electric field distribution is computed, using the system of partial differential equations consisting of the stationary continuity equation, Poisson's equation, and Ohm's law, i.e.,

$$\operatorname{div} \vec{J} = 0, \tag{1}$$

$$\operatorname{div}(\varepsilon \operatorname{grad} \varphi) = -\rho, \tag{2}$$

$$\vec{J} = \sigma \vec{E}, \tag{3}$$

where $\vec{J}$ is the current density inside the insulation, $\vec{E} = -\mathrm{grad}\,\varphi$ is the magnitude of the electric field, φ is the electric potential, ρ is the space charge density, σ is the electric conductivity, $\varepsilon = \varepsilon_0 \varepsilon_r$, where $\varepsilon_0 = 8.854 \times 10^{-12}$ As/(Vm) is the dielectric constant and ε_r is the relative permittivity [9]. The Equations (1)–(3) are solved using the finite-difference method in one dimension. Depending on the geometry, depicted in Figure 1, a uniform grid of spacing Δh in either radial direction or x-direction is utilized, respectively [10]. Simulation results of a 150 μm thick insulating material are given in [11], where a non-uniform grid of minimum nodal distance 0.1 μm is used. To provide a sufficiently accurate spatial discretization for both geometries in Figure 1, the distance between the anode and the cathode is discretized with $N = 1500$ equidistant grid points in this work.

The thickness of the planplanar insulation is L, the radius of the inner cable conductor is r_i, the radius of the outer sheath is r_a, the conductor temperature is T_i, the sheath temperature is T_a, and the applied voltage is U.

In literature, the conductivity of polymeric insulations is typically modeled with an Arrhenius-type conductivity-temperature relationship. The electric field shows a dependency with a "sinh" (see e.g., [12,13]). The bulk conductivity σ_B of polymeric insulations is given by

$$\sigma_B = (|\vec{J_0}|/|\vec{E}|)\exp(-E_a/(kT))\sinh(\gamma|\vec{E}|), \tag{4}$$

where $k = 1.38 \times 10^{-23}$ J/K is the Boltzmann constant, T is the temperature, $|J_0|$ is a current density constant, E_a is the activation energy, and γ is a constant to describe the dependency on the electric field [13]. These constants are determined by a fit of Equation (4) to measured data. The constants of XLPE insulations are approximately $|J_0| = 1 \times 10^{14}$ A/m^2, $E_a = 1.48$ eV and $\gamma = 2 \times 10^{-7}$ m/V (see [13]). For an LDPE insulation, the constants are $|J_0| = 0.04224$ A/m^2, $E_a = 0.84$ eV and $\gamma = 4.251 \times 10^{-7}$ m/V (see [14–16]).

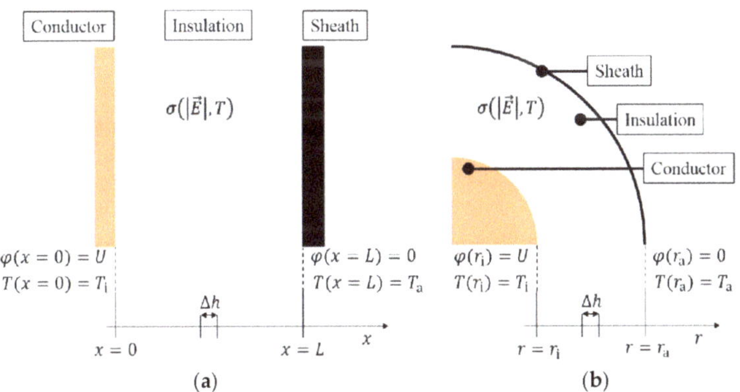

Figure 1. (**a**) geometry of a planplanar insulation; (**b**) geometry of a cylindrical insulation [10].

Analogously to [9], hetero charge accumulation in the vicinity of both electrodes is described by a conductivity gradient with two sigmoid functions. The total electric conductivity σ inside a cylindrical insulation is modeled by

$$\sigma = \sigma_B(K_1 - K_2), \tag{5}$$

$$K_1 = \frac{1}{1+\exp\left(-\frac{r-r_i-r_x}{\chi}\right)}, \tag{6}$$

$$K_2 = \frac{1}{1 + \exp\left(-\frac{r - r_a + r_x}{\chi}\right)}, \tag{7}$$

where K_1 describes the conductivity variations at the inner conductor (anode) and K_2 the conductivity variations at the outer sheath (cathode); in (6) and (7), r_x is the distance between r_i and the position of the highest gradient ($\sigma/\sigma_B(r = r_x) = 0.5$). If a planplanar insulation is considered, $r \rightarrow x$, $r_i = 0$ and $r_a = L$ holds.

The distance constant χ defines the conductivity gradient in the vicinity of both electrodes (see Figure 2a), which affects the magnitude and the shape of the hetero charges distribution (see Figure 2b). For example, in Figure 2 $\chi = 10$ μm and $\chi = 20$ μm with $r_x = 50$ μm are depicted. Increasing the distance constant χ results in a decreasing conductivity gradient and hetero charges distribution, with a slightly spread out charge shape. In Figure 2, a constant bulk conductivity (σ_B = const.) is used for the computation of the stationary space charge density ρ [10].

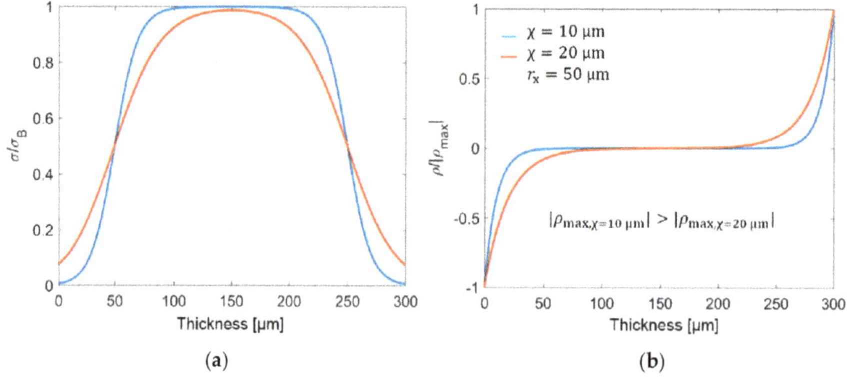

Figure 2. Influence of distance constant χ on conductivity gradient and resulting space charge distribution ρ. (**a**) normalized conductivity σ/σ_B; (**b**) normalized stationary space charge distribution [10].

A conductivity increase is described with K_1, resulting in negative charges at the anode and K_2 is describing a conductivity decrease, resulting in positive charges at the cathode. This relationship is obtained from the analytic solution of the stationary charge and electric field distribution, using a conductivity gradient.

Assuming a spatial varying conductivity $\sigma(r) = \sigma_0 \times K_1$, where σ_0 is a constant conductivity, and a planplanar insulation (see Figure 1a). Using (1), (3), and Gauss law, the stationary charge density is computed by

$$\mathrm{div}(\varepsilon \vec{E}) = \mathrm{div}\left(\varepsilon \frac{\vec{J}}{\sigma}\right) = \frac{\varepsilon}{\sigma}\mathrm{div}\left(\vec{J}\right) + \vec{J}\,\mathrm{grad}\left(\frac{\varepsilon}{\sigma}\right) = \sigma \vec{E}\,\mathrm{grad}\left(\frac{\varepsilon}{\sigma}\right) = \rho. \tag{8}$$

With (2) and $\vec{E} = -\mathrm{grad}\,\varphi$, the stationary electric field within a homogeneous (ε_r = constant) insulation is computed by

$$\varepsilon \frac{\partial}{\partial x}E(x) = \varepsilon \sigma E(x) \frac{\partial}{\partial x}\left(\frac{1}{\sigma}\right). \tag{9}$$

To obtain the stationary electric field, the solution of

$$\frac{\partial}{\partial x}E(x) - E(x)\left[\sigma \frac{\partial}{\partial x}\frac{1}{\sigma}\right] = 0 \tag{10}$$

is given by

$$E_{an}(x) = C \cdot \exp\left(\ln\left\{1 + \exp\left(-\frac{x - r_x}{\chi}\right)\right\}\right) = C \cdot \left\{1 + \exp\left(-\frac{x - r_x}{\chi}\right)\right\}, \tag{11}$$

where C is a constant, that is computed by

$$U = \int_0^L E(x)dx \rightarrow C = \frac{U}{\int_0^L 1 + \exp\left(-\frac{x - r_x}{\chi}\right)dx} = \frac{U}{L + \chi\exp\left(\frac{r_x}{\chi}\right) - \chi\exp\left(-\frac{L - r_x}{\chi}\right)}. \tag{12}$$

Measurements indicate hetero charges located near the electrodes and less charges are within the bulk. Thus, the conductivity gradient is also located near the electrodes and $L \gg r_x$. With $L \gg r_x$, the constant C is positive. With (11) and Gauss law, the stationary charge density is given by

$$\rho_{an}(x) = \frac{\partial}{\partial x}(\varepsilon E(x)) = -\frac{1}{\chi}\varepsilon C\exp\left(-\frac{x - r_x}{\chi}\right). \tag{13}$$

From (13), we see that the conductivity increases results in negative charges. Using a decreasing conductivity $\sigma(r) = \sigma_0 \times (1 - K_2)$ positive charges are modeled at the sheath, where the computation is analog to (8)–(10). The electric field is

$$E_{ca}(x) = C \cdot \exp\left(\ln\left\{1 + \exp\left(\frac{x - L_a + r_x}{\chi}\right)\right\}\right) = C \cdot \left\{1 + \exp\left(\frac{x - L + r_x}{\chi}\right)\right\}, \tag{14}$$

where C is equal to (12). The stationary charge density is

$$\rho_{ca}(x) = \frac{\partial}{\partial x}(\varepsilon E(x)) = \frac{1}{\chi}\varepsilon C\exp\left(\frac{x - L + r_x}{\chi}\right). \tag{15}$$

To determine the constants r_x and χ, the region of the conductivity gradient is defined by Δ. In Figure 3, (6) and (7) are depicted for a planplanar geometry [10]. To determine the dependency of r_x and χ on Δ, a straight line $f(x) = a(x - r_x) + b = (1/\Delta)(x - r_x) + 0.5$, depicted as the black line in Figure 3b, is used. With $f(x = r_x) = 0.5$, we define $r_x = \Delta/2$. As a first assumption to describe χ by Δ, we use the gradient of σ/σ_B at $x = r_x$ (equal to the gradient of σ/σ_B at $x = L - r_x$). For a planplanar geometry, the gradient of σ/σ_B is

$$\frac{\partial}{\partial r}(\sigma/\sigma_B) = \frac{\partial}{\partial r}(K_1 - K_2) = \frac{\exp\left(-\frac{r - r_x}{\chi}\right) \cdot \frac{1}{\chi}}{\left[1 + \exp\left(-\frac{r - r_x}{\chi}\right)\right]^2} - \frac{\exp\left(-\frac{r - L + r_x}{\chi}\right) \cdot \frac{1}{\chi}}{\left[1 + \exp\left(-\frac{r - L + r_x}{\chi}\right)\right]^2}. \tag{16}$$

If $L \gg r_x$, the gradient at $x = r_x$ or $x = L - r_x$ is equal to $1/(4\chi)$. Using $1/(4\chi)$ as the gradient a for a straight line $f(x) = a(x - r_x) + 0.5$, the green line in Figure 3b is obtained. Decreasing the gradient from $1/(4\chi)$ to approximately $1/(10\chi)$, depicted as the red dotted curve in Figure 3b, results in the best fit to describe the black line $f(x) = (1/\Delta)(x - r_x) + 0.5$. Thus, the constants r_x and χ are described by $r_x = \Delta/2$ and $10\chi \approx \Delta$, where now only Δ has to be determined by measurements.

In [17], hetero charge distributions in XLPE are discussed. The measured charge distribution shows an increasing hetero charge distribution from zero at the electrodes to approximately one third of the insulation thickness and then reducing to values close to zero around the center of the insulation. Thus, a first indication for the value of r_x is one third of the insulation and for the value of $\Delta = 10\chi$ is half of the insulation thickness. On the contrary, the analytic solution of the charge density (13) and (15) and Figure 2b show a maximum hetero charge density at the conductor, decreasing to zero within the range of Δ. The difference between measurements in [17] and (13) or (15) comes from filtered surface charges, which are considered in the measurements in [17]. The finite resolution of the measurement technique itself filters the surface charges, whereby they spread out and look like a Gaussian curve [8]. Consequently, a measured space charge distribution has its maximum hetero charge values in the vicinity of both electrodes, instead of a position immediately at the electrodes.

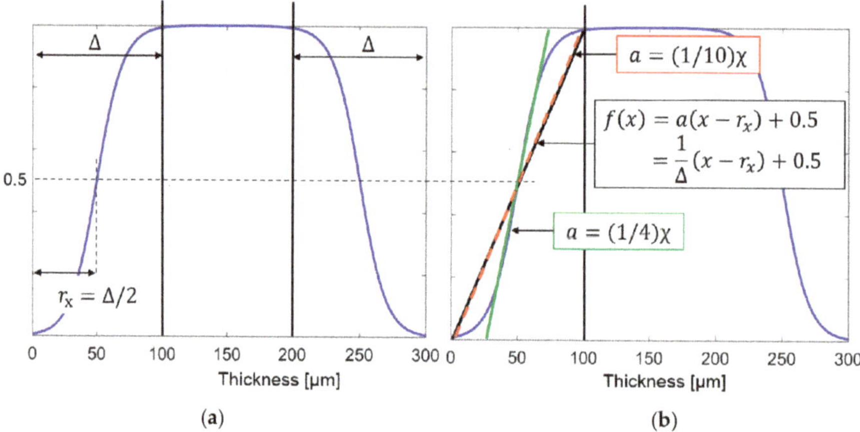

Figure 3. (**a**) Gradient region Δ, where the conductivity gradient is present. The position $r = r_x = \Delta/2$ has the highest gradient; (**b**) to determine χ by Δ, a straight line $f(x) = a(x - r_x) + 0.5 = (1/\Delta)(x - r_x) + 0.5$ is used, where the best approximation is shown by $\Delta = 10\chi$ [10].

To compare a measured and simulated space charge distribution, the surface charges have to be considered. With an applied voltage and accumulated space charges, surface charge (δ_+ and δ_-) accumulate at both electrodes. These charges are derived from [18] and are approximately described by

$$\delta_+ = -\int_{r_i}^{r_a} \frac{r_a - r}{r_a - r_i} \rho(r) \cdot dr + \varepsilon_0 \varepsilon_r(r_i) |\vec{E}(r_i)|, \tag{17}$$

$$\delta_- = -\int_{r_i}^{r_a} \frac{r - r_i}{r_a - r_i} \rho(r) \cdot dr + \varepsilon_0 \varepsilon_r(r_a) |\vec{E}(r_a)|. \tag{18}$$

The first term in (17) and (18) represents induced surface charges, by accumulated space charges, while the second term represents capacitance charges by the polarization of the insulation material.

The measurements are obtained, using the pulsed electro acoustic (PEA) method [6,19–21]. Utilizing an acoustic measurement technique, the resulting charge measurements are subjected to a Gaussian filtering process, giving inaccurate results of the measurements at the electrodes. Consequently, to compare the measurements against simulation results, the simulation results are filtered, using a Gaussian filter (see [18]).

3. Comparison between Simulated and Measured Space Charge Distribution

Simulation results of XLPE and LDPE insulation materials are depicted in Figures 4 and 5. The used parameters for the simulated results are summarized in Table 1 [10]. Comparing the simulation results in a planplanar insulation, the geometry in Figure 1a is used, while simulating a cylindrical insulation, the geometry in Figure 1b is used.

A description of the numerical implementation is found in [22], where the finite difference method is used for the one-dimensional problem. A brief description to setup the Gaussian filter is found in [22]. Space charge measurements and simulation results at the time $t \approx 0$ s are compared. At this time, (17) and (18) are reduced to $\varepsilon_0 \varepsilon_r(r_i) |\vec{E}(r_i)|$ and $\varepsilon_0 \varepsilon_r(r_a) |\vec{E}(r_a)|$, while no space charges are within the insulation material. The filter is calibrated by minimizing the difference between the measurements and the filtered simulation results at $t \approx 0$ s.

Table 1. Used constants for the simulation results in Figures 4 and 5 [10,19–21].

Figure 4a	Figure 4b	Figure 5
$U = 40$ kV	$U = 90$ kV	$U = 15$ kV
$L = 2$ mm	$r_i = 5$ mm $r_a = 9.5$ mm	$L = 0.3$ mm
$T = 27\,°\mathrm{C} = $ const.	$T_i = 65\,°\mathrm{C}$ $T_a = 50\,°\mathrm{C}$	$T = 27\,°\mathrm{C} = $ const.
$|\vec{J_0}| = 1 \times 10^{14}$ A/m^2	$|\vec{J_0}| = 1 \times 10^{14}$ A/m^2	$|\vec{J_0}| = 0.04224$ A/m^2
$E_a = 1.40$ eV	$E_a = 1.48$ eV	$E_a = 0.84$ eV
$\gamma = 2 \times 10^{-7}$ m/V	$\gamma = 2 \times 10^{-7}$ m/V	$\gamma = 4.251 \times 10^{-7}$ m/V
$\chi = 65.3\ \mu$m	$\chi = 0.15$ mm	$\chi = 10\ \mu$m
$r_x = 0.3$ mm	$r_x = 0.68$ mm	$r_x = 45\ \mu$m

The constant Δ is determined by minimizing the least squares problem

$$\eta = \sum_{i=1} \left(\rho_{M,i} - \rho_{S,i} \right)^2, \tag{19}$$

where the simulation results are ρ_S and the measurements are ρ_M.

3.1. Measurements of XLPE Insulation

Space charge measurements in planplanar and cylindrical XLPE insulations are found in references [6,19,20]. Here, the measured hetero charge distribution, including the surface charges, is positioned in about one third of the insulation thickness in the vicinity of both electrodes, while the magnitude varies with a varying temperature and electric field strength. With a constant magnitude of the applied voltage and the opposite polarity, the charge density also changes the polarity, but shape and magnitude remain the same. This so called "mirror image effect" is reported and discussed e.g., in [23]. The inversion of the equilibrium space charge distribution is explained e.g., with a spatially inhomogeneous polarization of the dielectric and the injection of charges at the electrodes [24,25].

Steady state charge distributions in a planplanar insulation configuration (Figure 4a) and in a cylindrical geometry (Figure 4b) are simulated and depicted in Figure 4 [10,19,20].

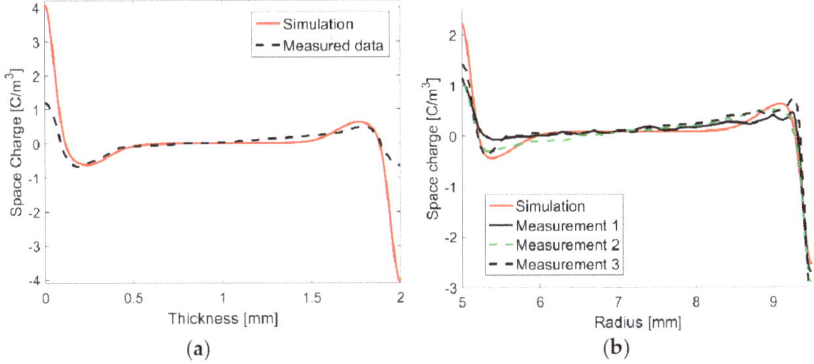

Figure 4. (a) Measured and simulated charge distribution in a planplanar cross-linked polyethylene (XLPE) insulation [20]; (b) Measured and simulated charge distribution in a cylindrical XLPE insulation. Three different XLPE cable measurements, labeled with numbers "1"–"3", are seen [10,19].

In Figure 4b, three different XLPE cable measurements, labeled with numbers "1"–"3", are seen. The cables differ for the semiconducting layer and the XLPE insulation type. The temperature in Figure 4a is constant ($T = 27$ °C), while in Figure 4b, a temperature gradient of 15 °C is used. The time independent temperature distribution is calculated by

$$T(r) \;=\; T_{\mathrm{i}} + \frac{T_{\mathrm{a}} - T_{\mathrm{i}}}{\ln(r_{\mathrm{a}}/r_{\mathrm{i}})} \cdot \ln(r/r_{\mathrm{i}}), \tag{20}$$

where the inner temperature is T_{i} and the outer temperature is T_{a}.

In Figure 4a, the hetero charges and surface charges are simulated and measured, both processes including a Gaussian filtering, in a region of about 0.52 mm (0.26·L) at each electrode. The used constants are $\chi = 0.052$ mm $= \Delta/10 \approx (0.26 \cdot L)/10$ and $r_{\mathrm{x}} = 0.25$ mm $= \Delta/2 \approx (0.26 \cdot L)/2$.

In Figure 4b, filtered hetero charges and surface charges are seen in a width of about 1 mm at each electrode, which is equal to $0.22 \times (r_{\mathrm{a}} - r_{\mathrm{i}})$. The used constants are $\chi = 0.12$ mm $= \Delta/10 \approx 0.22 \times (r_{\mathrm{a}} - r_{\mathrm{i}})/10$ and $r_{\mathrm{x}} = 0.6$ mm $= \Delta/2 \approx 0.22 \times (r_{\mathrm{a}} - r_{\mathrm{i}})/2$.

Comparing the simulation with "Measurement 1" indicate inaccurate results at the conductor and the sheath, where the simulation is overestimated. On the other hand, a comparison with "Measurement 2" and "Measurement 3" shows a sufficient accuracy of the simulated space charge distribution.

In Figure 4a, differences are observable especially at the anode and the cathode, which is a result of the filtering process. The computed surface charges (9) and (10) depend on the accumulated space charges and the local electric field. The local electric field depend on the space charges as well. If the simulated space charges differ in comparison to the measurement, the electric field and the surface charges also differ.

3.2. Measurements of LDPE Insulation

Measurements of a space charge distribution in a planplanar LDPE insulation are found e.g., in [6] and [21]. Simulation results and measurements of a 300 μm thick insulation are shown in Figure 5 [10,21]. The filtered hetero charges and surface charges accumulate in a region of approximately 80 μm at each electrode (0.267·L). The used constants are $\chi = 8$ μm $= \Delta/10 \approx 0.267 \cdot L/10$ and $r_{\mathrm{x}} = 40$ μm $= \Delta/2 \approx 0.267 \cdot L/2$.

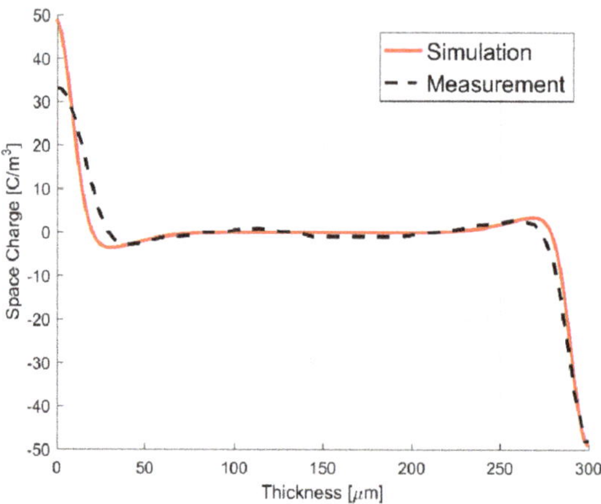

Figure 5. Measured and simulated charge distribution in a planplanar low-density polyethylene (LDPE) insulation [10,21].

In Figures 4 and 5, the defined region of accumulated space charges is equal to the region of the conductivity gradient in Figure 2. Differences between the charge distribution in Figures 4 and 5, compared to Figure 2b result from an assumed constant bulk conductivity in Figure 2. In Figures 4 and 5 a temperature and electric field dependent bulk conductivity is utilized. Accumulated hetero charges change the electric field and the conductivity at both electrodes, resulting in a spread-out charge shape compared to Figure 2b. Furthermore, in Figures 4 and 5, surface charges are considered and the simulation results are filtered, using a Gaussian filter [18].

Additional space charge simulations of XLPE and LDPE insulations are compared to measurements in [6]. The obtained results for the distance constant χ, the position r_x and the width of filtered surface and hetero charges Δ are seen in Table 2. A comparison between the simulation results and the measurements are seen in Figure 6 [10]. Accumulated hetero charges (including surface charges) are seen in approximately one quarter of the insulation thickness, whereby the width Δ is approximately independent of the geometry or the insulation material (see Table 2). Due to the "mirror image effect" of the charge distribution, the values for χ and r_x are constant, while changing the polarity of the voltage. It is not clear, why the hetero charges accumulate in one quarter of the insulation thickness. The resolution of the PEA method is 1.6 µm for a one dimensional planplanar insulation with a thickness of 25–27,000 µm and 0.1–1 mm for a cable insulation with a thickness of 3.5–20 mm [26]. The resolution is accurate enough to separate between hetero charges and bulk space charges.

Comparing, the constants χ and r_x in (6) and (7) with the width Δ (region of filtered surface and hetero charges), the approximation $\chi = (0.25 \cdot L)/10$ and $r_x = (0.25 \cdot L)/2$ for a planplanar insulation and $\chi = (0.25 \times (r_a - r_i))/10$ and $r_x = (0.25 \times (r_a - r_i))/2$ for a cylindrical insulation is defined.

Table 2. Distance constants χ and positions r_x for the simulation of space charge measurements in [6,19–21]. The absolute value of the applied voltage is $|U| = 20$ kV.

Ref.	χ	r_x	Insulation Thickness	Width of Charge Region Δ
[20], Figure 4a	0.052 mm	0.25 mm	2 mm	$0.26 \cdot L$
[19], Figure 4b	0.12 mm	0.60 mm	4.5 mm	$0.22 \times (r_a - r_i)$
[6], XLPE, planplanar, $+U$	0.052 mm	0.25 mm	2 mm	$0.26 \cdot L$
[6], XLPE, planplanar, $-U$	0.052 mm	0.25 mm	2 mm	$0.26 \cdot L$
[6], XLPE, cylindrical, $+U$	0.0875 mm	0.44 mm	3.5 mm	$0.28 \times (r_a - r_i)$
[6], XLPE, cylindrical, $-U$	0.0875 mm	0.44 mm	3.5 mm	$0.28 \times (r_a - r_i)$
[6], LDPE, planplanar, $+U$	0.052 mm	0.25 mm	2 mm	$0.26 \cdot L$
[6], LDPE, planplanar, $-U$	0.052 mm	0.25 mm	2 mm	$0.26 \cdot L$
[21], Figure 5	8 µm	40 µm	300 µm	$0.267 \cdot L$

Using a temperature and electric field dependent electric conductivity, e.g., (4), only charges within the bulk are simulated, which are mainly a result of the temperature gradient. Thus, comparing the measurements and the simulations, high differences are observable in the vicinity of the electrodes. For example, Figure 7 shows a space charge measurement and the simulating results using a commonly utilized conductivity Equation (4) and using (5). The applied constants are equal to the constants for Figure 4a,b and are seen in Table 1. Without a temperature gradient and electric field gradient, (4) is constant within the insulation and the simulated space charge density $\rho = 0$ C/m^3 (see (8)). The surface charges (17) and (18) are reduced to $\delta_+ = \varepsilon_0 \varepsilon_r(r_i)|\vec{E}(r_i)|$ and $\delta_- = \varepsilon_0 \varepsilon_r(r_a)|\vec{E}(r_a)|$ (see Figure 7a). A temperature gradient along the insulation results in accumulated bulk space charges, which are lower in the vicinity of the electrodes, compared to the measurements (see Figure 7b).

The conductivity of polymeric insulation materials differs with the sample preparation, which effects the constants $|J_0|$, E_a, and γ in (4). The resulting space charge distribution depends on many factors e.g., the conductivity, the local electric field, or the electrode material. As a result, it is very difficult to simulate the charge distribution of different references, even if it is the same material, like XLPE [27,28]. Differences between the measurements and the simulations in Figures 4–6 are small and the developed model yield results with good agreement to the measurements. The developed conductivity equation shows less differences to measurements compared to a commonly used conductivity equation and thus, the applicability of the formulation.

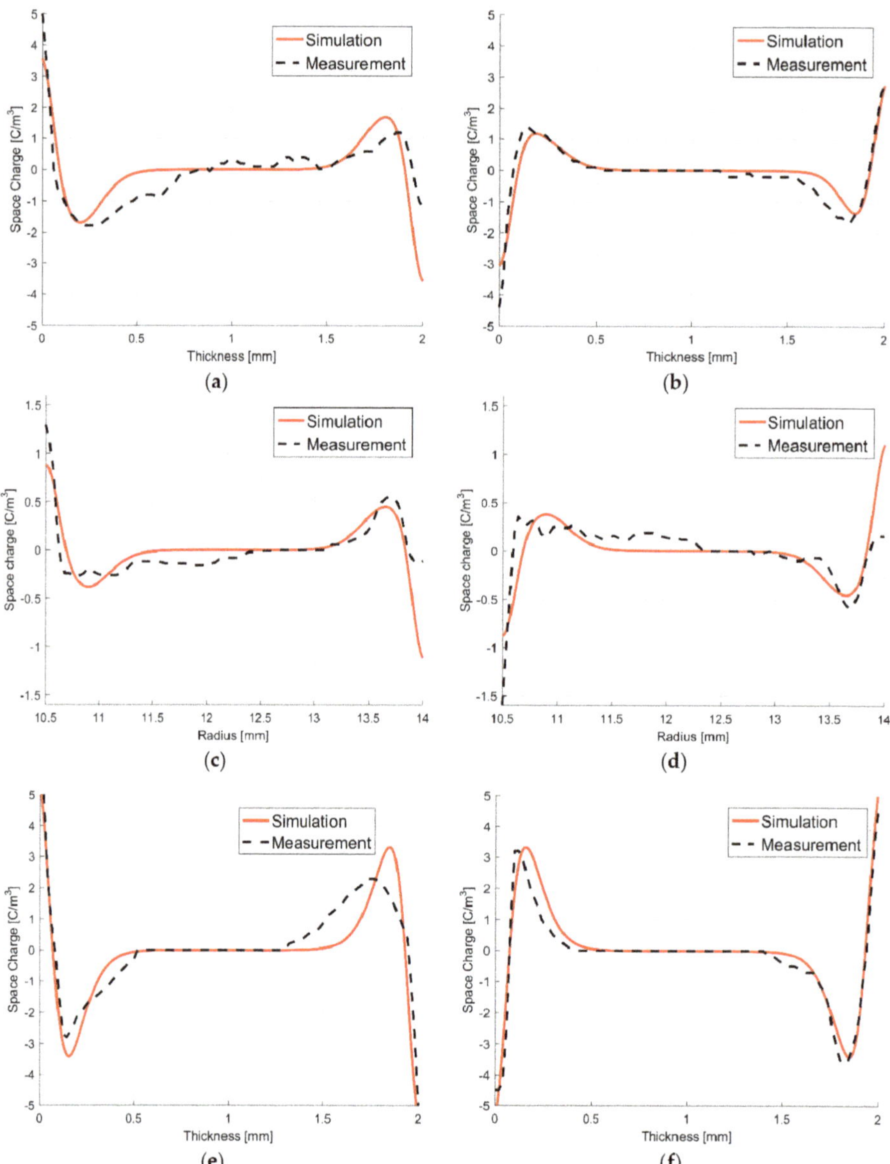

Figure 6. Measured and simulated charge distribution in a XLPE and LDPE insulation [6]. (**a**) XLPE, planplanar, $+U$; (**b**) XLPE, planplanar, $-U$; (**c**) XLPE, cylindrical, $+U$; (**d**) XLPE, cylindrical, $-U$; (**e**) LDPE, planplanar, $+U$; (**f**) LDPE, planplanar, $-U$. The absolute value of the applied voltage is $|U| = 20$ kV.

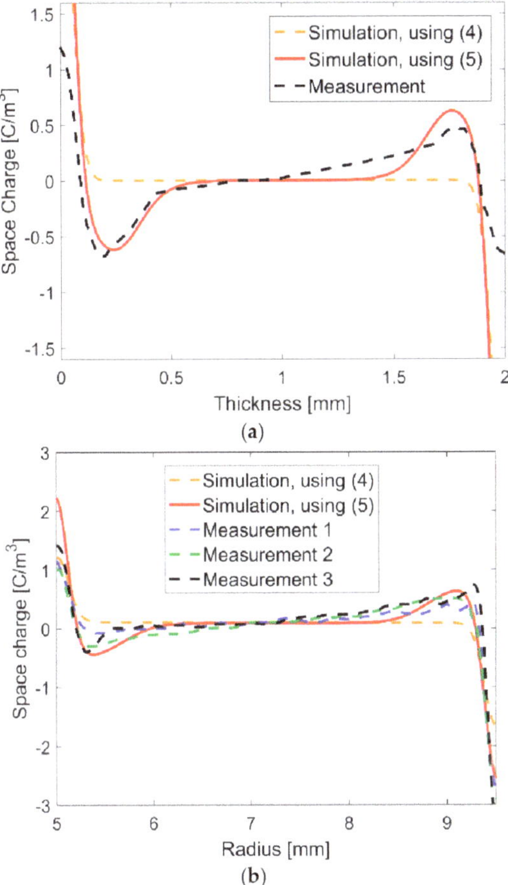

Figure 7. (**a**) Measured and simulated charge distribution, using a common conductivity equation in literature (4) and the developed conductivity equation (5) in a planplanar XLPE insulation [20]; (**b**) Measured and simulated charge distribution, using (4) and (5) in a cylindrical XLPE insulation [19].

4. Conclusions

An empirical conductivity model equation was developed for XLPE and LDPE insulation materials. As various space charge measurements in literature showed accumulated hetero charges in the vicinity of both electrodes, to simulate these charges, a commonly established, but insufficient accurate nonlinear field and temperature dependent conductivity model was used and improved by two sigmoid functions to create a conductivity gradient at both electrodes. The model was validated with its steady state space charge distribution and resulted in realistic accumulation of hetero charges typically positioned in about one quarter of the insulation thickness at each electrode. A comparison of simulation and measurement results validated the presented model, showing a good agreement and an improvement in comparison to those of previously established conductivity models.

Author Contributions: Conceptualization, C.J. and M.C.; Methodology, C.J. and M.C.; Software, C.J.; Validation, C.J.; Formal Analysis, C.J. and M.C.; Investigation, C.J. and M.C.; Resources, C.J. and M.C.; Data Curation, C.J. and M.C.; Writing-Original Draft Preparation, C.J. and M.C.; Writing-Review & Editing, C.J. and M.C.; Visualization, C.J. and M.C.; Supervision, M.C.; Project Administration, M.C.; Funding Acquisition, M.C.

Funding: This work was supported by the Deutsche Forschungsgemeinschaft (DFG) under the grant number CL143/17-1.

Acknowledgments: This work was supported by the Deutsche Forschungsgemeinschaft DFG (grant number CL143/17-1).

Conflicts of Interest: The authors declare no conflict of interest.

Nomenclature

$\vec{E}$	Electric field [V/m]
E_a	Constant for the temperature dependency of the bulk electric conductivity [eV]
$\vec{J}$	Current density [A/m^2]
$|\vec{J_0}|$	Constant for the bulk electric conductivity [A/m^2]
$k = 1.38 \times 10^{-23}$	Boltzmann constant [J/K]
K_1	Conductivity variations at the conductor
K_2	Conductivity variations at the sheath
L	Thickness of planplanar insulation [m]
N	Number of grid points
r	Coordinate for the cylindrical insulation [m]
r_a	Radius of the conductor in cylindrical coordinates [m]
r_i	Radius of the conductor in cylindrical coordinates [m]
r_x	Distance between the conductor (sheath) and the position of the highest gradient of K_1 (K_2) [m]
T	Temperature [°C]
T_a	Sheath temperature [°C]
T_i	Conductor temperature [°C]
U	Applied voltage [V]
x	Coordinate for the planplanar insulation [m]
γ	Constant for electric field dependency of the bulk electric conductivity [m/V]
Δ	Gradient region at the conductor and the sheath [m]
Δh	Distance between two grid points [m]
δ_+	Positive surface charges at the conductor [C/m^2]
δ_-	Negative surface charges at the sheath [C/m^2]
$\varepsilon_0 = 8.854 \times 10^{-12}$	Dielectric constant [As/(Vm)]
ε_r	Relative permittivity
η	Sum of the difference between ρ_s and ρ_M
ρ	Space charge density [C/m^3]
ρ_M	Measured space charge density [C/m^3]
ρ_s	Simulated and filtered space charge density [C/m^3]
σ	Total electric conductivity with hetero charges [S/m]
σ_B	Bulk electric conductivity without hetero charges [S/m]
φ	Electric potential [V]
χ	Distance constant to define the conductivity gradient in the vicinity of both electrodes [m]

References

1. Hanley, T.L.; Burford, R.P.; Fleming, R.J.; Barber, K.W. A general review of polymeric insulation for use in HVDC cables. *IEEE Electr. Insul. Mag.* **2003**, *19*, 14–24. [CrossRef]
2. Dissado, L.A. The origin and nature of 'charge packets': A short review. In Proceedings of the International Conference on Solid Dielectrics, Potsdam, Germany, 4–9 July 2010; pp. 1–6. [CrossRef]
3. Fleming, R.J. Space Charge in Polymers, Particularly Polyethylene. *Braz. J. Phys.* **1999**, *29*, 280–294. [CrossRef]
4. Küchler, A. *High Voltage Engineering–Fundamentals, Technology, Applications*, 5th ed.; Springer: Berlin/Heidelberg, Germany, 2018; pp. 583–590, 503–507. [CrossRef]
5. Fabiani, D.; Montanari, G.C.; Dissado, L.A.; Laurent, C.; Teyssedre, G. Fast and Slow Charge Packets in Polymeric Materials under DC Stress. *IEEE Trans. Dielectr. Electr. Insul.* **2009**, *16*, 241–250. [CrossRef]
6. Wang, X.; Yoshimura, N.; Murata, K.; Tanaka, Y.; Takada, T. Space-charge characteristics in polyethylene. *J. Phys. D Appl. Phys.* **1998**, *84*, 1546–1550. [CrossRef]

7. Takeda, T.; Hozumi, N.; Suzuki, H.; Okamoto, T. Factors of Hetero Space Charge Generation in XLPE under dc Electric Field of 20 kV/mm. *Electr. Eng. Jpn.* **1999**, *129*, 13–21. [CrossRef]
8. Jörgens, C.; Clemens, M. Modeling the Field in Polymeric Insulation Including Nonlinear Effects due to Temperature and Space Charge Distributions. In Proceedings of the Conference on Electrical Insulation and Dielectric Phenomena (CEIDP), Fort Worth, TX, USA, 22–25 October 2017; pp. 10–13. [CrossRef]
9. Hjerrild, J.; Holboll, J.; Henriksen, M.; Boggs, S. Effect of Semicon-Dielectric Interface on Conductivity and Electric Field Distribution. *IEEE Trans. Electr. Insul.* **2002**, *9*, 596–603. [CrossRef]
10. Jörgens, C.; Clemens, M. Empirical Conductivity Equation for the Simulation of Space Charges in Polymeric HVDC Cable Insulations. In Proceedings of the 2018 IEEE International Conference on High Voltage Engineering and Application (ICHVE), Athene, Greece, 10–13 September 2018. [CrossRef]
11. LeRoy, S.; Segur, P.; Teyssedre, G.; Laurent, C. Description of bipolar charge transport in polyethylene using a fluid model with constant mobility: Model predictions. *J. Phys. D Appl. Phys.* **2003**, *37*, 298–305.
12. Wintle, H.J. Charge Motion and Trapping in Insulators–Surface and Bulk Effects. *IEEE Trans. Electr. Insul.* **1999**, *6*, 1–10. [CrossRef]
13. Bodega, R. Space Charge Accumulation in Polymeric High Voltage DC Cable Systems. Ph.D. Thesis, Delft University of Technology, Delft, The Netherlands, 2006; pp. 78–88.
14. Kumara, J.R.S.S.; Serdyuk, Y.V.; Gubanski, S.M. Surface Potential Decay on LDPE and LDPE/Al$_2$O$_3$ Nano-Composites: Measurements and Modeling. *IEEE Trans. Dielectr. Electr. Insul.* **2016**, *23*, 3466–3475. [CrossRef]
15. Boudou, L.; Guastavino, J. Influence of temperature on low-density polyethylene films through conduction measurement. *J. Phys. D Appl. Phys.* **2002**, *35*, 1555–1561. [CrossRef]
16. Bambery, K.R.; Fleming, R.J.; Holboll, J.T. Space charge profiles in low density polyethylene samples containing a permittivity/conductivity gradient. *J. Phys. D Appl. Phys.* **2001**, *34*, 3071–3077. [CrossRef]
17. Fleming, R.J.; Henriksen, M.; Holboll, J.T. The Influence of Electrodes and Conditioning on Space Charge Accumulation in XLPE. *IEEE Trans. Electr. Insul.* **2000**, *7*, 561–571. [CrossRef]
18. Lv, Z.; Cao, J.; Wang, X.; Wang, H.; Wu, K.; Dissado, L.A. Mechanism of Space Charge Formation in Cross Linked Polyethylene (XLPE) under Temperature Gradient. *IEEE Trans. Dielectr. Electr. Insul.* **2015**, *22*, 3186–3196. [CrossRef]
19. Bodega, R.; Morshuis, P.H.F.; Straathof, E.J.D.; Nilsson, U.H.; Perego, G. Characterization of XLPE MV-size DC Cables by Means of Space Charge Measurements. In Proceedings of the Conference on Electrical Insulation and Dielectric Phenomena (CEIDP), Kansas City, MO, USA, 15–18 October 2006; pp. 11–14. [CrossRef]
20. Mizutani, T. Space Charge Measurement Techniques and Space Charge in Polyethylene. *IEEE Trans. Dielectr. Electr. Insul.* **1994**, *1*, 923–933. [CrossRef]
21. Wu, J.; Lan, L.; Li, Z.; Yin, Y. Simulation of Space Charge Behavior in LDPE with a Modified of Bipolar Charge Transport Model. In Proceedings of the International Symposium on Electrical Insulating Materials, Niigata, Japan, 1–5 June 2014; pp. 65–68. [CrossRef]
22. Jörgens, C.; Clemens, M. Conductivity-based model for the simulation of homocharges and heterocharges in XLPE high-voltage direct current cable insulation. *IET-SMT* **2019**. [CrossRef]
23. Fu, M.; Dissado, L.A.; Chen, G.; Fothergill, J.C. Space Charge Formation and its Modified Electric Field under Applied Voltage Reversal and Temperature Gradient in XLPE Cable. *IEEE Trans. Electr. Insul.* **2008**, *15*, 851–860. [CrossRef]
24. Bambery, K.R.; Fleming, R.J. Space Charge Accumulation in Two Power Cables Grades of XLPE. *IEEE Trans. Dielectr. Electr. Insul.* **1998**, *5*, 103–109. [CrossRef]
25. Lim, F.N.; Fleming, R.J.; Naybour, R.D. Space Charge Accumulation in Power Cable XLPE Insulation. *IEEE Trans. Dielectr. Electr. Insul.* **1999**, *6*, 273–281. [CrossRef]
26. Imburgia, A.; Miceli, R.; Sanseverino, E.R.; Romano, P.; Viola, F. Review of Space Charge Measurement Systems: Acoustic, Thermal and Optical Methods. *IEEE Trans. Dielectr. Electr. Insul.* **2015**, *23*, 3126–3142. [CrossRef]

27. Karlsson, M.; Xu, X.; Gaska, K.; Hillborg, H.; Gubanski, S.M.; Gedde, U.W. DC Conductivity Measurements of LDPE: Influence of Specimen Preparation Method and Polymer Morphology. In Proceedings of the 25th Nordic Insulation Symposium, Västerås, Sweden, 19–21 June 2017. [CrossRef]
28. Xu, X.; Gaska, K.; Karlsson, M.; Hillborg, H.; Gedde, U.W. Precision electric characterization of LDPE specimens made by different manufacturing processes. In Proceedings of the 2018 IEEE International Conference on High Voltage Engineering and Application (ICHVE), Athene, Greece, 10–13 September 2018. [CrossRef]

 *energies*

Article

Statistical Study on Space Charge effects and Stage Characteristics of Needle-Plate Corona Discharge under DC Voltage

Disheng Wang, Lin Du * and Chenguo Yao

State Key Laboratory of Power Transmission Equipment and System Security and New Technology, Chongqing University, Chongqing 400030, China
* Correspondence: dulin@cqu.edu.cn

Received: 30 June 2019; Accepted: 12 July 2019; Published: 17 July 2019

Abstract: The air's partial discharges (PD) under DC voltage are obviously affected by space charges. Discharge pulse parameters have statistical regularity, which can be applied to analyze the space charge effects and discharge characteristics during the discharge process. Paper studies air corona discharge under DC voltage with needle-plate model. Statistical rules of repetition rate (n), amplitude (V) and interval time (Δt) are extracted, and corresponding space charge effects and electric field distributions in PD process are analyzed. The discharge stages of corona discharge under DC voltage are divided. Furthermore, reflected space charge effects, electric field distributions and discharge characteristics of each stages are summarized to better explain the stage discharge mechanism. This research verifies that microcosmic process of PD under DC voltage can be described based on statistical method. It contributes to the microcosmic illustration of gas PD with space charges.

Keywords: partial discharge; needle-plate model; statistical rule; discharge stage; space charge

1. Introduction

Nowadays, the insulation requirements of DC equipment are further improved, with the wide application of DC system in high voltage direct current (HVDC) transmission and other aspects. As a reflection of insulation faults in DC system, PD under DC is also concerned in many researches.

Gas partial discharge (PD) under DC voltage is a process closely related to the generation, migration and dissipation of space charges. The main forms of gas PD in DC system are the surface discharge on dielectric materials and the corona discharges under non-uniform electric field [1,2]. Researches on gas corona discharges under DC mainly includes three directions. One is researching on microcosmic discharge theory and discharge mechanism explanation [3–5]. Another one is considering influencing factors of discharge [6–9], such as electrode structure [6,7], air pressure [8] and air flow [9]. The rest one is studying discharge parameters characteristics, and apply them to pattern recognition and stage division of discharges [10–14].

Different from the periodic generation and dissipation of space charges under alternate current (AC), many space charges will be retained in discharge region under DC because the polarity of electric field does not change. Therefore, one of the main points in the gas corona discharge researches under DC is the space charge effect and the electric field distortion caused by it. As yet, due to the lack of space charge measurement methods in gas, it is difficult to describe the space charge and the electric field distributions through testing. Reference [15] studies calibration of field-mill to measure DC electric field with space charges. However, the measurement scope cannot reach the distance of molecular level, so it still cannot precisely describe the distribution of space charge and electric field. Some researches apply simulation methods to describe the discharge mechanism [16,17]. In [17], applying corona discharge fluid model, space charge intensity and electric field variation during a single discharge process are described. But the simulation starts with none space charge situation and only lasts for the short time of a single discharge, which does not take the retained space charges into account. Space charge distribution are also studied in some experimental methods [18,19]. References [19] describes the corona layer morphology by gray value and thickness on the luminescent image. In this method, only space discharges distribution of ionization region can be obtained. In addition, the space charge behaviors and electric field variation of a single discharge cannot be described. Therefore, proposing new method to reflect the space charge effect and the electric field distortion is of significance.

It is obvious that the variation of electric field will affect PD phenomenon, and reflects in PD parameters. In theory, through the deep analysis of discharge parameter characteristics, the space charge and electric field distribution can be reflected. While existing researches on the discharge parameters characteristics are either focused on the differences of characteristics from other discharge stages and patterns [10–12], or focused on the characteristic extraction [13,14]. Few researches relate the statistical rules to the space charge effect and electric field distribution.

So in this paper, it is proposed to apply mathematical statistics methods and microcosmic discharge and space charge theories, to analyze space charge effect and electric field distribution on the basis of characteristic parameter rules of PD pulses.

In this paper, the air discharge of needle-plate model under DC is researched. Applying mathematical statistics, the stage characteristics of repetition rate (n), amplitude (V) and interval time (Δt) characteristic parameters are figured out to divide the discharge stages. The variation rule of characteristic parameters of each stage and transition process are explained by the space charge effects and electric field distribution. Furthermore, deduced space charge effects, electric field distributions and discharge characteristics of each stages are summarized to better explain the stage discharge mechanism.

2. Principle

2.1. Experiment Platform and Method

The schematic diagram of the experiment platform is shown in Figure 1. DC voltages are generated by the high voltage (HV) DC power supply Matsusada AU-30R2, whose output voltage range is ±30 kV. PD pulse acquisition adopts pulse current method. Oscilloscope is adapted to measure the voltage pulse signals on the 50 Ω non-inductive resistance. The needle-plate model is set in a cylindrical organic glass box (radius of 0.1 m), which is filled with air and blocks air flow. The curvature radius of needle tip r is 0.5 mm, and the distance between the needle and plate d is 15 mm. d/r is greater than 4, and the needle-plate electrical field is extremely uneven [20].

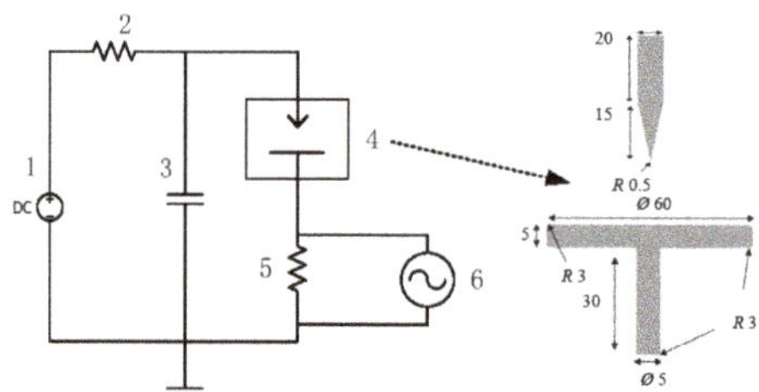

1- DC power supply 2- protective resistance 3- coupling capacitor
4- needle-plate model 5- non inductive resistance 6- oscilloscope

Figure 1. Schematic diagram of the experimental platform.

During the experiment process, applied voltage is raised with step-rise method. After initial discharge voltage are found, the voltage is increased to a certain voltage level with step of 0.25 kV (0.1 kV for some continuous voltage levels with marked PD signal change), and then sustains. Under each voltage level, when PD are relatively stabilized, PD signals in window width of 0.1 s are acquired with the sampling frequency of 500 MHz, which is repeated with 2 min interval. 5 groups of signals are recorded under each voltage level. To avoid occasionality, whole experiment process is repeated more than three times under each polarity. The temperature range is 293 K to 295 K, and humidity range is 75% to 79% during the experiment processes.

2.2. Relevant Space Charge Theory

According to the streamer theory, in a long gap, the PD will develop in form of streamer. When electric field intensity (E) on the head of streamer reaches a certain level, photons excited in strong ionizations induce photoelectric ionization, which produces new electrons and forms the secondary electron avalanches. Then the streamer can self-sustaining develop. The self-sustaining process of streamer is described in [21] under negative polarity needle-plate model. The ions on the head of streamer are considered to aggregate as a sphere, and self-sustaining condition can be expressed as:

$$\int_r^R \frac{2}{3} p f \mu e^{-\mu \rho} \frac{\sqrt{\rho} r}{(d - x_1)^{1/2}} \alpha e^{\int_\rho^r \alpha d\rho'} d\rho = 1,\tag{1}$$

where r is the radius of the sphere. ρ denotes radius of the photons from the center. R is the radius of the region where photons can induce photoionization, which is related to E. p denotes the probability of photoionization. d is the distance between the electrodes, and x_1 is the distance between the trigged electrons and the plate. The collision ionization coefficient α is related to E and pressure P, which can be expressed as:

$$\alpha/P = Ae^{-BP/E},\tag{2}$$

where A, B are coefficients. According to Equation (1) and Equation (2), E has obvious influence on the self-sustaining development of the streamer. When other variables are confirmed, the minima electric field intensity which meets the self-sustaining condition (E_0) can be determined.

During the formation process of electron avalanches, ionizations produce electrons and cations. Migration velocity v of ions is connected with the E. The charge–mass ratio of electron is much lower than ions, thus its migration rate is about two magnitudes higher. For needle-plate module, E in tiny region near the needle tip is significantly higher. Under high E, produced electrons mainly have migrated away from the discharge region when self-sustaining discharge stops. Meanwhile, high-speed electrons can hardly be trapped by molecules, thus merely small proportion of the electrons form anions in ionization region. Only in low E regions can more anions be formed gradually. Therefore, in the development path of streamers, the densities of retained electrons and anions are relatively low. Low-speed cations almost retain and play a major role.

2.3. PD Pulse Characteristic Parameters

The acquired PD pulses reflect the electric field variation of the gap in a discharge process, which is caused by the instantaneous change of electric charge quantity (q). Characteristic parameters amplitude (V), interval time (Δt) and repetition rate (n) of the PD pulses are selected for statistical study.

V can reflect the maximum change of q in a single discharge. From the beginning to the end of a streamer, space charges produced before are not completely dissipated by migration and recombination. Thus space charges retain and q rises. The maximum q should occur at the moment the self-sustaining discharge stops. Combined with the discussions above, amplitude of discharge pulse is related to E_0.

Δt is defined as the time interval from the end of a PD pulse to the beginning of the next. It is mainly composed of the electric field recovery time and the discharge delay of the next pulse. Discharge delay refers to the time delay of a pulse after applied voltage reaches the initial discharge voltage value. For streamer-type PD, discharge delay is about 10^{-8} s and can be ignored. The recovery time of electric field is related to the dissipation of space charges, which is influenced by q of the retained space charges and E of the space after a streamer. According to the analysis above, the maximum q in a discharge determines V. Therefore, Δt should have relationship with V if E is similar.

n is defined as the discharge repeating times per millisecond, which is related to Δt and mainly used for auxiliary analysis of discharge rules when Δt is too small to figure out its variation.

Acquired PD pulses under positive or negative polarity DC voltage are shown in Figure 2. In this research, the first peak value of a pulse is recorded as V. For positive polarity, the trigger voltage level is 0.2 V. For negative polarity, the trigger level is −0.1 V.

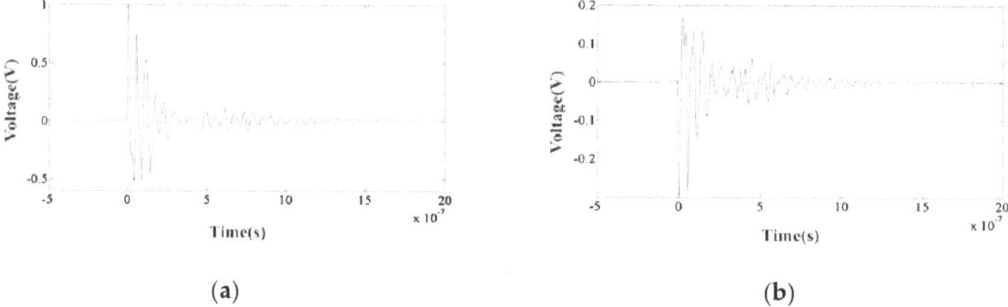

(a)   (b)

Figure 2. Single pulse under positive or negative polarity DC voltage. (**a**) Positive pulse; (**b**) Negative pulse.

3. Statistic Rules and Stage Characteristics of Positive PD

3.1. Formation of Positive PD Pulse

First of all, how space charge behaviors and electric field variation contribute to the formation of a PD pulse is deduced.

When E of the tiny region near the needle tip reaches the initial discharge value, electron avalanches occur here. When electron avalanche develops forward, produced ions will reduce E in the region where ionizations just happen. Thus ionizations mainly happen on the head regions. When ionizations on the head stop, so does the whole electron avalanche.

For positive polarity voltage, cations are concentrated in the head of electron avalanches. When the self-sustaining condition is satisfied, electron avalanche discharge transfers to streamer discharge. The cations produced by secondary electron avalanches further increase the density of cations on the head of the streamer. Cations can be approximately considered as motionless in the discharge path during the short time of discharge. While electrons generated by secondary electron avalanches are injected into the body of the streamer, and most are transformed to anions because of the low E. Then, the anions and the cations form plasma, which is approximately electrically neutral inside. Thus the streamer shows positive polarity on the whole affected by the high-density cations on the head during the discharge process. Because the charge–mass ratio of cations is relatively high, the interaction effect between cations is low. The paths of streamers under positive polarity DC voltage are relatively concentrated. An instantaneous, enhanced positive polarity electric field will be generated by the cations near the needle tip. The E near the plate will also increase. A growing number of electrons flow toward the gap from the plate, thus forming the rising edge of the pulse current on the sampling resistance.

The streamer continues to develop towards the plate until E is insufficient to maintain self-sustaining discharges on the head. Then under the action of electric field, retained cations migrate toward the plate and anions migrate toward the needle. The electric field intensity gradually returns to the initial value, and the enhanced electric field decreases. The number of electrons flowing from the plate decreases gradually, and the falling edge of PD pulse occurs. Because the E on the discharge path at the falling stage is lower than before, the falling time is longer than the rising time. When the electric field intensity of the region resumes to the initial discharge value, a new PD can be generated.

3.2. Discharge Stages

Air corona discharge of the needle-plate module under DC has obvious space charge effect. When E of discharge paths resumes to the initial discharge value, part of ions still retain. After multiple discharges, ion densities reaches an equilibrium state. The cation density near the needle tip should be higher than the average. When the density reaches a certain level, the ionic electric field will distort to the original electric field.

The minimum applied voltage under which the PD pulse can be detected is defined as the initial discharge voltage. First, the average value of parameters n, V and Δt are extract. The expression of the average value is:

$$\overline{y} = \sum_{1}^{N} y_i / N, \tag{3}$$

where N denotes the number of data, and y_i denotes the i th data.

In this research, for each voltage level, five or more groups of acquired PD data (0.1 s each) are chosen from different measurements, ensuring that inside which are at least one hundred pulses. Then characteristic parameters are extracted from these PD pulse data.

Figure 3 shows the average value of PD parameters under each voltages. According to rules of $\overline{n}$, $\overline{V}$ and $\overline{\Delta t}$, the discharge stages are divided. The discharge process can be divided into three stages: initial streamer stage, glow-like discharge stage and breakdown streamer stage.

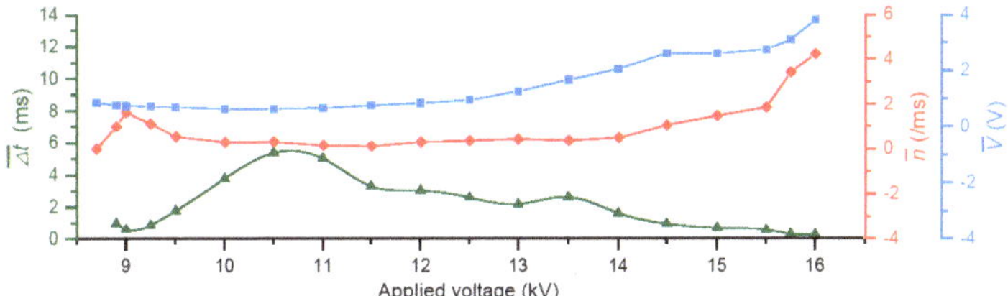

Figure 3. Average value diagram of n, V and Δt on positive polarity.

The initial streamer stage is 8.8 kV to 9 kV. The density of retained cations at this stage is relatively low and is close to the needle tip, so the effect of cation electric field is weak. E of outer regions is below E_0. Thus the development degrees of streamers are similar, and the generated space charge amounts of each discharge are close. Thus, $\overline{V}$ basically retains unchanged. Under initial discharge voltage, the high-intensity electric field region is tiny. The discharges is random, thus $\overline{\Delta t}$ is large (84.44 ms, not shown in Figure 3). With the increase of applied voltage, amount of retained cations increases. Because the cations group is close to the tip, thus their main effect is enhancing original E near the needle tip. The amount of produced cations and anions is relatively small, so when E increases, the electric field recovers quickly, thus $\overline{n}$ rises.

Nine kV to 9.5 kV is the transition stage, where the cation group gradually becomes intensive and moves away from the needle tip. For the electric field between cations and the needle tip, the cation group generates an opposite electric field and weakens the original one. In some regions, E is weakened to the value below E_0, so some streamers are weakened and stopped inside cation group. While in some regions, the lowest E in cation group is higher than E_0, and even stronger more outside, streamers are enhanced. So on average, $\overline{V}$ changes little. Weakened E inside results in the increasing of electric field recovery time. So $\overline{\Delta t}$ increases.

The glow-like discharge stage is about 9.5 kV to 13.5 kV. At this stage, the development of streamer is significantly inhibited by the cations group. The cations group stays at a certain distance from the tip and the opposite electric field is more remarkable. Majority of streamers are stopped inside the cation group. In this situation, the equipotential line of E where E is equal to the E_0 is basically unchanged, so the development degrees of the streamers at the initial stage are similar. $\overline{V}$ is basically unchanged. Then at the later transition stage, the distance of the cation group from the needle expands under the action of the needle-tip electric field. The cation electric field cannot offset the original electric field increases. So E gradually increases and streamers develop further. $\overline{V}$ rises slightly. At the initial stage, due to the increasing amount of generated space charges and the decreasing E, the discharge recovery time increases and $\overline{\Delta t}$ is large. In 10 kV to 11.5 kV, E inside decreases significantly and $\overline{\Delta t}$ reaches the maximum value. Then with the expansion of the cation group and the slightly increase of E, recovery time slightly decreases, and so does $\overline{\Delta t}$.

As E between the needle tip and the cation group is weakened, electrons are more likely to form anions and accumulate near the needle tip under the action of the electric field. If the curvature radius of the positive polarity electrode is relatively large, the density of accumulated anions can reach a certain level and anion group can be formed. The electric field between the anion group and the needle tip will be enhanced, while the electric field between anion group and the cation group will be further weakened. Then, the stable glow discharge may occur between the needle tip and the cations. There is no obvious glow discharge phenomenon under positive polarity voltage in this needle–plate gap.

At the breakdown streamer stage (13.5 kV to 16 kV), the cations group is far away from the needle tip and its density is also relatively reduced. Its weakening effect to the electric field is reduced, and the streamers can develop through the cation region. As a result, the streamers are stronger than the stage before. Raising the applied voltage, more cations are generated and $\overline{V}$ rises. E in the discharge regions increases. The recovery time decreases significantly and so does $\overline{\Delta t}$.

When it close to breakdown (after 15.5 kV), the negative polarity plate begins to affect the discharge process. At this time, the head of streamers will be close to the negative polarity plate. Due to the high conductivity of the plasma, the electric field between the streamer head and the plate electrode is significantly enhanced, which is conducive to ionization. The streamers will be strongly enhanced here. Therefore, V increases with obviously rising trend. After a discharge, E of the discharge path is still in an enhanced state. So Δt decreases a lot (about 0.23 ms at 16 kV). Continue to increase the applied voltage, streamers link the gap and breakdown occurs quickly. Due to the randomness of the breakdown streamer, the breakdown may occur under a certain applied voltage within the scope of 16.5 kV to 17.5 kV.

3.3. Derived Parameter Rule

The equivalent charge quantity q is defined as the charge quantity of electrons passing through the sampling resistance in unit time, which can indirectly reflect the changes of the space electric field and discharge energy. Regarding single PD pulse as a triangular wave approximately, the expression of q is:

$$q = \overline{n} \cdot \frac{\overline{V} t_w}{2R}. \tag{4}$$

where t_w denotes single PD pulse width, which is set as 50 ns. R denotes the sampling resistance value.

Equivalent charge quantities of positive PD are shown in Figure 4. At the initial streamer stage, there is an obvious q, while at the glow-like discharge stage q is close to 0. Then in the breakdown streamer stage, q increases obviously, especially near breakdown. It indicates that the discharges of glow-like stage are obviously inhibited under the action of space charges, which also means that the electric field variation is little, and the discharges consume less energy. q is obvious at breakdown streamer stage.

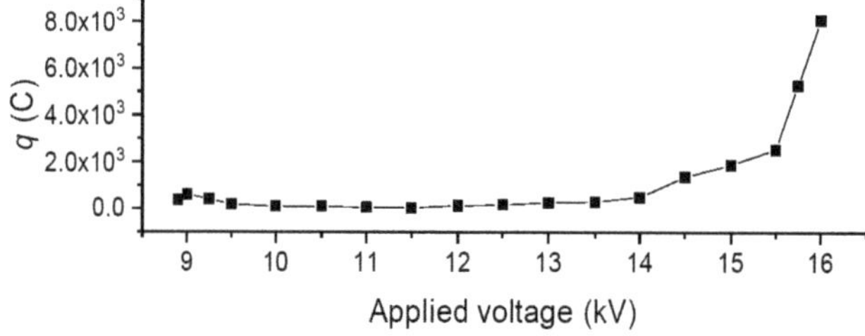

Figure 4. Equivalent charge quantity diagram of positive PD.

The variation coefficient (*VC*) is adopted to reflect the fluctuation of PD parameters under different voltages, furthermore, to judge the unevenness of electric field distribution. The expression of variation coefficient is:

$$VC = \frac{\sqrt{\sum_{1}^{N} (y_i - \overline{y})^2 / N}}{\overline{y}}. \tag{5}$$

The *VC* of *V* and Δ*t* under different voltages are shown in Figure 5. *VC* for *V* basically remains unchanged at the initial streamer stage. Then *VC* decreases at the early stage of glow-like discharge and reaches the lowest value. It rises at the end of glow-like discharge stage, and reaches a higher value at the breakdown streamer stage and then retains. When approaching breakdown, *VC* has a small increase. Amplitude fluctuation in the glow-like discharge stage are relatively smaller, indicating that the streamers are relatively regular, and the electric field distribution of discharge regions is even at this stage because of the cation group. Especially at the initial stage of glow-like discharge, the space charge electric field plays an obvious role.

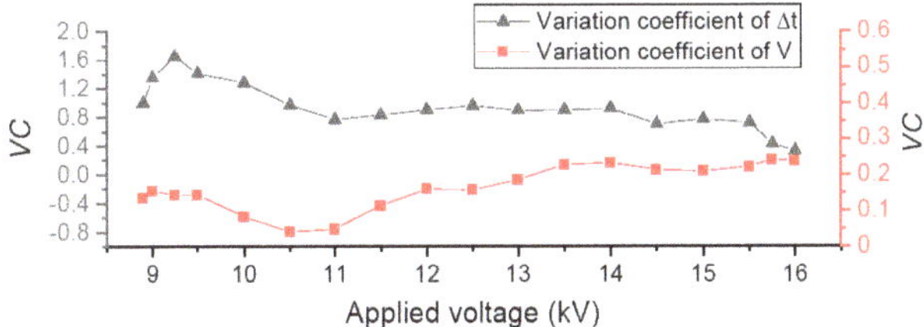

Figure 5. Variation coefficients diagram of positive polarity.

The *VC* for Δ*t* increases at the initial streamer stage where space charges are few, and decreases at the transition stage with the increasing density of cations. Then *VC* sustains in glow-like discharge stage. At last, it decreases again near breakdown. It can be seen that the fluctuation of Δ*t* will be lower with certain cation density, because space charges retained after a single discharge have less effect on the electric field. The cation group contributes to a relatively stable electric field in discharge regions, thus ions migrate regularly and the fluctuation of discharge recovery time is low.

3.4. Δ*t*–*V* Distribution Rules

Discharges of initial streamer stage are very few and relatively random, whose Δ*t*–*V* distribution is not researched. Instead, typical Δ*t*–*V* distribution of transition stage after initial streamer stage is described. In Figure 6, the Δ*t*–*V* points is concentrated in two regions, which forms two "sharp peaks". As discussed before, the points in high-amplitude peak represent the pulses generated by the streamers passing through the cation group, while points in low-amplitude peak represent the pulses generated by the streamers stopped inside. Figure 7 shows *V* and Δ*t* proportion distribution respectively. Each Δ*t* proportion curves has two peaks. While in each Δ*t* proportion curve, there is only one peak, and low-value Δ*t* makes up the majority. Combined with Figure 6, it can be concluded that the difference of the two forms of discharges is mainly reflected in the amplitude. Δ*t* distributions of this two forms of discharges are basically the same. It may because the streamers corresponding to the high-amplitude pulses generate more ions, but after the discharge, the recovery electric field for ion migration has a higher *E*.

The typical Δ*t*–*V* scatter diagram of glow-like discharge stage is shown in Figure 8a,b. At the initial stage, the discharges are restrained on weaken electric field inside action group. *V* changes little, so Δ*t*–*V* scatters basically form a horizontal line. Notice that the central value of *V* in 10 kV is close to that of the low-value peaks in 9 kV (about 0.65 V), which demonstrates the two discharge forms are similar. Then at the later stage of glow-like discharge, the range of *V* enhances. Δ*t*–*V* points still distribute in a horizontal ribbon basically. Figure 9a,b show the proportion distributions of V and Δt. At the beginning, V is relatively concentrated (10 kV–11 kV). But at later stage, V gradually distributes dispersedly. The abscissa of the peak point is larger. For Δt, the abscissa of the highest proportion point and the abscissa ranges of each proportion curve are basically the same. It can be observed that the distribution curves of Δt can be divided into two parts. One part is of peak shape in abscissa of 0 to 0.005 s. Another part is of flat shape in 0.005 s to 0.01 s. The flat part may represent the Δt after the streamers that pass though the cation group region, since the ions take a long time to migrate because of the low E in this region. For Δt, their distribution are dispersed at the early stage. The abscissa of the highest proportion point decreases with applied voltage rises, reaching the lowest value under 11 kV. The Δt points then become more concentrated at the later stage, and the highest proportion value rises.

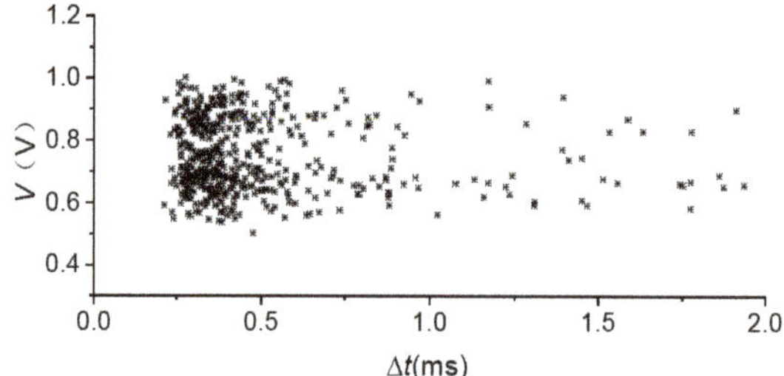

Figure 6. Δt–V distribution of 9 kV at the transition stage.

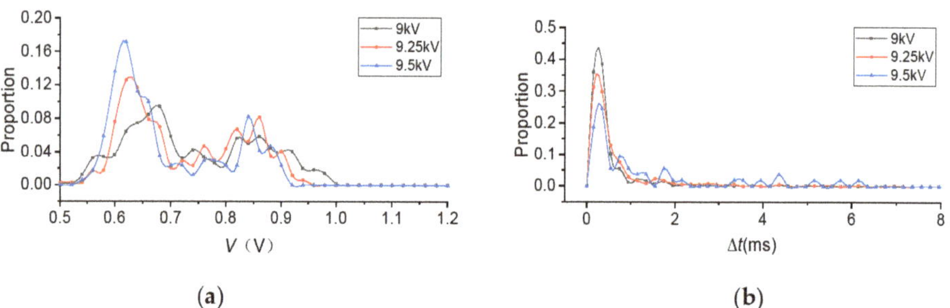

(a) (b)

Figure 7. Proportion distribution diagrams at the transition stage. (**a**) Proportion distribution of V; (**b**) Proportion distribution of Δt.

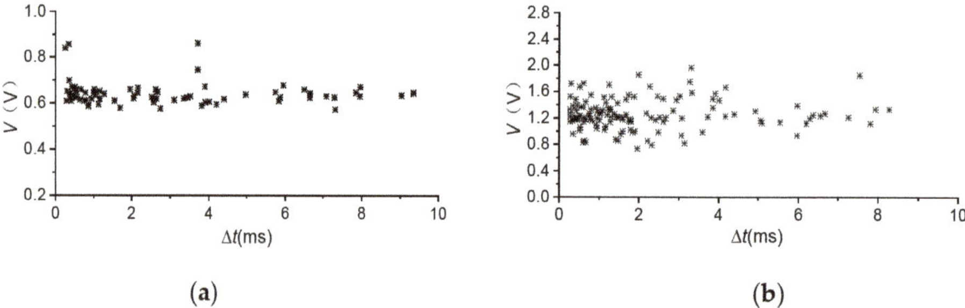

(a) (b)

Figure 8. Δt–V distribution of glow-like discharge stage. (**a**) Δt–V distribution of 10 kV; (**b**) Δt–V distribution of 13 kV.

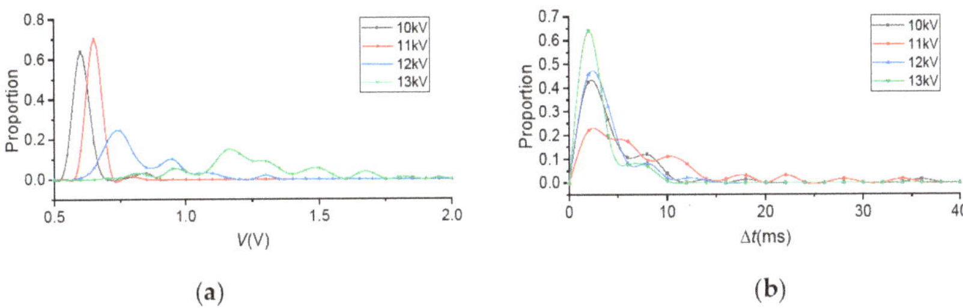

Figure 9. Proportion distribution diagrams at the glow-like discharge stage. (**a**) Proportion distribution of V; (**b**) Proportion distribution of Δt.

At breakdown streamer stage, E in the cation group region can satisfy self-sustaining condition and majority of streamers can pass through the cation region. As shown in Figure 10a, the main Δt–V points form a triangular. When Δt abscissa is small, the density of points is large and V range is wide. With the increase of Δt, density of point and V range are decreased. Compared Figure 10a to Figure 10b, with the increase of applied voltage, the range of V increases, and the Δt decreases. Figure 11a,b show the V and Δt distribution proportions. For V, the shapes of the proportion curves are similar in the beginning, and the abscissa of the highest proportion point gradually increases. While V overall increases, and V range expands significantly near breakdown. For Δt, with the increase of applied voltage, the concentration degree of Δt is significantly increased. The highest proposition of the curve increases, and the abscissa of highest proposition point decreases. Similarly, the variations are especially obvious near the breakdown. It can be concluded that breakdown streamers in positive polarity are strong and intensive.

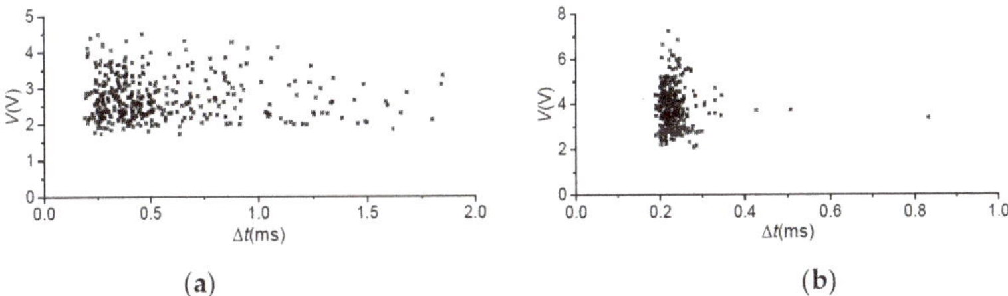

Figure 10. Δt–V distribution of breakdown streamer stage. (**a**) Δt–V distribution of 15 kV; (**b**) Δt–V distribution of 16 kV.

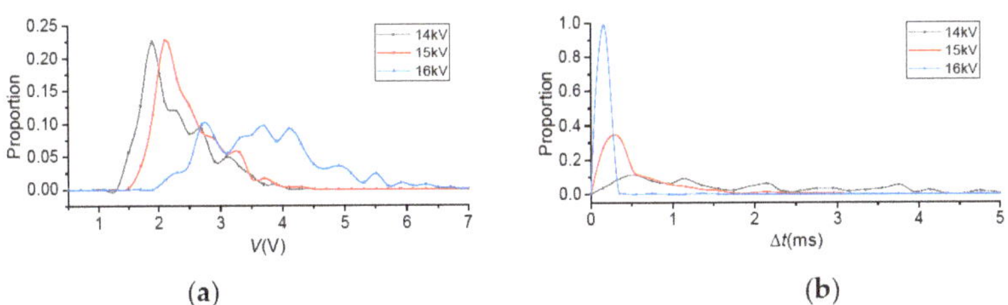

Figure 11. Proportion distribution diagrams at breakdown streamer. (**a**) Proportion distribution of V; (**b**) Proportion distribution of Δt.

4. Statistic Rules and Stage Characteristics of Negative PD

4.1. Formation of Negative PD Pulse

For negative polarity PD, there are many electrons in the head of streamers, which is different from the positive polarity situation. Due to the large charge–mass ratio, interaction forces between electrons play a role. As electrons migrate away from the needle tip dispersively, the discharge paths of the streamers are also dispersive. Retained space charges will distribute with a wider radius.

When streamers stop, the generated electrons are repelled by the electric field and leave the discharge regions quickly. Cations are regarded as immobile during the discharge process.

When the electron avalanches are transformed to streamers, the secondary collapses produce more cations, which gather in the head of the streamer. The ion density in the head of the streamer is high, and E in this region is low. Many electrons are transformed to anions. Cations main form plasma with the anions in the body of the streamer. Since the plasma is electrical neutrality, the main electrical property displayed by the streamer is negative polarity during the discharge process, which is caused by the electrons and anions in the front of streamer head. Thus an instantaneous negative polarity electric field toward the plate electrode will be generated, and E near the plate also increases instantaneously. Growing number of electrons flow towards plate, forming the rising edge of pulse. When streamer stops, the ions migrate to the electrodes, and E gradually recovers. The enhanced negative polarity electric field decreases, forming the falling edge of pulse. When the space charge dissipates, E near the needle tip returns to initial discharge value, and a new PD may be generated.

4.2. Discharge Stages

Due to the slow migration velocity, cations may retain in discharge regions. Meanwhile, the density of anions outside will increase. Under the influence of ion electric field, different forms of discharge are also formed under negative polarity voltage.

Similarly, average values of n, V and Δt are extracted and are shown in Figure 12. According to the rules of n, V and Δt, process of negative polarity corona discharge can be distinguished as four stages: initial streamer stage, Trichel discharge stage, glow-discharge stage and breakdown streamer stage.

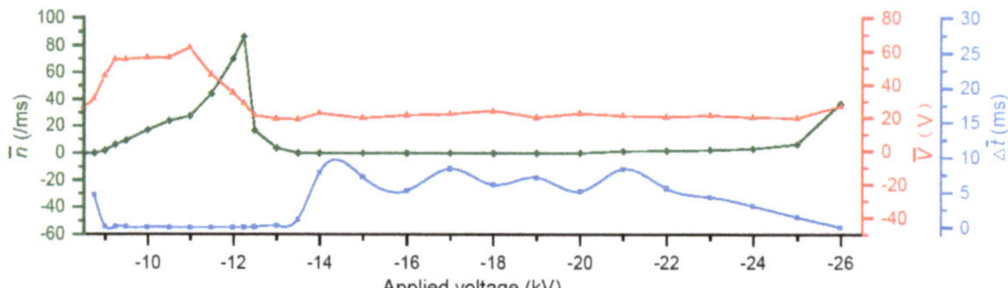

Figure 12. Average value diagram of n, V and Δt on negative polarity.

The initial streamer stage is -8.5 kV to -9.0 kV. At the initial discharge voltage, E is low and the streamers are weak. V is low and Δt is large. With the increase of the applied voltage, the amount of cations increases, and E between needle tip and cations increased slightly. Recovery time of the electric field becomes shorter; thus n rises. Streamers are strengthened, and the space charges generated by a single discharge are also slightly increased, so V increased slightly.

The Trichel discharge stage is -9.0 kV to -12.25 kV. At this stage, the density of cations near the needle tip reaches a certain level, forming a cation layer because of the dispersive development streamers. Cation layer greatly enhances the electric field between the needle tip and the cation layer. Outside the cation layer, E toward the plate is weakened, and anions in space also play a weaken role.

The Trichel discharge stage can be divided into initial and later stages. At initial stage, the streamers stop in the weakened periphery region of cation region. Since the weaken effect of cations layer basically offsets the increase of E produced by of applied voltage, streamers almost stop at same distance from the needle tip, and the development degrees of the stream are approximately the same. V is almost unchanged. E between the cation layer and the needle tip is flat and steady increases, so that n basically presents a linear growth trend.

At the later stage, electrons begin to form anions very close to the outer edge of cation layer, and plasma layer is formed when cations and anions are both very intensive. Variation of E in the plasma layer is very low. As the anions and cations get closer and closer, plasma layer gradually takes place the peripheral cation layer. E inside plasma layer increases. It is easier for E to recover to E_0 after a streamer, and ion variation generated by a single discharge is relatively less. Therefore, the recovery time of electric field decreases. n increases significantly, reaching the maximum value under -12.25 kV. Ion variation generated by a single streamer is less and density of original ions grows, thus single streamer affect the electric field less and less, instantaneous variation of space charge intensity is low. So V decreases.

After that, with the increase of applied voltage, E between the needle electrode and the cation layer is so high that it will not drop blow E_0 even after a streamer. So stable glow discharges occur (-12.5 kV to -24 kV). At this time, the generation and dissipation of ions basically reach a dynamic balance. Almost no instantaneous space charge changes occur, so there is no PD pulse. E is very low on the outer edge of the plasma layer, where streamers can hardly be produced. PD pulses will be generated only in a few of the peripheral regions of the cation layer when the E is weakened and may drop below E_0 after a streamer. These streamers are soon stopped by the plasma layer, so V is small. n is also small at this stage. Within a certain voltage range, the increase of applied voltage only causes the expanding of cation layer. The periphery weakened electric field still cannot supports streamers, so the negative polarity glow discharge stage lasts long.

The breakdown streamer stage is -24 kV to breakdown voltage. Glow discharges still happen inside. E on the outer edge of the plasma layer has risen high enough. So streamers are generated from the plasma region and continues to develop outward. Streamers are stronger and E in discharge regions increases. So both n and V increases. When breakdown streamer occurs, breakdown will happen soon when applied voltage increases a little. The breakdown voltage is approximately -27 kV to -28 kV.

4.3. Derived Parameter Rule

The q diagram of negative polarity discharge pulse is shown in Figure 13. At initial streamer stage, q is small. Then, in the Trichel discharge stage, q increases linearly throughout the stage, which is verified in Figure 14b. The linear increase of q in the Trichel discharge stage also indicates that the electric field inside the cations layer is relatively flat. Subsequently, in glow discharge stage, q is approximately 0. When it comes to the breakdown streamer stage, q rises again.

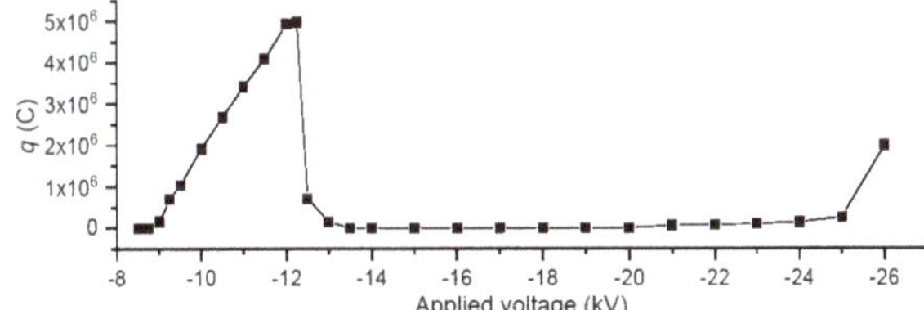

Figure 13. Equivalent charge quantity diagram of negative PD.

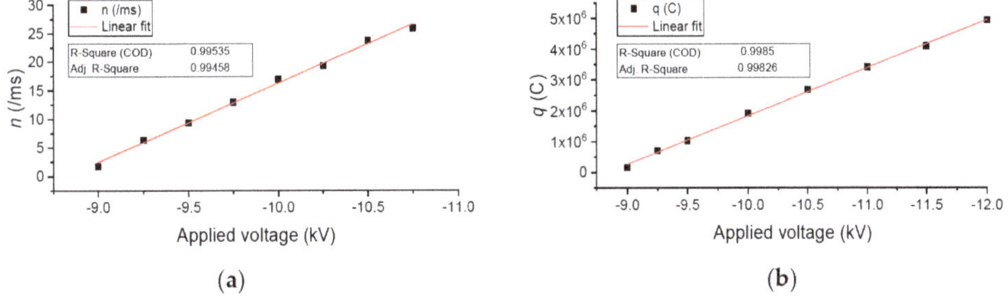

(a)

(b)

Figure 14. Linear fit analysis of n, q at Trichel discharge stage. (a) Linear fit of n; (b) Linear fit of q.

PD fluctuation is researched by *VC*. As shown in Figure 15, for *V*, *VC* increases at initial streamer stage, has a sudden rise in −9 kV and then stable at the initial stage of Trichel discharge. It indicates that the gradually formed cation layer makes the electric field distribution uniform. At the later stage of Trichel discharge, *VC* remains stable after rising to a certain extent. It is because with the increase of plasma, *E* between the needle tip and the plasma layer is quicker strengthened. The electric field intensity in a certain region is all higher than E_0, so the streamer may develop in many paths within the region. The development degree of the streamer varies, and the fluctuation of *V* is relatively large. At the glow discharge stage, only few PD happen in peripheral regions of plasma. Few PD pulses with small *V* are achieved in tiny peripheral regions. So *VC* basically remains stable at glow discharge stage, and it slightly decreases at the streamer breakdown stage.

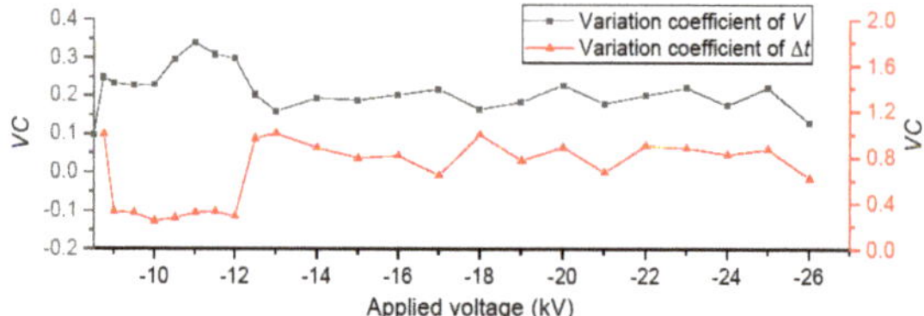

Figure 15. Variation coefficients diagram of negative polarity.

For Δt, due to the generation of the cation layer and the plasma layer, the fluctuation is relatively small. The remarkable characteristic is that *VC* remains low throughout the Trichel discharge stage. It indicates that the space charge migration after a single discharge is under a relatively stable electric field, and the recovery times of the electric field are similar. Subsequently, *VC* in glow discharge stage maintains in a larger value, and decreases a little in the breakdown streamer stage. It means discharges in the breakdown streamer stage are more regular than that in the glow-discharge stage.

4.4. Δt–V Distribution Rules

The typical Δt–V scatter diagram of the initial streamer stage is shown in Figure 16. When discharges are relatively regular, the scatter points of Δt–V basically form a horizontal line. It indicates that the discharge region is tiny. The development degree of streamers are similar, so *V* of PD pulses are unchanged. Figure 17 shows proportion distribution of *V*. When applied voltage increases, abscissas of *V* become bigger and the abscissa range is enhanced. Highest proportion value decreases. It indicates that the effect of ion electric field is weak, and the electric field generated by the needle tip mainly plays a role. So the change of applied voltage has a greater impact on the *V*. At this stage, Δt is large and decreases sharply with the increase of applied voltage. The scatter diagram of Δt is not listed.

The typical Δt–V scatter diagram of the Trichel discharge stage is shown in Figure 18. Some Δt–V points form an inclined spindle. It indicates there is an approximate proportional relationship between *V* and Δt, which confirms the analysis in Section 2.3. In Figure 18a, at the initial stage, the points in high Δt abscissas are relatively intensive. In Figure 18b, the points in low Δt abscissas are intensive. The density of the points outside the spindle rises because plasma layers form in some regions. The weaken effect of cation layer for electric field outside decrease because of the plasma. So regions where *E* is higher than E_0 were enhanced in some directions. Some Δt–V points in low-Δt abscissas have higher *V*, and their distribution does not follow the original proportional relationship.

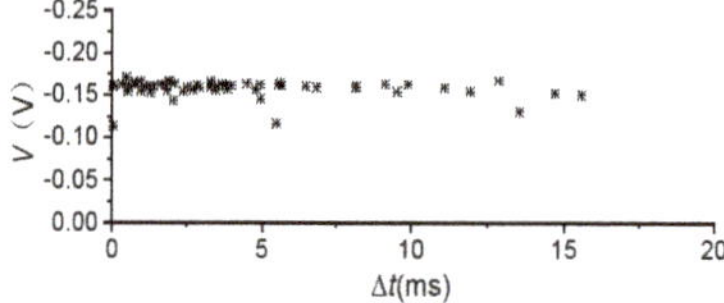

Figure 16. Δt–V distribution diagram of −8.75 kV at initial streamer stage.

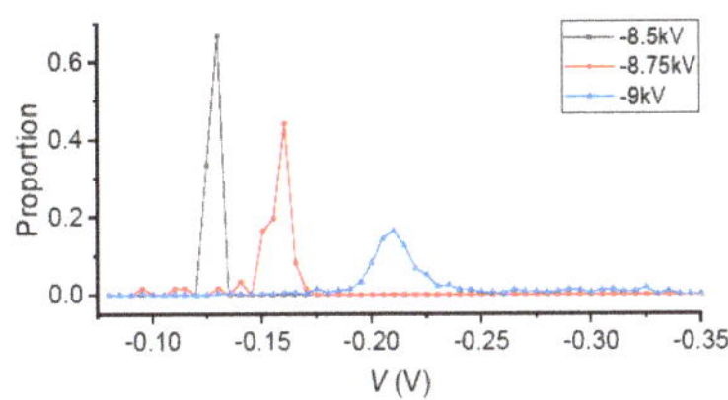

Figure 17. Proportion distribution diagrams of *V* at the initial streamer stage.

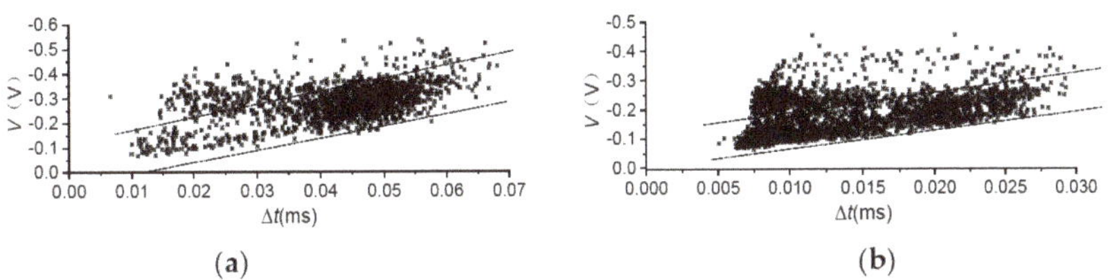

(**a**)

(**b**)

Figure 18. Δt–V distribution diagrams at Trichel discharge stage. (**a**) Δt–V distribution of −10 kV; (**b**) Δt–V distribution of −12 kV.

At the initial stage, the proportion distribution of *V* under different applied voltage are approximately similar, as shown in Figure 19a. At the later stage, the distribution curve of *V* overall moves to lower *V* abscissas gradually, and the highest proportion gradually increases, as shown in Figure 20a.

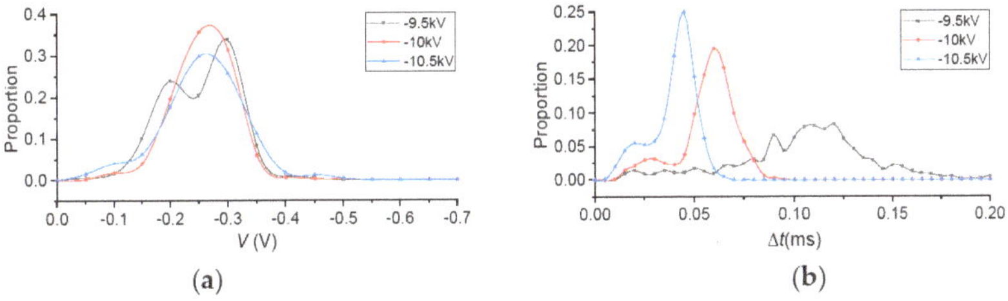

(**a**)

(**b**)

Figure 19. Proportion distribution diagrams at the initial stage of Trichel discharge. (**a**) Proportion distribution of *V*; (**b**) Proportion distribution of Δt.

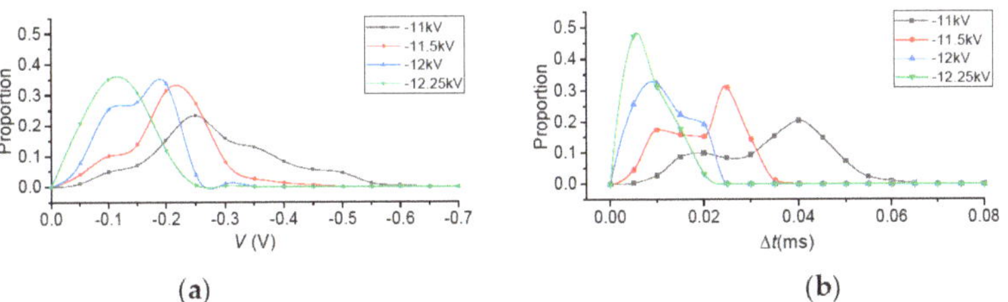

(**a**)

(**b**)

Figure 20. Proportion distribution diagrams at the last stage of Trichel discharge. (**a**) Proportion distribution of *V*; (**b**) Proportion distribution of Δt.

As shown in Figure 19b, the range of Δt decreases overall when applied voltage rises. Notice that Δt curves can be divided into two parts. One part is of flat shape and low proportion in low Δt abscissas, the other one is of peak shape and high proportion in high Δt abscissas. The flat shape part is caused by the plasma in some regions, because part of cation layer is transformed to plasma layer, and high *E* regions expand. At the later stage, as shown Figure 20b, with the increase of applied voltage, the proportion of the flat shape part increases, at last flat shape part becomes the main body. It indicates that diffuse plasma layer is formed, which almost covers all discharge regions.

Since PD in the glow-discharge stage are very few, only Δt–V distribution of the transition stages before and after glow-discharge stage are analyzed. The typical Δt–V scatter diagrams are shown in Figure 21a,b. The Δt–V points basically distribute in a horizontal band, which means V changes little, because PD can be measured are only generated in tiny regions. Δt–V distribution is similar to that of initial streamer stage.

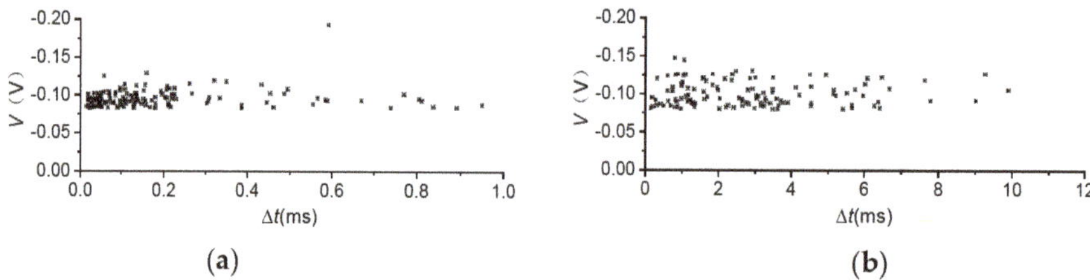

Figure 21. Δt–V distribution diagrams of glow discharge stage. (**a**) Δt–V distribution of −13 kV; (**b**) Δt–V distribution of −24 kV.

Proportion distributions of V and Δt at glow discharge stage are shown in Figure 22. In Figure 22a, it can be found that the V proportion curves at transition stage after glow discharge have double peak. The points in low-V abscissas peak should link to PD in few peripheral regions of cation layer where E drops below E_0 after a streamer. So the low V abscissas peak in proportion curves of transition stage after the glow discharge, are similar to the single peak in proportion curves of transition stage before the glow discharge. The points in high V peak link to PD starting from the outer edge of plasma layer. The two forms of discharge streamers have obvious difference, thus form the double peaks.

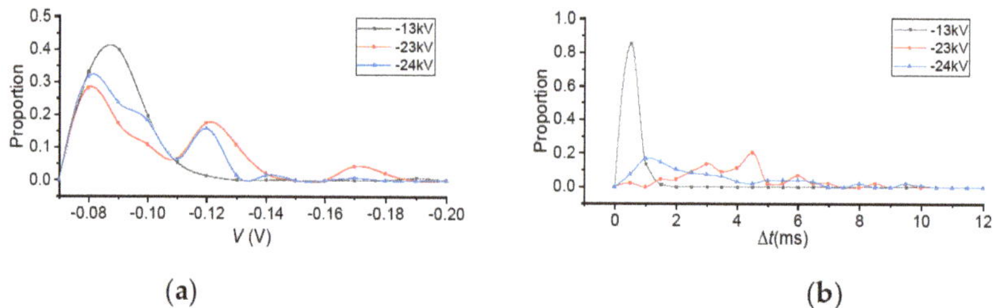

Figure 22. Proportion distribution diagrams at glow discharge stage. (**a**) Proportion distribution of V; (**b**) Proportion distribution of Δt.

At breakdown streamer stage, Δt significantly decreases. The typical Δt–V scatter diagram is shown in Figure 23. The distribution of Δt–V points form a sharp cone like shape. In this situation, the discharge region is close to the plate, and the axial electric field is stronger than other directions. In axial direction streamers are continuously produced. Therefore, low Δt and high V discharge pulses account for the main proportion. Proportion distribution characteristics of V and Δt at breakdown discharge stage are also shown in Figure 24. In Figure 24b, it can be observed that the concentration degree of Δt is significantly increased as it approaches breakdown.

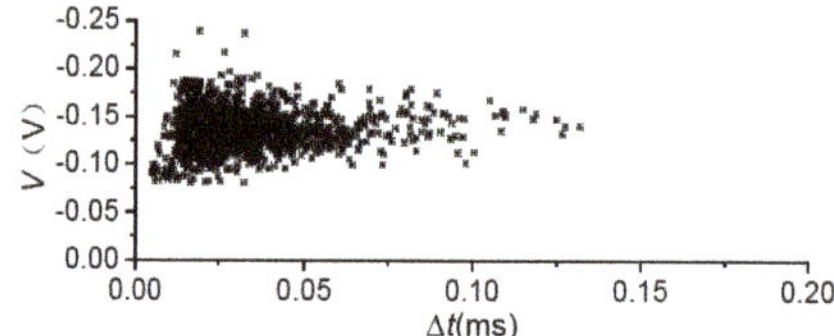

Figure 23. Δt–V distribution of −26 kV at breakdown streamer stage.

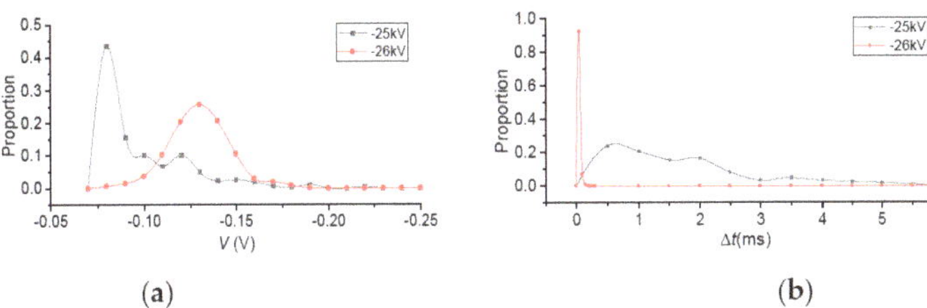

Figure 24. Proportion distribution diagrams at breakdown streamer stage. (**a**) Proportion distribution of V; (**b**) Proportion distribution of Δt.

5. Discussion

The deduced space charge effect in the corona discharge development process of both polarities are concluded and compared, to further discuss the PD mechanisms.

For initial streamer stage, space charge effect is weak. Streamers develops when E of tiny region reaches E_0. The development degree of streamers is low, so V is low and Δt is short. Discharges are relatively random.

Then space charges begin to retain, and actions play a main role. At transition stage of positive PD, the action group weaken the internal electric field while strengthen the external one. But E of the weaken regions are not even. In central axial region, electric field is the strongest and still higher than E_0 after being weakened. While E of regions around is not, so streams develop toward these regions are stopped earlier. Thus, two forms of streamer occur.

At Trichel discharge stage of negative PD, the action layer strengthen the internal electric field while weaken the external one. Actions are more dispersive and intensive, and are closer to the needle tip. Because of the strengthen effect, from needle tip to action layer, electric field decreases little, or even increase. While the E's outside cation layers are basically lower than E_0, which stops streamers in a similar degree and causes unchanged V. The electric field inside cation layer is high, flat, and linear, and grows with the increase of applied voltage, so some Δt–V has an approximate proportional relationship. Recovery times are short and stable, so n rises in a linear fashion. Discharges inside the cation layer are strong. More anions are transformed and gradually become close to cations at the later stage.

Because actions are more intensive and closer to the needle tip on negative polarity, its effects are more obvious, which reflects on the variations of parameters, especially n, q.

Next, retained space charges increase to a certain level, electric effects are obvious. At glow-like discharge stage of positive PD, the effect of actions are more obvious because of higher density. The streamers are basically limited inside action group and cause unchanged V. Notice that PD parameter rules of this stage are similar to that of Trichel discharges in negative polarity. So it can be concluded that the electric field distribution of discharge regions are similar at two stages. The phenomenon delay and are not so obvious on negative polarity because of the different effects of actions. At the later transition stage, because cation group expands and becomes sparse, its effect is weakened and streamers in some regions are not limited, discharge form is transforming to breakdown streamer.

At the glow-discharge stage of negative PD, intensive anions occur close to the outer edge of cation layer, and plasma layer is formed. Stable glow discharges occur inside plasma layer, which does not produce a PD pulse without instantaneous space charge variations. In a few regions of the cation layer and plasma layer, some intermittent streamers occur. Because the regions are tiny and E is relatively weak, the PD is weak. Within a certain voltage range, the increase of applied voltage only causes the expanding of cation layer, so negative polarity glow discharge stage lasts long.

At last, streamers with few limitation occur gradually and cause breakdown. At breakdown streamer stage of positive PD, the cations group is far from the needle tip and its density decreases. Its effects are reduced, and the streamers are almost not limited and become strong. While at breakdown streamer stage of negative PD, cation layer and plasma layer still work. When the plasma layer is close to the plate, the high E regions in peripheral regions of plasma expand, the negative streamers 'restart' and develop without much limitation. As such, the negative streamers are relatively weak, and cause a breakdown sooner because they start near the plate.

The statistic rules of PD parameters in this paper and others support each other in some aspects [10–14]. The mechanism explanations are based on the classical theories and well-founded. The deduced result can be interpreted in perspective of previous studies, for example, space charge and plasma distribution at some stages can be verified by luminescent image [19]. In follow-up research, we are hopeful the deduced space charge effects and electric field distributions will be verified by a simulation on the basis of the corona discharge fluid model mentioned previously [17].

The statistical analysis method and some deduced rules may also be effective in other gas PD under an uneven electric field, with DC voltage and AC voltage with low frequency, where space charges have enough time to retain.

6. Conclusions

In this paper, a 15-mm needle-plate model is adopted to study the air corona discharge under positive and negative polarity DC voltage. The statistical rules of characteristic parameters of each stage and transition process are explained by the space charge effects and electric field distributions. Furthermore, the discharge characteristics, the reflected space charge effects of each stage are summarized and compared. Discharge mechanisms are explained in microcosmic angle. Microcosmic process of PD under DC voltage can be described based on statistical methods, and several conclusions can be drawn.

Space charges obviously distort electric fields, and the most effective influence to PD development is decreasing E down to E_0 somewhere. For positive corona discharge, the space charge effects are most obvious at the glow-like discharge stage, where cations play a main role. The internal electric field of cation group is weakened but its distribution is relatively flat. The cation group inhibits the development of streamer, but inhibition effect is weakened later. At the breakdown flow stage, the influence of the action group is almost not reflected. For negative corona discharge, at Trichel discharge stage, cations work obviously, producing an electric field distribution with high and flat E inside. Then anions take part in and plasma layer forms in peripheral regions of cation layer, and further weakens the periphery electric field. At the glow discharge stage, E inside is high enough to sustain stable glow discharges inside. While E outside plasma layer is low, until in the breakdown streamer stage, E in outer edge of plasma layer reaches E_0 and streamers 'restart'.

Space charge effects are connected with PD parameter by electric field, and some electrical field distributions contribute to obviously PD parameter rules. When discharge regions are tiny and E is relatively weak, it may cause unchanged V, and Δt–V points distribute horizontally, which are shown at initial discharge stages of both polarities and transition stages of negative-glow discharge. When the electric field is high and relatively flat, Δt–V may have an approximate proportional relationship, and Δt–V points distribute inclined, which is shown at Trichel discharge stage of negative polarity.

Different forms of discharges may happen under a certain voltage, especially at transition stages, and different forms of discharges can be distinguished on PD parameters. For example, the V proportion distribution at glow-like stage of positive polarity, and Δt proportion distribution at Trichel discharge stage of negative polarity. The phenomenon arises in that E varies significantly in different discharge regions, which, to a large extent, are caused by the existence of cations and plasmas.

Author Contributions: Conceptualization, D.W. and L.D.; Methodology, D.W.; Software, D.W.; Validation, D.W.; Formal Analysis, D.W.; Investigation, D.W.; Resources, L.D. and C.Y.; Writing-Original Draft Preparation, D.W.; Writing-Review & Editing, L.D. and C.Y.; Visualization, D.W.; Project Administration, L.D.; Funding Acquisition, L.D.

Funding: This research was funded by "the National Key R&D Program" grant number "2017YFB0902400".

Acknowledgments: Many thanks to Deming Zhan and Han Yan for their contributions in experiments and investigations.

Conflicts of Interest: The authors declare no conflict of interest.

References

1. Lutz, B.; Kindersberger, J. Surface charge accumulation on cylindrical polymeric model insulators in air: Simulation and measurement. *IEEE Trans. Electr. Insul.* **2011**, *18*, 2040–2048. [CrossRef]
2. Zhang, Z.J.; Zhang, D.D.; Zhang, W.; Yang, C.; Jiang, X.L.; Hu, J.L. DC flashover performance of insulator string with fan-shaped non-uniform pollution. *IEEE Trans. Electr. Insul.* **2015**, *22*, 177–184. [CrossRef]
3. Thomson, J.J.; Thomson, G.P. *Conduction of Electricity through Gases*; Cambridge University: London, UK, 1933.
4. Trichel, G.W. Mechanism of the negative point-to-plate corona near onset. *Phys. Rev.* **1938**, *54*, 1078–1084. [CrossRef]
5. Morshuis, P.H.F.; Smit, J.J. Partial discharge at DC voltage: their mechanism, detection and analysis. *IEEE Trans. Dielectr. Electr. Insul.* **2005**, *12*, 328–340. [CrossRef]
6. Kachi, M.; Nadjem, A.; Moussaoui, A. Corona discharge as affected by the presence of various dielectric materials on the surface of a grounded electrode. *IEEE Trans. Dielectr. Electr. Insul.* **2018**, *25*, 390–395. [CrossRef]
7. Piccin, R.; Mor, A.R.; Morshuis, P. Partial discharge analysis of gas insulated systems at high voltage AC and DC. *IEEE Trans. Dielectr. Electr. Insul.* **2015**, *22*, 218–228. [CrossRef]
8. Marek, F.; Barbara, F.; Pawel, Z. Partial discharge forms for DC insulating systems at higher air pressure. *IET Sci. Meas. Technol.* **2016**, *10*, 150–157.
9. Stephan, V.; Joachim, H. Experimental evaluation of discharge characteristics in inhomogeneous fields under air flow. *IEEE Trans. Dielectr. Electr. Insul.* **2018**, *25*, 721–728.
10. Si, W.R.; Li, J.H.; Yuan, P.; Li, Y.M. Digital detection, grouping and classification of partial discharge signals at DC voltage. *IEEE Trans. Electr. Insul.* **2008**, *15*, 1663–1674.
11. Tang, J.; Liu, F.; Zhang, X.X.; Meng, Q.H.; Zhou, J.B. Partial discharge recognition through an analysis of SF6 decomposition product part 1 decomposition characteristics of SF6 under four different partial discharges. *IEEE Trans. Electr. Insul.* **2012**, *19*, 19–36. [CrossRef]
12. Liu, M.; Tang, J.; Pan, C. Development processes of positive and negative DC corona under needle-plate electrode in the air. *High Voltage Eng.* **2016**, *42*, 1018–1027.
13. Wei, G.; Tang, J.; Zhang, X.X.; Lin, J.Y. Gray intensity image feature extraction of partial discharge in high-voltage cross-linked polyethylene power cable joint. *IEEE Trans. Electr. Insul.* **2016**, *23*, 1076–1087. [CrossRef]
14. Zhang, S.Q.; Li, C.R.; Wang, K.; Li, J.Z.; Liao, R.J.; Zhou, T.C.; Zhang, Y.Y. Improving recognition accuracy of partial discharge patterns by image-oriented feature extraction and selection technique. *IEEE Trans. Dielectr. Electr. Insul.* **2016**, *23*, 1076–1087. [CrossRef]
15. Zhang, B.; Wang, W.; He, J. Impact factors in calibration and application of field mill for measurement of DC electric field with space charges. *CSEE J. Power Energy Syst.* **2015**, *1*, 31–36. [CrossRef]
16. Abdel-Salam, M.; Wiitanen, D. Calculation of corona onset voltage for duct-type precipitators. *Ind. Appl.* **1993**, *29*, 274–280. [CrossRef]
17. Wu, F. Numerical analysis on microcosmic process of corona discharge and ionized filed of HVDC transmission lines. Ph.D. Thesis, Chongqing University, Chongqing, China, 2014.
18. Liu, K. Study on the effect of corona discharge space charge background on gap discharge. Master's Thesis, Huazhong University of Science and Technology, Wuhan, China, 2013.
19. Liu, Z.; Liu, T.; Miao, X.; Guo, W. Research on corona layer of needle-plate discharge in atmosphere. *Sci. Technol. Eng.* **2013**, *13*, 1553–1556.
20. Yang, J. *Gas Discharge*, 1st ed.; Science Press: Beijing, China, 1983; pp. 154–196.
21. Yan, Z.; Zhu, D. *High Voltage Insulation Technology*, 2nd ed.; China electric power press: Beijing, China, 2007; pp. 47–76.

Article

Design of Cable Termination for AC Breakdown Voltage Tests [†]

Arthur F. Andrade [1,*], Edson G. Costa [2], Filipe L.M. Andrade [1], Clarice S.H. Soares [1] and George R.S. Lira [2]

[1]   Center of Electrical Engineering and Informatics, Federal University of Campina Grande, Campina Grande 58429900, Brazil
[2]   Department of Electrical Engineering, Federal University of Campina Grande, Campina Grande 58429900, Brazil
*   Correspondence: arthur.andrade@ee.ufcg.edu.br; Tel.: +55-83-2101-1140
†   This paper is an extended version of our paper published in 2018 IEEE International Conference on High Voltage Engineering and Application (ICHVE 2018), Athens, Greece, 10–13 September 2018; pp. 1–4.

Received: 3 July 2019; Accepted: 30 July 2019; Published: 9 August 2019

Abstract: International standards prescribe overvoltage tests to evaluate the insulating material performance of high-voltage cables. However, it is difficult to manage the electric fields at the cable ends when laboratory measurements are carried out because surface and external discharges occur at the cable termination. Therefore, this paper presents a procedure for designing cable terminations to reduce the electric field at the cable ends to appropriate levels even in the case of overvoltage tests. For this purpose, computer simulations of electric field distribution using the finite element method (FEM) were performed. A 35 kV cable model was employed as a sample. An voltage with RMS (root mean square) value of 300 kV was used as an overestimate of breakdown voltage for the internal insulating material. The cable termination model obtained through the proposed methodology allows an electric field reduction in air, preventing the occurrence of external discharges, and thus permitting the breakdown voltage measurement of the cable's inner insulation.

Keywords: cable termination; electric field; high-voltage test; stress relief cone

---

1. Introduction

In engineering, an ideal solution or project must be established due to economic, technical, practical, and environmental limitations. In the design and construction of power transmission or distribution lines, two types of cable can be used, namely overhead or underground cables [1]. High-voltage underground cables have been widely used as electrical conductors in different applications such as medium-voltage industrial facilities, underground and submarine transmission connections, renewable energy plants [1], the feeding of residential buildings and urban centers, as well as in facilities where environmental and visual aspects require the use of underground cables. However, the installation of underground cables represents a significant financial expenditure when compared with the application of overhead cables [1,2].

In addition, the complex manufacturing process of underground cables and the diversity of products and manufacturers can lead to the commercialization of low-performance cables. The use of good performance cables is essential for underground applications, since the cables are subjected to several stresses during their lifetime, such as electrical (due to operation voltages, overvoltage surges, and others), thermal (since the cables are subjected to abnormal temperature rises, thermal expansion, and contraction), mechanical (such as external damages, lateral impact, and pressure abnormalities), and environmental (due to humidity, oxidation, solar radiation, and other phenomena) [1].

Therefore, in order to ensure operation under the aforementioned stresses and improve the power supply reliability and continuity, cables must be exposed to routine and type electrical tests to guarantee, essentially, the dielectric performance of the insulation material and reduce, consequently, the financial losses for power utilities and industries. One of the most widely used materials is cross-linked polyethylene (XLPE) [2–5].

Although most failures in a power cable occur at its junctions and terminations, the evaluation of the cable insulation material is extremely necessary. One of the tests required for this evaluation is the cable breakdown voltage determination. International standards IEC 60229 and IEC 60520-2 [6,7] establish the test requirements for high-voltage cables. In Brazil, the standards NBR 10299 and NBR 16132 [8,9] provide the specifications in regard to the tests for the statistical distribution of puncture electric field strength in cables for systems with a voltage above 15 kV. NBR 10299 aims to set the minimum failure rate based on the installed cable length. A generally accepted value is 6.7×10^{-4} failure/(year $\times$ km) [8].

The prescribed tests should be performed using a sample of at least 3 m effective length, that is, without considering the terminations on both sides. The external shielding is grounded and a rising AC voltage is applied over the cable until the internal breakdown is reached. The experimental setup should ensure the breakdown occurrence on the effective length cable. Considering that the tests can subject the cable to overvoltages with values of 5 to 10 times greater than the normal operation voltage, the main problem faced during the tests is the electric field distortion at the cable ends, which causes external rupture and prevents the evaluation of the inner insulating material.

To carry out the installation in field or breakdown voltage tests in insulated cables, it is necessary to remove a part of the cable insulation. While the electric field inside the effective length has a predictable distribution, with a radial direction and logarithmical behavior [1,2,10], there is a great field intensification at the cable ends. The electric field in the vicinity of the shield end is illustrated in Figure 1. Thus, such a predictable distribution should be guaranteed to allow successful tests.

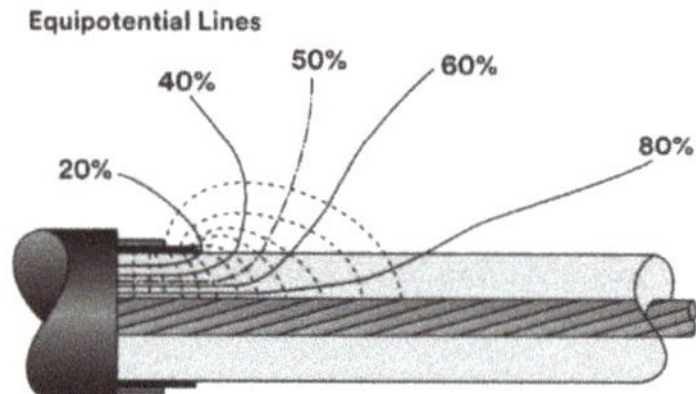

Figure 1. Electric field in the vicinity of the shield discontinuity [11].

The cable end consists of a conductor, semiconductor layers, insulating layer, and conductive tape for the shielding, in addition to air. The diversity of materials with different electrical characteristics, dielectric strength, and relative permittivity provides highly non-uniform electric fields with both axial and tangential field components. The tangential electric field is one of the main reasons for failures in terminals [12,13]. The field enhancement at the cable end produces surface and external discharges in the air, which can be prevented by using properly designed terminations [1,9,10,14]. In this sense, arrangements with different characteristics were studied by [12–19].

Most of the studies that have dealt with the electric field distribution in high-voltage cable terminations apply commercial software based on the finite element method (FEM) for the purpose of analysis and/or design of terminations, joints, or stress relief cones. In [12,15,16], simulations were used to assist in the design of stress cones based on high-temperature superconductors (HTS). The authors of [16] proposed the use of an epoxy/ZnO conductive layer to improve the electric field distribution. Other studies compared different materials and field grading options for cables. In [13], for example, different types of field grading options for 36 kV paper-insulated lead cables (PILC) and cross-linked polyethylene (XLPE) cables were compared. In [17], the impact of defects on cable joints was analyzed, and [18] studied the electric field in a cable termination undergoing transient stresses. During the project stage of a termination, another possible objective is to estimate areas more susceptible to defects and thus improve prototypes. In this regard, [19] calculated the electric field in a sleeve for a 110 kV cable using FEM, in order to estimate zones more susceptible to dielectric breakdown and thus improve a sleeve prototype.

Some of the aforementioned studies presented termination projects based on the stress cone concept, and analyzed electric field distribution or the influence of different materials. However, there is a lack of studies related to termination performance during overvoltage tests or breakdown voltage tests. In addition, some proposed prototypes require expensive materials.

Therefore, a methodology for the conception, drawing, and electrostatic simulation of a feasible termination is reported in this paper. The termination must be able to ensure the completion of overvoltage tests on cables. The proposed procedure can also be used for the design of optimized cable terminations. Such terminations are usually responsible, for example, for the linking between different transmission lines, functioning as connectors. For the analyses of the electrical stresses at the termination, computational simulations were performed using a commercial software based on the finite element method and a single-phase 35 kV cable model, which was used as a sample. Conventional materials were considered in the project, which represents a potential cost reduction.

2. Materials and Methods

Sections of an XLPE 35 kV insulated cable designed for application in underground and underwater environments were used as design samples. As can be seen in Figure 2, the cable consists of a copper conductor suitable for the electric current conduction; a thin semiconductor layer for the reduction of electric field distortions; a thick layer of insulation material, the XLPE being used for electrical insulation; another semiconductor layer; a conductive layer applied for the shielding or uniformization of the electric field inside the cable; and, finally, a layer of insulating rubber for mechanical protection. The layers can be seen from the (right) outermost to the (left) innermost part.

Figure 2. Cable layers.

Initially, 5 m wide cable sections were used as samples in the laboratory tests. The samples were composed of 3 m of intact cable and two bare ends on both sides. An AC voltage of 300 kV was applied to the high-voltage cable to perform a dielectric withstand test on the internal insulation material. Although several termination configurations were implemented, the breakdown occurred externally to the cable before it was possible to puncture the test portion, resulting in an incorrect application of the insulation material evaluation test.

In order to identify the cable termination points most susceptible to discharges and to design an optimized termination in which the electric field intensity in those points is minimal, the COMSOL Multiphysics® software (version 5.4, COMSOL AB, Stockholm, Sweden) was used to perform 2D axis-symmetric simulations of the cable terminations. This software implements FEM to solve electrostatic problems [20] through three main steps—geometry modeling, assignment of material properties to each part of the modeled geometry, and choice of a mathematical model to describe the phenomenon.

For the studies presented in this paper, the electrostatic physics module was used. Using this model, it is possible to solve an electrostatic problem using Gauss's Law:

$$\nabla \cdot \mathbf{D} = \rho \tag{1}$$

associated with the definition:

$$\mathbf{E} = -\nabla V \tag{2}$$

and the constitutive equation:

$$\mathbf{D} = \varepsilon_0 \varepsilon_r \mathbf{E}. \tag{3}$$

In Equations (1)–(3), the electric displacement field ($\mathbf{D}$) symbolizes the way in which the electric field ($\mathbf{E}$) will affect the organization of electric charges in the surroundings. $\mathbf{D}$ (C/m^2) and the electric field itself (V/m) represent the interaction between charged objects. In Equation (1), ρ represents the free electric charge density [20]. The operator nabla (∇) represents the vector differential operator.

By definition, ε_0 is the permittivity of free space and its value is 8.8542×10^{-12} F/m. ε_r represents the relative permittivity of the material. The relative permittivity values of the materials used in the simulations can be seen in Table 1.

Table 1. Termination and cable materials.

Material	Copper	Rubber	XLPE	Semiconductor
Relative Permittivity	-	3	3.2	200

With the selected physics and material properties, the initial geometry of the cable termination that presented external discharges was modeled and simulated in order to identify the points where the electric field exceeded the breakdown's stipulated limits for the air. Figure 3 shows that the intensification of the electric field occurred at the cable termination. Due to the high-intensity electric field at the external shielding end, it was necessary to build a termination model for the cable to accomplish successful overvoltage tests.

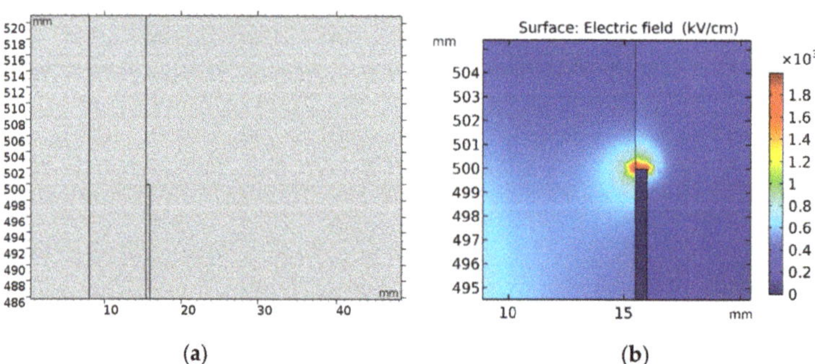

(a)   (b)

Figure 3. (**a**) Detail of the cable termination and (**b**) field intensification on the same area.

By applying the steps presented for preparing the simulation environment, we simulated successive cable termination geometries from the concept of the stress relief cone. Several models were created and tested until a final configuration was established. The main objectives of the simulations were to minimize the electric field near the cutting region of the semiconductor layer and in the air region between the termination and the high-voltage electrode, in order to minimize corona and reduce the possibility of discharges between the high-voltage electrode and the external grounded shield. Thus, in an overvoltage test, the insulation evaluated was guaranteed to be the internal one, since it was subjected to greater electrical stresses.

The simulation tests consisted of evaluating internal and external maximum electric field strength, as well the average electric field strength between the conductor terminal and the grounding end. In order to make a conservative estimation of the termination performance, a value of 424 kV was used as boundary condition for the applied potential on the simulations. This value corresponds to the peak value for a 300 kV RMS voltage. Due to the axisymmetric geometry, only a half longitudinal sectional view of the model was used in the analysis.

3. Results

In this section, the main termination designs created and analyzed during the test stage are presented, as well as the optimized final termination design. The performance of the prototypes was evaluated by the distribution and maximum value of the calculated electric field.

3.1. Intermediate Stages of the Termination Design

Initially, the main intermediate stages of the termination design are shown in Figure 4.

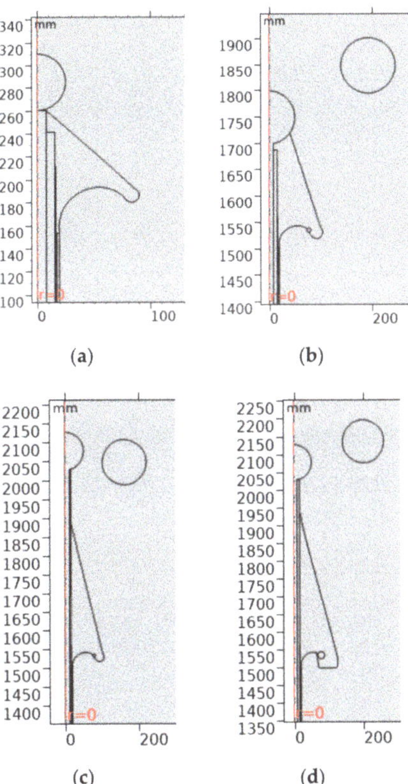

Figure 4. Main intermediate stages of the termination design: (**a**) initial stress cone; (**b**) enhanced stress cone and corona ring inserted; (**c**) elongated stress cone with terminal away and (**d**) improvement in external shield and corona ring position.

Successive improvements were made between the initial design step and the final step, aiming to improve the electric field distribution and reduce the maximum electric field on the termination surface and in the air region around the termination. The initial stress relief configuration is shown in Figure 4a. In Figure 4b, the shape of the stress relief cone was modified, having been elongated, and a conductive material with a metal ring for improving the field distribution was applied to the shield termination on the outer surface of the cone. A corona ring was also added to the high-voltage electrode. In the Figure 4c configuration, the cone length was increased, the geometry was improved, and the corona ring position was modified. In Figure 4d, the portion of the stress cone in which the external shield ends was modified again, and the position of the corona ring was optimized.

In Figures 5 and 6, the potential distribution and maximum electric field around the termination are shown in the same order as the four intermediate stages presented in Figure 4.

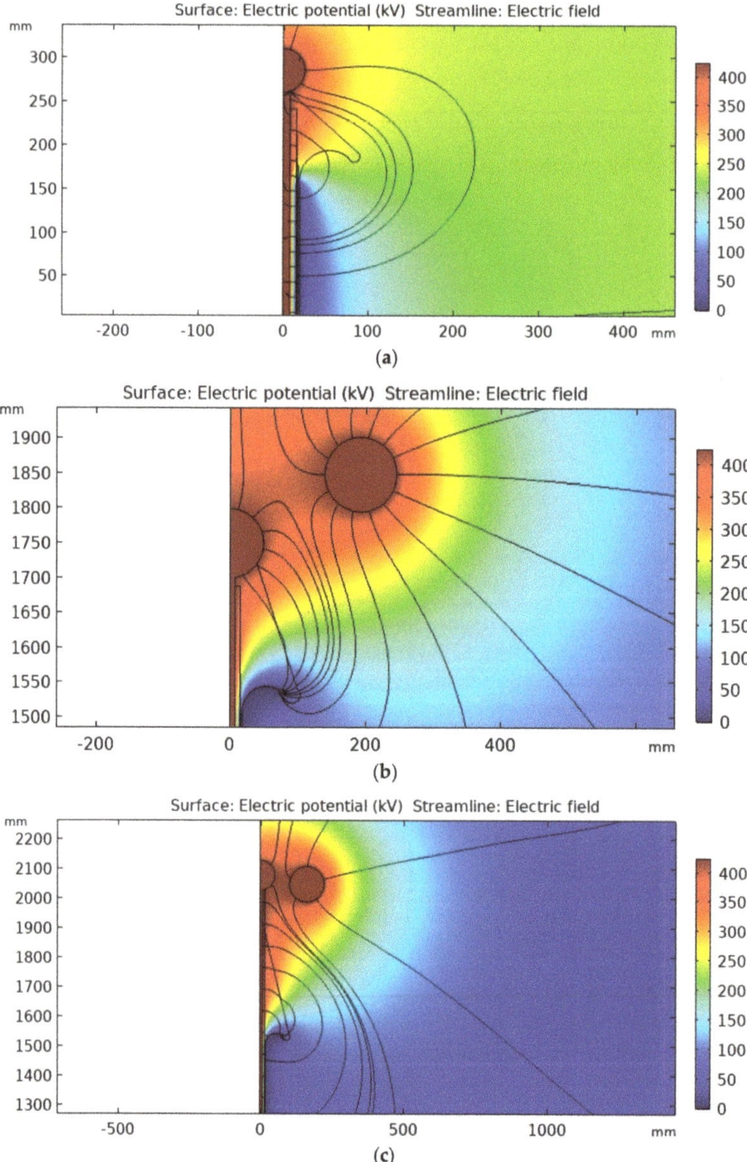

Figure 5. *Cont.*

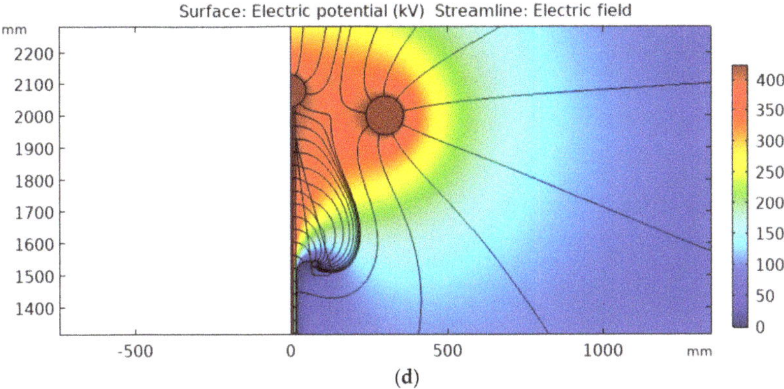

(d)

Figure 5. Potential distribution and electric field lines for the configurations of cable termination shown in Figure 4: (**a**) initial stress cone, (**b**) enhanced stress cone and corona ring inserted, (**c**) elongated stress cone with terminal away and (**d**) improvement in external shield and corona ring position.

As can be observed in Figure 5, a progressive change in the potential distribution was achieved due to modifications in the geometry. Moreover, as indicated in Figure 6, there was a progressive decrease of the maximum electric field in the air as the termination was being improved.

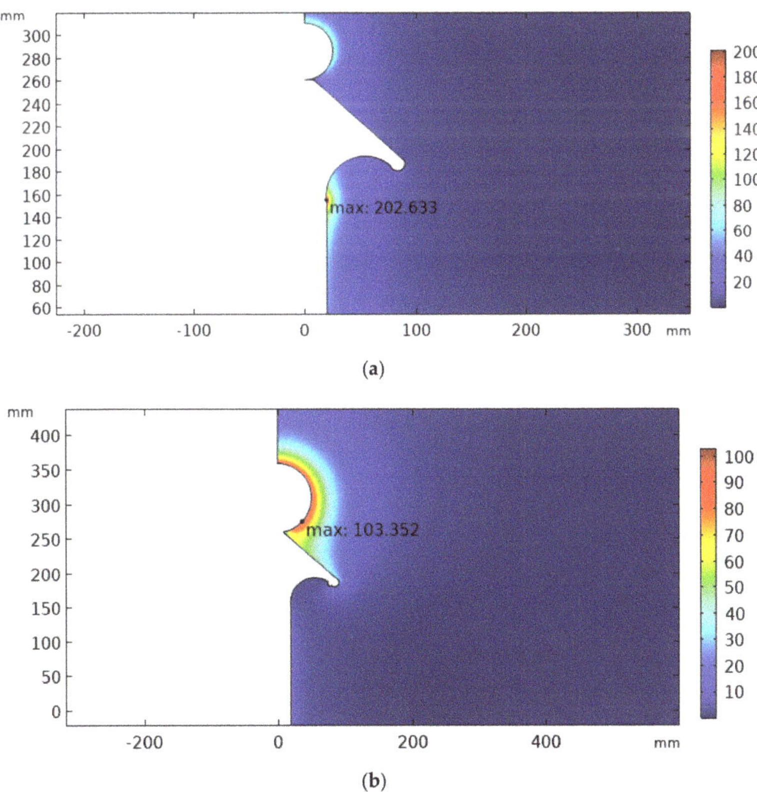

Figure 6. *Cont.*

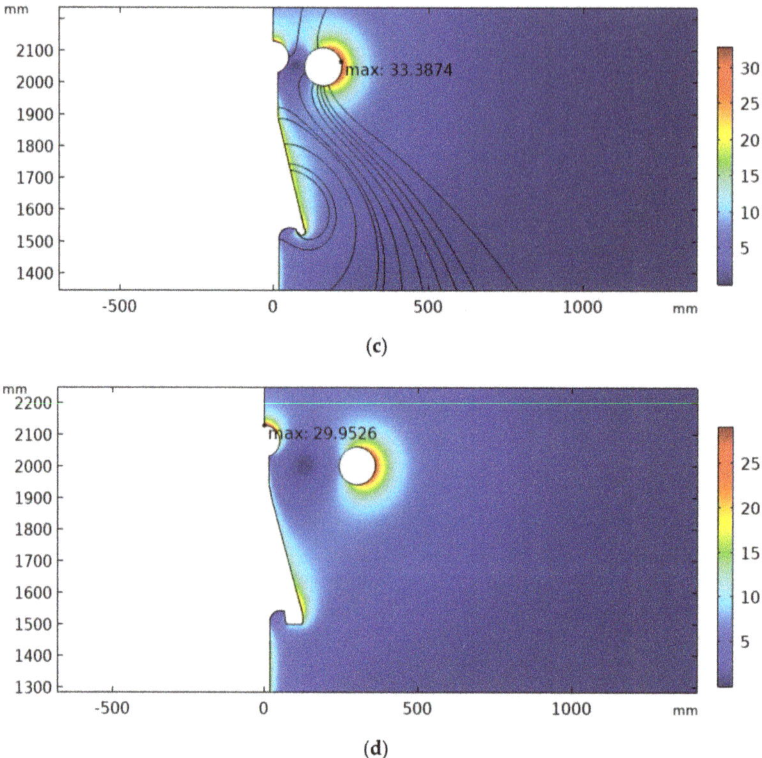

(c)

(d)

Figure 6. Electric field distribution and position of its maximum value for the configurations of cable termination shown in Figure 4: (**a**) initial stress cone; (**b**) enhanced stress cone and corona ring inserted; (**c**) elongated stress cone with terminal away and (**d**) improvement in external shield and corona ring position.

3.2. Final Termination Design

In the final version of the proposed termination, a rubber-composed stress relief cone and a termination electrode with a round end were designed. Near the cone, the termination also had rubber sheds to prevent the formation of a superficial discharge. At the end of the termination itself, a sphere-shaped corona ring was used for decreasing the electric field enhancement at the cable end.

A drawing of the final termination design is shown in Figure 7.

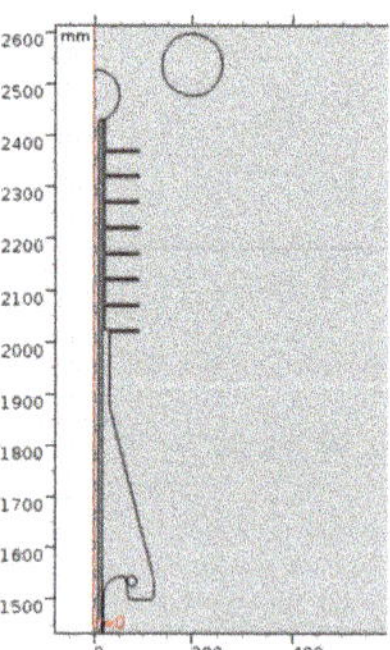

Figure 7. Drawing of the proposed model geometry.

After the design definition, its dimensions were obtained, and an 2D-revolution form of the termination was drawn by using AutoCAD® computer-aided design software (2018 student version, AutoDesk, Inc., San Rafael, CA, USA). 3D drawing views of the designed termination are shown in Figure 8.

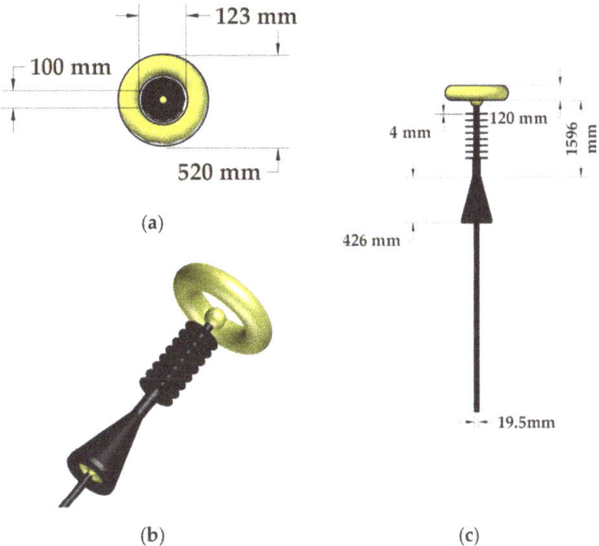

Figure 8. 3D drawing views of the designed termination.

Similar to the results regarding the intermediate versions of the termination, the potential and electric field distribution obtained for the final version of the termination are shown in Figure 9.

From the streamlines in Figure 9, field intensification points can be observed. They are located on the stress relief cone and at the torus-shaped electrode. As can be seen in Figure 9b, the maximum electric field achieved its minimum value for the last configuration. In addition, unlike in the initial design stages, the point of maximum external electric field in the final design is not in the region between the high-voltage terminal and stress cone, but on the external shield, away from the region susceptible to breakdown. This fact results in a better performance of the termination. In Table 2, the maximum electric field calculated for each of the presented termination designs are presented and compared.

Table 2. Maximum electric field calculated for the analyzed termination designs.

Termination Design	Maximum Calculated Electric Field (kV/cm)
Stage a	202.6
Stage b	103.4
Stage c	33.4
Stage d	30.0
Final design	27.7

The results shown hereafter refer to the final termination design. The obtained result for the inner electric field on the termination is presented in the color plot in Figure 10.

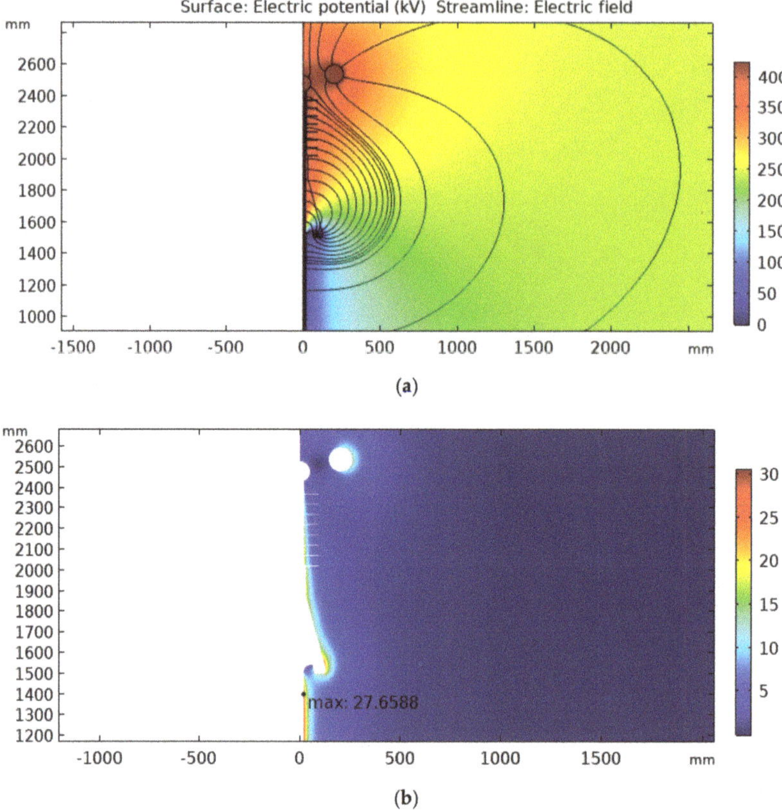

Figure 9. (**a**) Potential distribution and (**b**) external electric field distribution for the final version of the proposed cable termination.

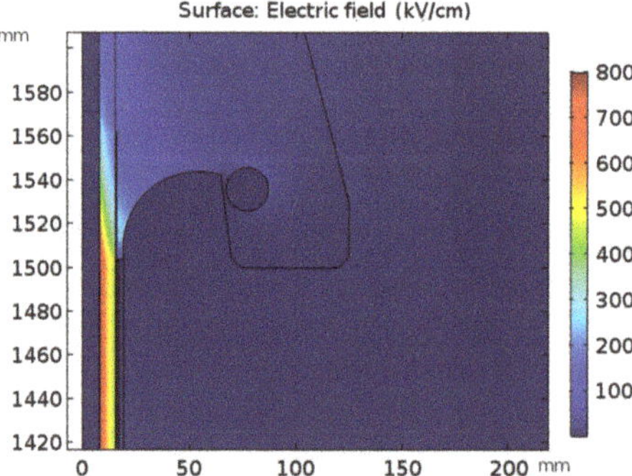

Figure 10. Electric field on the effective length end.

As shown in Figure 10, the maximum electric field is of the order of 800 kV/cm, exhibiting a reduction in the maximum electric field of 2.25 times when compared to the preliminary computer simulations shown in Figure 3.

As a way to verify the electric field distribution on the XLPE dielectric, the electric field was investigated over two paths of the insulation material. These paths are shown in Figure 11.

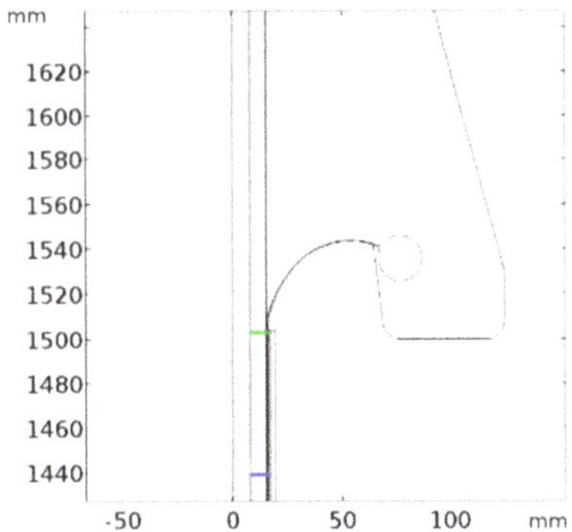

Figure 11. Paths for analysis of the electric field distribution.

In Figure 12, the distribution of the electric field over the points of Figure 11 is presented. Figure 12 curves represent the electric field decay along the insulating material radius. It is expected that a discharge must begin from a point where the field intensification is maximum. According to the standards, this point shall be inside the effective length of the cable, far enough from the external shield end.

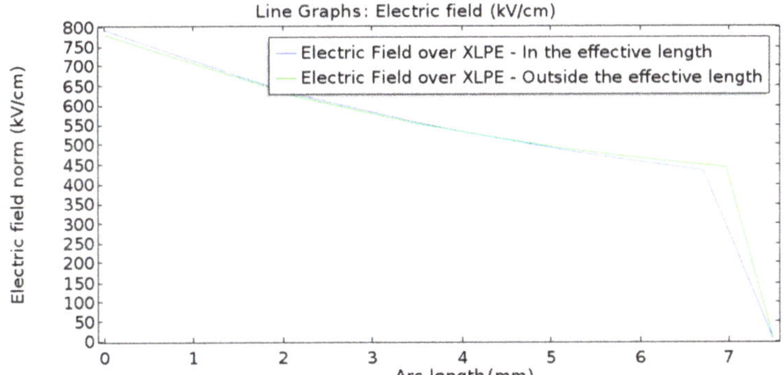

Figure 12. Distribution of electric fields over the termination.

Figure 12 curves show that the maximum electric field occurs on the blue curve, inside the effective length of the cable, even though it is also observed that both curves are close. In addition, as shown in the figure, the maximum field in the XLPE decreases with radius, thus indicating that breakdown will begin next the conductor surface and occur inside the effective length.

In Figure 13, the reduction of the maximum electric field along the cable length is shown considering a line adjacent to the inner conductor. The electric field intensity starts decreasing at the point where the external shield ends, which occurs at about 1504 mm.

In order to check the termination performance against superficial discharges as well as corona and arc formation, the electric field at the rubber termination surface was analyzed. Therefore, the defined path is the one exhibited as a red line in Figure 14, which unites high-voltage and zero potentials.

The electric field strength along the red line highlighted in Figure 11 can be seen in Figure 15. The length count starts from the termination electrode and ends at the surface of the high-voltage electrode.

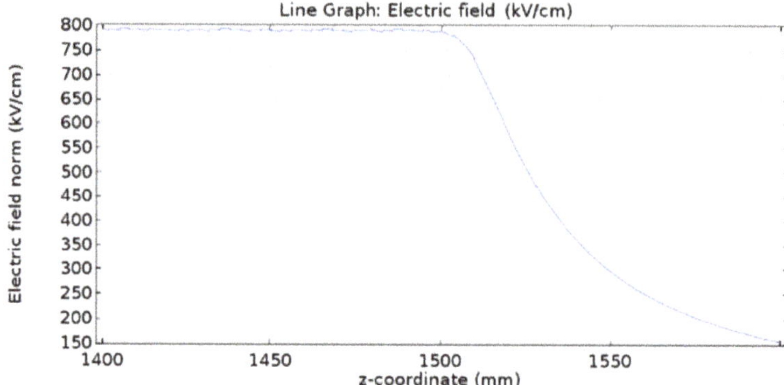

Figure 13. Electric field along the inner conductor length.

Figure 14. Outer path for external electric field analysis.

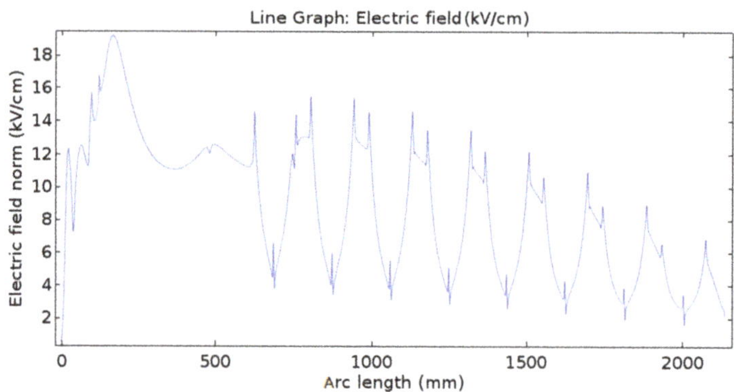

Figure 15. Simulated electric field along a path adjacent to the termination surface, highlighted in red in Figure 14.

The maximum electric field strength obtained on the termination surface was about 19 kV/cm near the stress relief cone. The effect of the insulating sheds can also be observed. The mean electric field throughout the superficial path is about 8.8 kV/cm.

4. Summary and Conclusions

The main contribution of this work is the proposition of a methodology that allows the design of adequate terminations that can withstand voltage tests and breakdown voltage tests in cables. Therefore, it was necessary to develop a termination that guaranteed that the greater electrical stress was applied to the internal insulation.

The optimized cable termination ensured that the electric field presented a greater probability of electrical breakdown inside the cable, in the cable part represented by the 3 m effective length. The achieved termination model minimized the probability of surface discharges and electrical breakdown between the high-voltage electrode and the external ground. In this context, the electric field distribution in the air and termination surface was also analyzed. As shown in Figure 15, the proposed termination allows the control of the surface electric field.

The proposed methodology was tested using a 35 kV XLPE cable model. The designed termination model contained an additional insulator to avoid tangential discharges and increase creepage distance, a metal sphere at the end of the structure, and a ring-shaped electrode to improve the electric field distribution in the surrounding air. As a result, the average electric field throughout the superficial path was about 8.8 kV/cm, and the maximum electric field strength was about 19 kV/cm, near the stress relief cone surface.

From the analysis of the results obtained in the simulations, it was observed that the objective was reached, given that, when applying the termination created, dielectric breakdown is more likely to occur in the effective length (inner part) of the cable. The performed simulations showed that a termination with the proposed dimensions and materials decreased the probability of electrical discharge at the terminations of the cable to be tested. The proposed termination eliminated the field intensification at the termination of the cable.

The proposed methodology benefits from advanced simulations to perform initial tests and to avoid wasting resources with unsatisfactory termination models. A termination was effectively developed for the cable considered so that overvoltage tests could be performed. Furthermore, conventional materials were considered in the proposed design, which represents a potential cost reduction. Future research may be carried out by developing new models in search of an improved result. The proposed termination should be built to carry out experimental tests in relation to its functionality. This study may also contribute as a basis for future studies in the HVDC area, especially the ones related to cable terminations.

Author Contributions: Conceptualization, E.G.C., F.L.M.A., and A.F.A.; methodology, C.S.H.S.; software, F.L.M.A., A.F.A. and C.S.H.S.; formal analysis, F.L.M.A. and A.F.A.; resources, E.G.C. and G.R.S.L.; writing—original draft preparation, F.L.M.A., A.F.A. and C.S.H.S.; writing—review and editing, A.F.A., E.G.C. and C.S.H.S.; visualization, supervision, E.G.C. and G.R.S.L.

Funding: This research received no external funding.

Acknowledgments: The authors acknowledge the UFCG Graduate Program for Electrical Engineering (COPELE), the Coordination of Improvement of Higher Education Personnel (CAPES), and the National Council for Scientific and Technological Development (CNPq) for granting scholarships.

References

1. Moore, G.F. *Electric Cables Handbook*, 3rd ed.; Blackwell Science: Bristol, UK, 1997.
2. Fothergill, J.C.; Hampton, R.N. Polymer Insulated Power Cable. In *Advances in High Voltage Engineering*, 1st ed.; Haddad, A., Warne, B., Eds.; The Institution of Engineering and Technology: Stevenage, UK, 2004; pp. 477–510.
3. Katahoire, A.M.S.; Raghuveer, M.R.; Kuffel, E. Determination of stress cone profiles for termination of high voltage XLPE cables. *IEEE Trans. Power Appar. Syst.* **1982**, *PAS-101*, 3804–3809. [CrossRef]
4. Illias, H.A.; Ng, Q.L.; Bakar, A.H.A.; Mokhlis, H.; Ariffin, A.M. Electric field distribution in 132 kV XLPE cable termination model from finite element method. In Proceedings of the IEEE International Conference on Condition Monitoring and Diagnosis, Bali, Indonesia, 23–27 September 2012; pp. 80–83.
5. Metwally, I.A. Reduction of Electric-Field Intensification and Hot-Spot Formation inside Cable Terminations. In Proceedings of the MELECON 2014 17th IEEE Mediterranean Electrotechnical Conference, Beirut, Lebanon, 13–16 April 2014.
6. International Electrotechnical Commission. *Electric Cables—Tests on Extruded Oversheaths with a Special Protective Function*; IEC 60229; IEC: Geneva, Switzerland, 2007.
7. International Electrotechnical Commission. *Power Cables with Extruded Insulation and their Accessories for Rated Voltages from 1 kV (Um = 1,2 kV) Up to 30 kV (Um = 36 kV)—Part 2: Cables for Rated Voltages from 6 kV (Um = 7.2 kV) Up to 30 kV (Um = 36 kV)*; IEC 60502-2; IEC: Geneva, Switzerland, 2014.

8. Brazilian Association of Technical Standards. *Electrical Cables in Alternating Current and Impulse—Statistical Analysis of Dielectric Strength*; NBR 10299; Brazilian Association of Technical Standards: Rio de Janeiro, Brazil, 2011.

9. Brazilian Association of Technical Standards. *Halogen Free, Low Smoke Insulated and Sheathed Power Cables for Rated Voltages from 3 kV Up to 35 kV—Performance Requirements*; NBR 16132; Brazilian Association of Technical Standards: Rio de Janeiro, Brazil, 2012.

10. Malik, N.H.; Al-Arainy, A.A.; Qureshi, M.I.; Pazheri, F.R. Calculation of Electric Field Distribution at High Voltage Cable Terminations. In Proceedings of the International Conference on High Voltage Engineering and Application, New Orleans, LA, USA, 11–14 October 2010.

11. 3M. Power Cable Splicing and Termination Guide. Austin, TX, USA. Available online: https://media.distributordatasolutions.com/3M/2018q1/5a9ec41a8b4865eb70059fe3670116e4d63a4078.pdf (accessed on 25 March 2018).

12. Lu, K.K.; Fang, J.; Huang, X.H.; Fang, X.Y.; Shen, Z.; Song, W.J.; Zhang, H.J.; Qiu, M. On Prefabricated Stress Cone for HTS Cable Termination. *IEEE Trans. Appl. Supercond.* **2015**, *25*, 1–6. [CrossRef]

13. Lewarkar, A.; Bergsma, D.; Conradie, J.; Buddhawar, S. Study of Electric Field Grading Types on Medium Voltage Polymeric and Paper Insulated Cables. In Proceedings of the 2018 Electrical Insulation Conference (EIC), San Antonio, TX, USA, 17–20 June 2018.

14. Andrade, A.F.; Costa, E.G.; Andrade, F.L.M.; Soares, C.S.H.; Lira, G.R.S. Design of Termination for an AC Disruptive Voltage Test on a 35 kV Cable. In Proceedings of the 2018 IEEE International Conference on High Voltage Engineering and Application (ICHVE), Athens, Greece, 10–13 September 2018.

15. Wu, C.Y.; Fang, J.; Huang, X.H.; Lu, W.J.; Li, D.; Guo, L.J. The study on stress-cone based on HTS cable terminal. *Physica C* **2013**, *484*, 229–233. [CrossRef]

16. Li, Z.; Yang, Z.; Xing, Y.; Zhu, W.; Su, J.; Kong, X.; Jiang, J.; Du, B. Improving the Electric Field Distribution in Stress Cone of HTS DC Cable Terminals by Nonlinear Conductive Epoxy/ZnO Composites. *IEEE Trans. Appl. Supercond.* **2019**, *29*, 1–5. [CrossRef]

17. Yang, H.; Liu, L.; Sun, K.; Li, J. Impacts of different defects on electrical field distribution in cable joint. *J. Eng.* **2019**, *16*, 3184–3187. [CrossRef]

18. Bhattacharyya, S.; Chakraborty, A.; Saha, B.; Chatterjee, S. Electric Stress Analysis of a Medium Voltage Cable Termination Subjected to Standard and Non-Standard Lightning Impulse Voltages. In Proceedings of the 2016 International Conference on Intelligent Control Power and Instrumentation (ICICPI), Kolkata, India, 21–23 October 2016.

19. Seleznev, D.A.; Obraztsov, N.V.; Kiesewetter, D.V. Numerical Simulation of the High Voltage Cable Sleeve Operation for 110 kV. In Proceedings of the 2018 IEEE Conference of Russian Young Researchers in Electrical and Electronic Engineering (EIConRus), Moscow, Russia, 29 January–1 February 2018.

20. COMSOL AB. AC/DC Module User's Guide. 2018. Available online: http://bit.do/comsol-ac-dc-m (accessed on 25 March 2018).

Article

Static Voltage Sharing Design of a Sextuple-Break 363 kV Vacuum Circuit Breaker [†]

Xiao Yu [1,*], Fan Yang [1], Xing Li [1], Shaogui Ai [2], Yongning Huang [2], Yiping Fan [2] and Wei Du [3]

[1]   State Key Laboratory of Power Transmission Equipment & System Security and New Technology, School of Electrical Engineering, Chongqing University, Chongqing 400044, China

[2]   Electric Power Research Institute of Ningxia Electric Power Company of State Grid Corporation of China, Yinchuan 750001, China

[3]   Wuhan NARI Limited Company of State Grid Electric Power Research Institute, Wuhan 430074, China

*   Correspondence: yuxiao@cqu.edu.cn; Tel.: +86-136-6761-8048

†   This paper is an extended version of our paper published in 2018 IEEE International Conference on High Voltage Engineering and Application (ICHVE 2018), Athens, Greece, 10–13 September 2018; pp. 1–4.

Received: 22 May 2019; Accepted: 26 June 2019; Published: 29 June 2019

Abstract: A balanced voltage distribution for each break is required for normal operation of a multi-break vacuum circuit breaker (VCB) This paper presented a novel 363 kV/5000 A/63 kA sextuple-break VCB with a series-parallel structure. To determine the static voltage distribution of each break, a 3D finite element method (FEM) model was established to calculate the voltage distribution and the electric field of each break at the fully open state. Our results showed that the applied voltage was unevenly distributed at each break, and that the first break shared the most voltage, about 86.3%. The maximum electric field of the first break was 18.9 kV/mm, which contributed to the reduction of the breaking capacity. The distributed and stray capacitance parameters of the proposed structure were calculated based on the FEM model. According to the distributed capacitance parameters, the equivalent circuit simulation model of the static voltage distribution of this 363 kV VCB was established in PSCAD. Subsequently, the influence of the grading capacitor on the voltage distribution of each break was investigated, and the best value of the grading capacitors for the 363 kV sextuple-break VCB was confirmed to be 10 nF. Finally, the breaking tests of a single-phase unit was conducted both in a minor loop and a major loop. The 363 kV VCB prototype broke both the 63 kA and the 80 kA short circuit currents successfully, which confirmed the validity of the voltage sharing design.

Keywords: vacuum circuit breaker; multi-break; voltage distribution; FEM; stray capacitance; grading capacitor

---

1. Introduction

Modern power systems have a high requirement for switching appliances, which are essential for the safe and reliable operation of power systems. As an important part of electrical switchgear, the vacuum switch plays an important role in controlling and protecting electrical power systems [1]. Vacuum interrupters have the advantages of environmental friendliness, fine extinction capability, and long lifespan [2]. Vacuum circuit breakers (VCBs) are widely applied in 40.5 kV and lower voltage power systems, while for 126 kV and higher voltage power systems SF_6 circuit breakers are mainly used. However, modern power systems have larger loads, which have higher requirements of power quality and safety. Hence, short circuit currents should be interrupted instantly to reduce losses [3]. To date, the commonly used SF_6 circuit breakers in 330 kV and 500 kV AC power systems have a long remove fault time of approximately 50 ms [4]. During this period, the fault current reaches its peak several times [5]. According to the dielectric strength, SF_6 behaves better than VCBs; hence, VCBs have higher dielectric strength restoration after the current is turned to zero, in comparison with other types of circuit breakers. Nowadays, VCBs with controlled switching technology, such as permanent magnetic mechanisms, spring type mechanisms, and electromagnetic repulsion mechanisms, are able to achieve fast interruption during a rising slope within 2–7 ms [6–9]. However, single-break VCBs are limited to 126 kV due to the saturation effect of the vacuum gap and the overheating problem [10].

There are two ways to develop high-voltage VCBs: higher-voltage single-break VCBs and multi-break VCBs. Due to the saturation characteristic of the vacuum gap, single VCBs are limited to 126 kV. Conversely, multi-break VCBs, which are several vacuum interrupters connected in series, can eliminate the saturation effect and increase the breaking capacity [10]. Hence, the multi-break VCB has drawn more attention [11–15]. The total interruption time of VCBs with an ultra-fast electromagnetic repulsion mechanism and a fault current detection system can be reduced to 10–20 ms [13]. Based on this technology, several features of double- and triple-break VCBs have been investigated, including the open velocity, voltage distribution, and synchronization of each VCB; additionally, a breaking test has been conducted on prototypes [13,16,17]. Vertical series connected structures of double- and triple-break VCBs have been investigated with the purpose of analyzing their static voltage distribution. According to these results, the stray capacitance and post-arc characteristic of uneven voltage distribution limit the breaker capacity and the grading capacitors can improve the balance of voltage distribution and breaking capacity.

However, multi-break VCBs are still limited to under 126 kV. Compared with 126 kV VCBs, higher-voltage VCBs are quite different in structure due to their insulation requirements. Therefore, to extend normal VCBs to 363 kV power systems, two key points should be taken into consideration: (i) the uneven voltage distribution of each break and (ii) the synchronized switching of each break. The voltage distribution of each break can be improved by installing a grading capacitor at each break in parallel to balance the voltage distribution equally [18,19]. As for synchronized switching, a phase-controlled algorithm is always used to detect the fault current changing rate to determine the open time for each break, which ensures the synchronization of the multiple breaks. Furthermore, stray capacitances of multi-break VCBs are mainly decided by the number of breaks, topology, structure, and dimensions. Therefore, for a new 363 kV multi-break VCB, it is necessary to investigate the voltage distribution and stray capacitance in order to design a suitable grading capacitor.

This paper is an extension of our previous work published in ICHVE2018 [20]. Our goal was to determine the value of the grading capacitor for a novel 363 kV sextuple-break VCB taking into account its gas-insulated scheme and series-parallel layout. In order to obtain the static voltage distribution and distributed capacitances of the VCB, a three-dimensional finite element model based on the actual dimension parameters of the single-phase unit was established. An equivalent capacitance network model in PSCAD was also set up to analyze the influence of the grading capacitor on the voltage distribution of each break. Finally, a breaking test was conducted to verify the performance of the grading capacitors.

2. Structure of the 363 kV Circuit Breaker

This section gives the topological structure, dimensions, and parameters of the proposed sextuple-break 363 kV VCB.

2.1. Determination of the Number of Breaks

The vacuum gap is able to withstand a strong electric field and voltage level, and its breakdown voltage is directly proportional to its open distance. However, the large vacuum gap exhibits a saturation phenomenon. The voltage level of VCBs varies greatly, ranging from 3.6 to 72 kV. For a 363 kV VCB, in an open state it must be able to sustain the nominal voltage and the surge voltage. If low-voltage VCBs such as 12 kV VCBs are adopted, many breaks are needed, which would lead to the need for more complex structures and greater costs. The most widely used commercial VCBs are 40.5 kV VCBs, and these are considered to be the best choice. However, the rated current of a

40.5 kV VCB is 2500 A, so a parallel structure of two branches is needed for a rated current of 5000 A and a rated breaking current of 63 kA. To allow for sufficient breaking capacity and sufficient margin, sextuple-break with two branches of parallel structure was adopted. Consequently, the physical structure of the proposed 363 kV VCB was formed by 12 40.5 kV VCBs in a series-parallel structure with grading capacitors and inductors (see Figure 1). Series inductors were used to share the current for each phase and grading capacitors were used to share the voltage for each break.

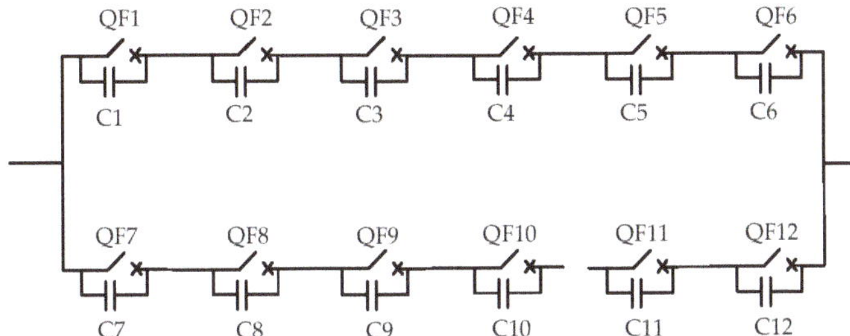

Figure 1. Topology of a 363 kV sextuple-break vacuum circuit breaker (VCB). Notes: QF means VCB; C means grading capacitor.

2.2. Structure Design

When designing the structure of a high-voltage multi-break VCB, insulation, size, maintenance, and total cost are the key factors for consideration. A gas-insulated high-voltage apparatus, such as a gas-insulated switchgear, has the advantage of being compact and maintenance-free with a long mechanical life, while other outdoor apparatuses requires tall structures to satisfy their requirement for insulation against ground. Hence, the gas-insulated apparatus was the best choice. The prototype design and connection structure are shown in Figure 2. The VCBs were totally enclosed in the aluminum tank filled with 0.4 MPa SF_6 with inlet busbar and outlet busbar. The 40.5 kV VCBs were also connected to terminals and other VCBs by a concentric cylindrical busbar. The dimensions and parameters of the 363 kV VCB and 40.5 kV VCB are shown in Tables 1 and 2, respectively.

Table 1. Dimension of the 363 kV VCB.

Parameter Name	Value
Tank radius	410 mm
Tank height	2047 mm
Busbar external radius	80 mm
Upper and lower terminal radius	156 mm
Transfer flange radius	245 mm

Table 2. Parameters of the 40.5 kV VCB.

Part Name	Size
Moving contact	Radius = 25 mm, length = 190 mm
Static contact	Radius = 25 mm, length = 105 mm
Contact	Radius = 40 mm, thickness = 30 mm, round radius = 4 mm
Ceramic envelope	Radius = 35.5 mm, thickness = 8.5 mm
Ceramic envelope	External radius = 240 mm, internal radius = 200 mm, height = 378 mm

The vacuum interrupter worked with a fast short circuit current prediction algorithm and a phase control system. The moving contact was driven by an ultrafast electromagnetic repulsion mechanism, so the opening time (500 μs) and opening velocity (up to 5 m/s) could be achieved [6,15]. It adopted a fiber-controlled system to synchronize the VCBs. The open dispersibility (i.e., the time delay of the operation of all VCBs) of these VCBs was shorter than 0.1 ms, and the closed dispersibility was shorter than 0.2 ms. Thus, combined with the above components, a quick current interruption within a short arcing time (3 ms) was achieved. Compared to typical high-voltage circuit breakers 30-60 ms, the total interruption time was reduced to 10–20 ms.

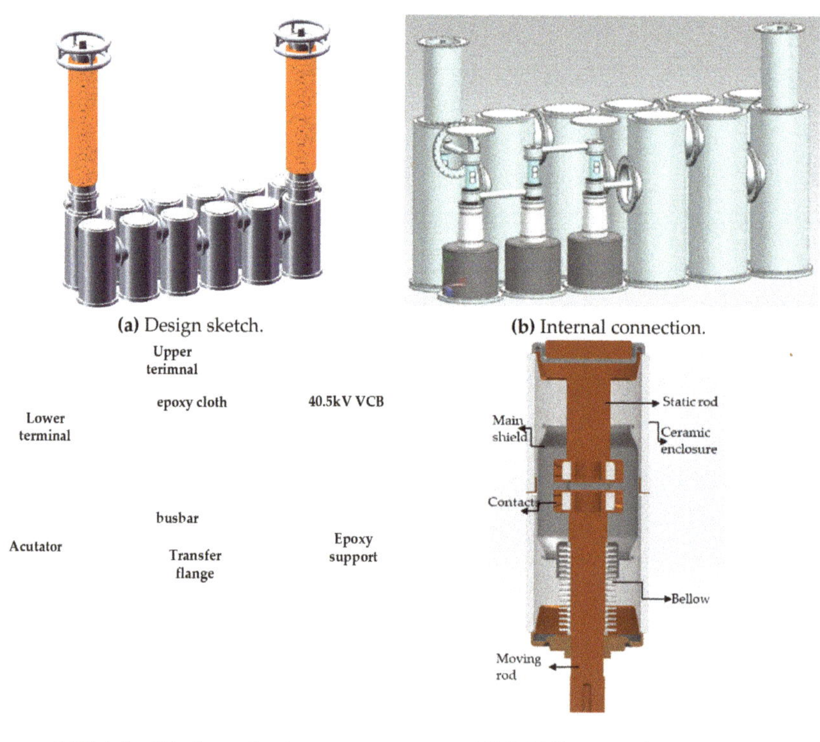

(a) Design sketch.

(b) Internal connection.

(c) Details of the inner structure.

(d) 40.5 kV vacuum interrupter.

Figure 2. Single phase structure of the 363 kV VCB.

3. Finite Element Method (FEM) Model and Voltage Distribution Calculation

As the 363 kV sextuple-break VCB has a complex structure, analytical methods cannot be used to calculate its voltage distribution. As a consequence, taking into account its symmetry, an FEM model of the single unit was established, and a numerical calculation was performed. In establishing the FEM model, some components were simplified as follows: (1) the end shield, bellows, and contacts were simplified to a metal cylinder; (2) the control system, actuator, disc spring, screw holes, and bolts were removed. The entire opening distance was 20 mm. In the FEM model, the relative permittivity of the insulators was set at 4.95, SF_6 at 1.002, and epoxy at 4. The applied voltage was 300 kV on the surface of the static rod of the first break and its connected busbar. The outside tank and the moving rod of the last break and its connected busbar were grounded. All the main shields of the vacuum interrupters were set to floating potential. The voltage freedom degrees of other components were coupled using COMSOL software. Tetrahedral mesh was adopted to deal with the irregular geometry. The total mesh elements, as shown in Figure 3, were 39,442,046. The voltage distribution profile and electric field distribution profile are shown in Figure 4.

Figure 3. The mesh of the simulation model.

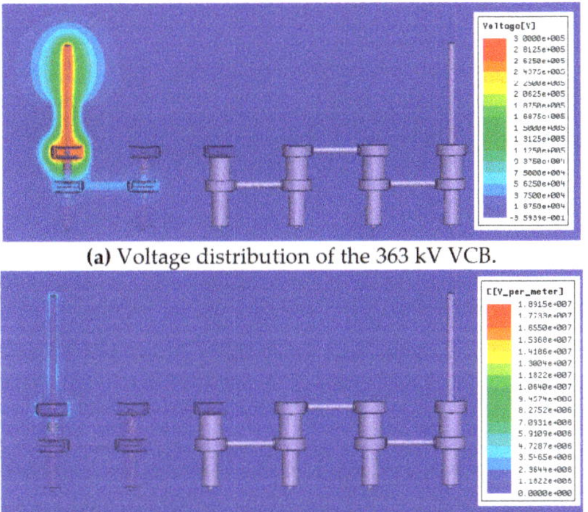

(a) Voltage distribution of the 363 kV VCB.

(b) Electric field distribution of the 363 kV VCB.

Figure 4. The static voltage and electric field distribution of the 363 kV VCB.

According to the calculation results, the voltage distribution was exceedingly uneven due to the stray capacitance. The first break shared most of the total applied voltage, reaching 259 kV and accounting for 86.3% of the voltage, while the sixth break shared only 0.0153 kV. The maximum electric field of the first break was 18.5 kV/mm, which easily gave rise to arc reigniting and reduced the interrupting capacity of the VCB. Table 3 shows the voltage of each break.

Table 3. Voltage sharing of each break.

Break	Voltage	Voltage Distribution Ratio (%)
V1	256 kV	86.3%
V2	35.18 kV	11.73%
V3	5.03 kV	1.68%
V4	0.684 kV	0.23%
V5	0.0982 kV	0.032%
V6	0.0153 kV	0.00%

Note: V1 to V6 denote the first to sixth breaks.

4. Grading Capacitor Design

When designing the grading capacitor, we needed to know the distributed capacitance and stray capacitance of the 363 kV VCB, which can be regarded as a multi-conductor system considering the influence of the grounded tank. Under AC voltage, the voltage distribution of a multi-conductor system is determined by the self-capacitance and mutual capacitance of these conductors.

The commercial COMSOL software can easily obtain the distributed capacitance and extract a capacitance matrix by defining the conductors successively. As shown in Figure 5, all of the 13 conductors were defined, where the incoming busbar, the upper terminal of the first break, and the static contact were defined as conductor 1; the main shield, which is the floating potential, of the first VCB as conductor 2; the lower terminal of the first break, lower busbar, and the lower terminal of the second break as conductor 3; and so forth, and the exterior tank was grounded. Among the conductors, conductor 13 was composed of the last connecting terminal, last busbar, and outside tank. The capacitances between the moving contact and the static contact, and the capacitance between the conductor and the tank made a greater difference. Therefore, other stray capacitances between the conductors that were at a great distance from each other, such as the capacitance between the main shields of different VCBs, were ignored.

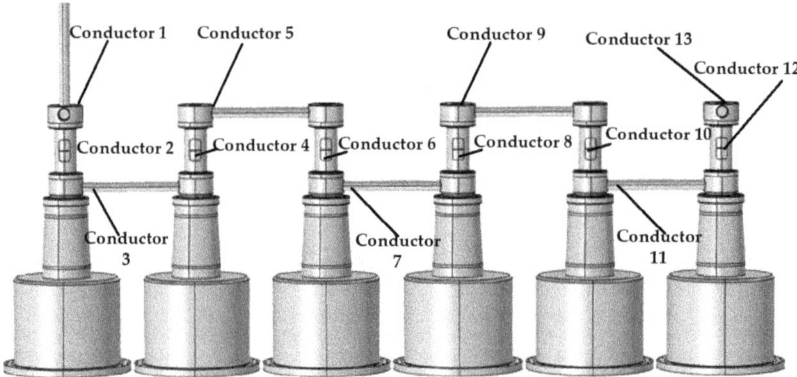

Figure 5. The conductor definitions.

According to the calculated capacitance matrix, the stray capacitances to the earth were greater than the capacitance of the break of the VCB, which makes the voltage of each break unequal, indicating the need for a voltage sharing design. A grading capacitor was an effective and low-cost measure to balance the voltage of each break. To analyze the influence of the grading capacitor, the equivalent circuit model can be used. Figure 6a illustrates the equivalent capacitance network and the simulation model with a grading capacitor in PSCAD of the 363 kV VCB, where C1 was the capacitance between the moving contact and the static contact of the first break, and C10 was the capacitance between conductor 1 and the tank. Cg was the grading capacitor. The capacitance between the main shield and nearby conductors, which was very small, was attributed to the capacitance nearby to simplify the equivalent capacitor network under the criterion that the voltage of each break equals the results of the FEM simulation. Finally, parameters for the equivalent circuit simulation model were as follows: C1 = C2 = C3 = C4 = C5 = C6 = 14.7 pF; C20 = C40 = C60 = 80 pF; C30 = C50 = 76 pF.

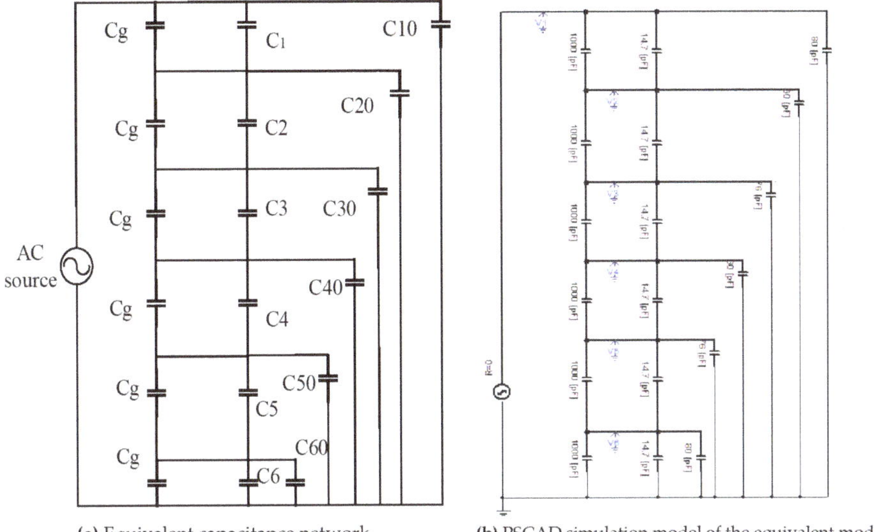

(a) Equivalent capacitance network.   (b) PSCAD simulation model of the equivalent model.

Figure 6. The equivalent circuit model of the 363 kV VCB with a grading capacitor.

By assigning a different value to the grading capacitors, the change in the voltage of the first break along with the grading capacitors is shown in Figure 7. Table 4 shows the voltage variation of the first break at different grading capacitor values. The values of the grading capacitors were set in the range of 700 to 12,000 pF. The authors of [13] indicated that grading capacitors of 1000 pF can meet the opening requirements of a 126 kV triple-break VCB. However, for the 363 kV VCB, the first break shared 23.3% of the applied voltage when the grading capacitor was 1000 pF. When the grading capacitor was 6000 pF, the voltage distribution ratio of the six breaks from one to six was 18.4%, 17.5%, 16.3%, 15.9%, 15.5%, and 15.1%, respectively. The voltage distribution ratio of the first break was 3.2% higher than the sixth. Furthermore, when the grading capacitor exceeded 10,000 pF, the rate of improvement slowed down.

Table 4. Voltage of the first break at different grading capacitor values.

Grading Capacitor (pF)	Voltage of the First Break (kV)
700	89.4
900	83
1100	78
1500	72
2000	67
4000	56
6000	55
7000	55
8000	54.5
10,000	53.5
12,000	52.7

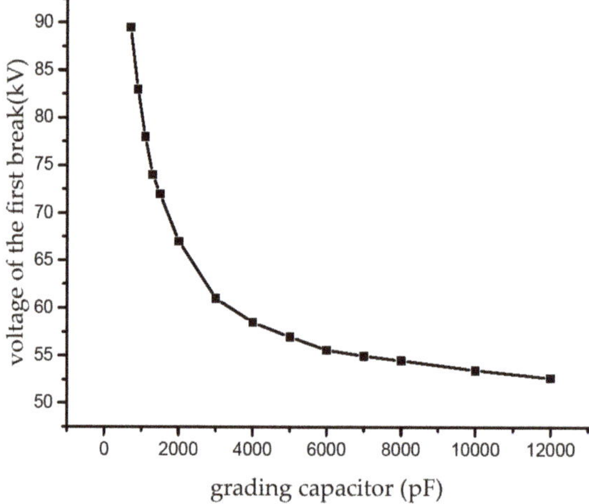

Figure 7. Influence of grading capacitors for the 363 kV VCB.

The voltage unbalance coefficient (K) is defined as:

$$K = \text{Max}(\frac{V_i - \overline{V}}{\overline{V}})$$ (1)

where V_i is the voltage of the ith break and $\overline{V}$ is the average voltage of each break.

When K > 1, the voltage was unevenly distributed. When the grading capacitance equaled 6000 pF, K = 1.1; when the grading capacitance equaled 10,000 pF, K = 1.07. In addition, the accepted standard for selecting grading capacitors in China is that the voltage unbalanced coefficient should be below 1.1. However, according to China's national standard GB 4787-2010 [21], the preferred minimum for a grading capacitor is 1000 pF in consideration of the function of reducing the steepness of the transient recovery voltage (TRV) [17]. Other researchers have adopted 400 pF grading capacitors for double-break VCBs [19] and 1000 pF grading capacitors for 126 kV triple-break VCBs [17]. Greater grading capacitor values lead to a decrease in the breaking capacitor value because of the post arc current. Hence, the value of grading capacitors for the sextuple-break 363 kV VCB was set at 10 nF in this study.

5. Breaking Test of the 363 kV VCB

The synthetic test circuit, as recommended by IEC standards, was used to impose the appropriate stresses on the test circuit breaker (CB) during the breaking test, as shown in Figure 8. These circuits were essential to verify the short circuit current making and breaking capabilities of the 363 kV VCB. The supply circuit capacity of the synthetic test circuit was 480 kV/80 kA. The grading capacitors were connected in parallel with the VCB. The test was carried out on a single phase only, and in type T100s and T100a both for 63 kA (minor loop) and 80 kA (major loop). The first-pole-to-clear factor was 1.3.

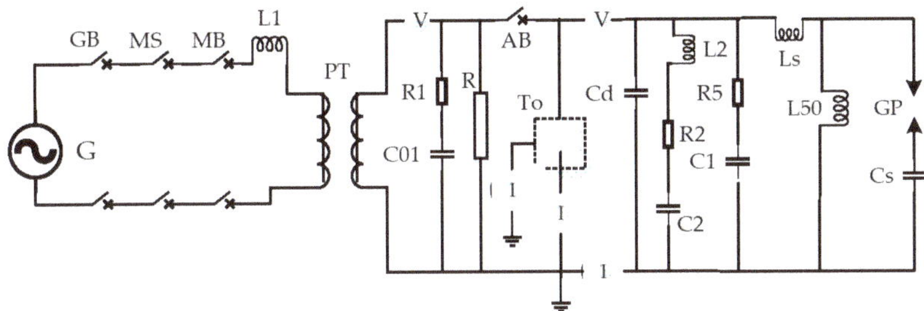

Figure 8. Breaking test circuit for the 363 kV VCB. G, generator; GB, generator breaker; MS, make switch; MB, master breaker; L, reactor; PT, power transformer; R, resistor; C, capacitor; To, test object; R, arc prolonging circuit; V, voltage measurement; GP, gap; I, current measurement; AB, auxiliary breaker.

The prototype of the 363 kV sextuple-break VCB interrupted the short circuit current (symmetrical and asymmetrical) successfully in a high-voltage apparatus quality supervision and inspection test center as a preliminary research test. The key information is summarized in Table 5, and all recorded waveforms of the breaking current and the TRV are shown in Figures 9 and 10. In Figures 9 and 10, I is the test current, Ih is the small zero crossing current of synthetic test circuit, Ur is the total transient recovery voltage, Ucs is the voltage of the current source, HFUcs is the high-frequency voltage of the current source, and HFUr is the high-frequency transient recovery voltage. The recovery voltage is less than 258 kV and the TRV peak is less than 544 kV, which indicates that the grading capacitors of 10 nF can meet the requirement of the voltage sharing design and breaking capacity.

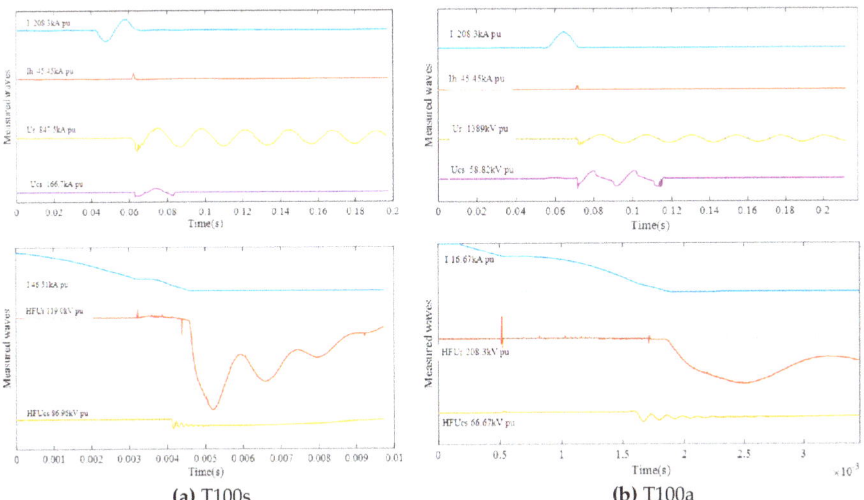

(a) T100s

(b) T100a

Figure 9. Test waveforms of the synthetic breaking test at 63 kA.

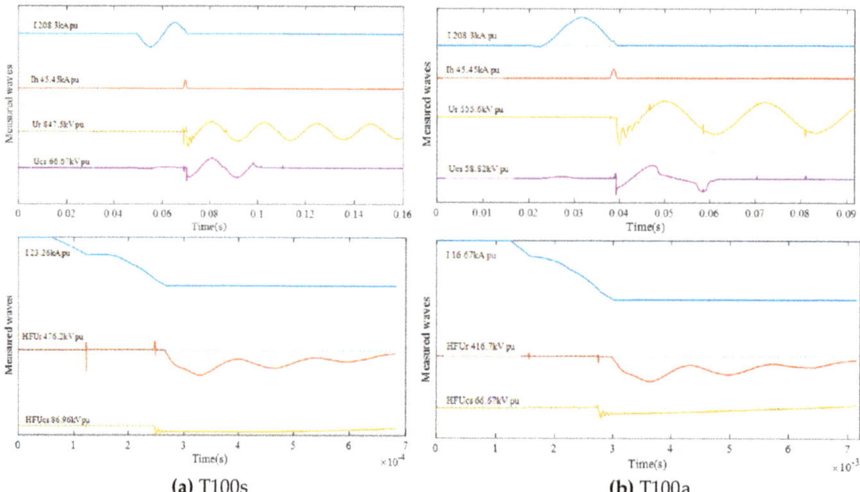

(a) T100s   **(b)** T100a

Figure 10. Test waveforms of the synthetic breaking test at 80 kA.

Table 5. Results of the breaking test.

Test Parameters	63 kA		80 kA	
	T100s	**T100a**	**T100s**	**T100a**
Charging voltage	512 kV	460 kV	487 kV	471 kV
Breaking current	64 kA	63.7 kA	73.5 kA	80.2 kA
di/dt	28 A/us	24.4 A/us	35.5 A/us	35.5 A/us
Recovery voltage	258 kV	228 kV	254 kV	246 KV
TRV peak	–544 kV	–484 kV	–542 kV	–526 kV
Th	461 us	230 us	200 us	233 us

6. Summary and Conclusions

This paper presented a novel sextuple-break 363 kV VCB that can remove a short circuit fault in an extremely short time. An FEM model based on actual dimensions was established to calculate the static voltage distribution and the stray capacitance to determine its grading capacitor design. At the fully open state, the voltage distribution of the proposed structure was uneven, and the first break undertook 259 kV, accounting for 86.3% of the total applied voltage. Subsequently, an equivalent capacitor network model was set up in PSCAD based on the parameters from the FEM model. Along with the increase of the grading capacitor value, the voltage of the first break decreased rapidly, and the voltage distribution of each break became more even. The best value of the grading capacitor was 10 nF, and its corresponding voltage unbalance coefficient was 1.07. Finally, T100s and T100a breaking tests were conducted at the short current of 63 kA and 80 kA both in a minor loop and a major loop. The prototype 363 kV VCB passed the test at the maximum TRV of 544 kV, which demonstrated that the performance of grading capacitors can satisfy the requirements of the voltage sharing design. In our future work, we will focus on the sextuple-break 363 kV VCB's dynamic opening mechanism to gain a deeper insight into the dynamic voltage distribution of the sextuple-break 363 kV VCB and put forward better grading measures.

Author Contributions: Conception, modelling, writing, editing: X.Y; supervision: F.Y.; funding acquisition: S.A., Y.H., Y.F.; experiment: W.D.; review and editing: X.Y, X.L.

Funding: This research was founded by Science and Technology Project of State Grid Corporation of China (grant 5229DK160005), National Key R&D Program of China (grant 2017YFB0902703) Chongqing Postdoctoral Fund Special Grant Project (grant Xm2017195).

Conflicts of Interest: The authors declare no conflict of interest.

References

1. Chen, H.; Liu, X.; Li, L.; Liu, Y.; Zhang, Y.; Huang, Y. Analysis of Dynamic Arc Parameters for Vacuum Circuit Breaker Under Short-Circuit Current Breaking. *IEEE Trans. Appl. Supercond.* **2019**, *29*, 1–5. [CrossRef]
2. Yao, X.; Wang, J.; Geng, Y.; Yan, J.; Liu, Z.; Yao, J.; Liu, P. Development and type test of a single-break 126-kV/40-kA–2500-A vacuum circuit breaker. *IEEE Trans. Power Deliv.* **2016**, *31*, 182–190. [CrossRef]
3. Liu, H.; Wang, Z.; Yang, J.; Li, B.; Ren, A. Circuit Breaker Rate-of-Rise Recovery Voltage in Ultra-High Voltage Lines with Hybrid Reactive Power Compensation. *Energies* **2018**, *11*, 100. [CrossRef]
4. Franck, C.M. HVDC circuit breakers: A review identifying future research needs. *IEEE Trans. Power Deliv.* **2011**, *26*, 998–1007. [CrossRef]
5. Jankowski, P.; Mindykowski, J. Study on the Hazard Limitation of Hybrid Circuit Breaker Actuator Operation. *Energies* **2018**, *11*, 416. [CrossRef]
6. Zhang, B.; Ren, L.; Ding, J.G.; Wang, J.; Liu, Z.; Geng, Y.; Yanabu, S. A Relationship between Minimum Arcing Interrupting Capability and Opening Velocity of Vacuum Interrupters in Short-circuit Current Interruption. *IEEE Trans. Power Deliv.* **2018**, *33*, 2822–2828. [CrossRef]
7. Tan, Y.; Kun, Y.; Xiang, B.; Wang, J.; Liu, Z.; Geng, Y.; Yanabu, S. Repulsion mechanism applied in resistive-type superconducting fault current limiter. *IEEE Trans. Appl. Supercond.* **2016**, *26*, 1–9. [CrossRef]
8. He, Z.; Wang, S. Design of the electromagnetic repulsion mechanism and the low-inductive coil used in the resistive-type superconducting fault current limiter. *IEEE Trans. Appl. Supercond.* **2014**, *24*, 1–4. [CrossRef]
9. Hou, C.; Yu, X.; Cao, Y.; Lai, C.; Cao, Y. Prediction of synchronous closing time of permanent magnetic actuator for vacuum circuit breaker based on PSO-BP. *IEEE Trans. Dielectr. Electr. Insul.* **2017**, *24*, 3321–3326. [CrossRef]
10. Homma, M.; Sakaki, M.; Kaneko, E.; Yanabu, S. History of vacuum circuit breakers and recent developments in Japan. *IEEE Trans. Dielectr. Electr. Insul.* **2006**, *13*, 85–92. [CrossRef]
11. Cheng, X.; Chen, S.; Ge, G.; Wang, H.; Wu, Q. Investigating on Synergy Effect of Series-Connected Vacuum Arcs in Multi-break VCBs. In Proceedings of the 2018 28th International Symposium on Discharges and Electrical Insulation in Vacuum (ISDEIV), Greifswald, Germany, 23–28 September 2018; Volume 2, pp. 671–674.
12. Ge, G.; Liao, M.; Duan, X.; Cheng, X.; Zhao, Y.; Liu, Z.; Zou, J. Experimental investigation into the synergy of vacuum circuit breaker with double-break. *IEEE Trans. Plasma Sci.* **2016**, *44*, 79–84. [CrossRef]
13. Huang, D.; Wu, G.; Ruan, J. Study on Static and Dynamic Voltage Distribution Characteristics and Voltage Sharing Design of a 126-kV Modular Triple-Break Vacuum Circuit Breaker. *IEEE Trans. Plasma Sci.* **2015**, *43*, 2694–2702. [CrossRef]
14. Horn, A.; Lindmayer, M. Investigations on the series connection of two switching gaps in one tube in vacuum. *IEEE Trans. Plasma Sci.* **2005**, *33*, 1594–1599. [CrossRef]
15. Zhang, B.; Tan, Y.; Ren, L.; Wang, J.; Geng, Y.; Liu, Z.; Yanabu, S. Interruption Capability of a Fast Vacuum Circuit Breaker with a Short Arcing Time. In Proceedings of the XXVIIth International Symposium on Discharges and Electrical Insulation in Vacuum, Suzhou, China, 18–23 September 2016; pp. 1–4.
16. Huang, D.; Shu, S.; Ruan, J. Transient recovery voltage distribution ratio and voltage sharing measure of double-and triple-break vacuum circuit breakers. *IEEE Trans. Compon. Packag. Manuf. Technol.* **2016**, *6*, 545–552. [CrossRef]
17. Ge, G.; Cheng, X.; Liao, M.; Huang, Z.; Zou, J. Mechanism of dynamic voltage distribution in series-connected vacuum interrupters. *IEEE Trans. Plasma Sci.* **2018**, *46*, 3083–3089. [CrossRef]
18. Fugel, T.; Koenig, D. Influence of grading capacitors on the breaking performance of a 24-kV vacuum breaker series design. *IEEE Trans. Dielectr. Electr. Insul.* **2003**, *10*, 569–575. [CrossRef]
19. Sugita, M.; Igarashi, T.; Kasuya, H.; Okabe, S.; Matsui, Y.; van Lanen, E.; Yanabu, S. Relationship between the voltage distribution ratio and the post arc current in double-break vacuum circuit breakers. *IEEE Trans. Plasma Sci.* **2009**, *37*, 1438–1445. [CrossRef]

20. Shaogui, A.; Xiao, Y.; Yongning, H.; Fan, Y.; Yiping, F.; Xing, L. Voltage Distribution Design of a Novel 363 kV Vacuum Circuit Breaker. In Proceedings of the 2018 IEEE International Conference on High Voltage Engineering and Application (ICHVE), ATHENS, Greece, 10–13 September 2018; pp. 1–4.
21. Grading Capacitors for High-Voltage Alternating Current Circuit Breakers. Available online: https://infostore.saiglobal.com/preview/is/en/2014/i.s.en62146-1-2014%2Ba1-2016.pdf?sku=1723996 (accessed on 21 May 2019).

Article

Accuracy and Reliability of Switching Transients Measurement with Open-Air Capacitive Sensors

Fani Barakou [1], Fred Steennis [1,2] and Peter Wouters [1,*]

[1]   Department of Electrical Engineering, Eindhoven University of Technology, P.O. Box 513,
      5600MB Eindhoven, The Netherlands; fani.barakou@npl.co.uk (F.B.); fred.steennis@dnvgl.com (F.S.)
[2]   DNV GL, P.O. Box 9035, 6800ET Arnhem, The Netherlands
*   Correspondence: p.a.a.f.wouters@tue.nl; Tel.: +31-40-2474-244

Received: 21 March 2019; Accepted: 10 April 2019; Published: 11 April 2019

Abstract: Contactless capacitive (open-air) sensors are applied to monitor overvoltages near overhead line terminations at a substation or at the transition from underground cables to overhead lines. It is shown that these sensors, applied in a differentiating/integrating measuring concept, can result in excellent characteristics in terms of electromagnetic compatibility. The inherent cross-coupling from open-air sensors to other phases is dealt with. The paper describes a method to calibrate the sensor to line coupling matrix based on assumed 50 Hz symmetric phase voltages and in particular focuses on uncertainty analysis of assumptions made. Network simulation shows that predicted maximum overvoltages agree within typically 7% compared to reconstructed values from measurement, also with significant cross-coupling. Transient voltages from energization of an (extra-)high voltage connection can cause large and steep rising ground currents near the line terminations. Comparison with results obtained by a capacitive divider confirms the intrinsic capability in interference rejection by the differentiating/integrating measurement methodology.

Keywords: air capacitive sensors; power system transients; high-voltage measurements; high-voltage monitoring

---

1. Introduction

Knowledge on actual overvoltages from transients in electrical transmission systems can be very helpful to estimate optimal measures for overvoltage protection, to understand the background of problems (if any) and/or to estimate related degradation of electrical insulation. Such transients arise e.g., from line energization and can cover a wide frequency spectrum [1]. The installation of equipment such as (R) C-dividers capable to capture high-frequency signals is costly, leaves a relatively big footprint and should be planned ahead. Dividers can also be prone to EMC (electromagnetic compatibility) disturbance [2], e.g., arising from ground currents upon a switching event. A non-invasive, cost-efficient, easy to (re-)install and EMC-proof measurement system would therefore be appealing to occasionally perform voltage monitoring.

This paper describes applications based on open-air capacitive sensors as part of a D/I (differentiating/integrating) measurement concept. The signal at the measurement cable end, characteristically terminated with 50 Ω, is basically the time-derivative of the phase-voltage as picked up with the open-air capacitive sensors. These sensors can be mounted in a short time without adaptation to the EHV (extra-high voltage) system. In other words, it is a non-invasive measuring system. The original waveform is restored by means of an analogue integrator. One of the earlier papers describing the method for application on high-voltage measurement is reference [3]. As the sensors basically consist of metal plates which form capacitances to high-voltage conductors nearby, the requirements related to costs, installation effort and being non-invasive are fulfilled. A drawback of the method relates to the prior unknown coupling strengths of open-air sensors and the limited selectivity as they couple to any phase conductor in the vicinity. A matrix containing all couplings between sensors and phases needs to be established to entangle the cross-coupling.

A number of open-air sensor applications with solutions for the selectivity problem can be found in literature. In [4], line energization was studied. Decoupling could be achieved since different phases had distinct contact moments. This allowed to determine the responses of the sensors upon the three distinct energization moments. Each phase energization provided three sensor responses, enabling to establish the complete coupling matrix. In [5] the capacitive sensors were inside a GIS (gas insulated system) and in prior tests the phases could be energized separately. A harmonics study was published in [6] where the sensors were positioned halfway between two overhead line pylons. The capacitances were derived from equations for cylindrical line conductors above perfect conductive earth. Unknowns in distances (e.g., line heights) could be resolved by assuming symmetric power frequency voltages on the overhead lines and fitting the measured and calculated waveforms. For the transient overvoltage study in [7], a similar approach was taken. However, the sensors were placed near the terminations for high sensitivity. Due to the complex termination design, analytical field calculation was not possible, leaving the coupling matrix not fully determined. It was shown that with a few assumptions based on symmetry in the configuration the reconstruction could be realized.

This paper develops a methodology to analyse the consequence of uncertainties in the assumptions made to reconstruct the phase voltages. Section 2 briefly describes the methodology. Confidence bounds are needed to judge the accuracy of network simulations, which are implemented in PSCAD [8]. References [7] and [9] provide information on the modelled network detail and extension, respectively. In Section 3.1, these simulations are compared with measurements in relation to the experimental uncertainty. In [2] the merits of the D/I method in terms of EMC immunity are discussed. To verify the claims, simultaneous measurement of switching transients by the D/I system and a high voltage divider is performed. The reliability of the recordings from both systems is compared in Section 3.2. Section 4 reflects on the results in terms of accuracy as well as reliability and Section 5 concludes with situations for which D/I measurement is particularly of interest.

2. Measurement Analysis and Uncertainty

Measured sensor signals u_i relate to the phase voltages U_j according to a coupling matrix **M**: $u = MU$. Reconstruction of the phase voltages from recorded signals can be achieved by inverting this relation. The matrix elements M_{ij} are obtained by selecting the part from the recorded signals containing only the power frequency waveform (with angular frequency ω_0). Sinusoidal fits $u_{pf,i}$ are made and related with an assumed base of symmetric (per unit) phase voltages, i.e., equal amplitudes and 120° phase angle differences:

$$\begin{pmatrix} u_{pf,1}(t;\varphi_0) \\ u_{pf,2}(t;\varphi_0) \\ u_{pf,3}(t;\varphi_0) \end{pmatrix} = \begin{pmatrix} M_{11} & M_{12} & M_{13} \\ M_{21} & M_{22} & M_{23} \\ M_{31} & M_{32} & M_{33} \end{pmatrix} \begin{pmatrix} \cos(\omega_0 t + 2/3\pi) \\ \cos(\omega_0 t) \\ \cos(\omega_0 t - 2/3\pi) \end{pmatrix} \tag{1}$$

$$\text{with } u_{pf,i}(t;\varphi_0) = a_i \cos(\omega_0 t + \varphi_0) + b_i \sin(\omega_0 t + \varphi_0)$$

The reference time $t = 0$ in (1) corresponds with the central phase reaching its maximum value. If the trigger moment of the measurement device is not synchronized with (one of) the phase voltages, it can be implemented as a phase shift ϕ_0. This additional unknown can be omitted by having the measurement time base related to a simultaneous recording of the phase voltage information, e.g., from a voltage transformer (synchronized), or it can be obtained from the measured waveforms themselves (unsynchronized). As (1) relates sinusoidal waves of a single frequency, each row equation determines only two parameters and the set of three equations is underdetermined. With ten unknowns (for unsynchronized measurements) and six fitted parameters also other sources of information must be employed. Reduction of the number of independent components M_{ij} can be achieved by neglecting components when these are very small, or by having a symmetric positioning of sensors with respect to the phase conductors meaning that various matrix components are equal [7,10]. Also parameter estimates can be made and their uncertainties can be retrieved, e.g., from comparison with numerical electrostatic field simulations on the system. Which assumptions are to be made and what the effects will be on the uncertainty margins are specific for the configuration at the measurement site. Measurements of energization transients are obtained from a double circuit 380 kV line-cable-line connection [7]. More specifically, measurements were taken at the underground cable to overhead line transition points (at both cable ends) and near an overhead line termination at one of the substations.

Uncertainty analysis concerns the determination on how input margins propagate into end result deviations according the designed measurement chain. In addition, reliability of the measurement can be impeded when also signal coupling along an unintended route takes place. This can be a serious problem when recording switching transients in (E)HV networks. Line energization causes steep switching fronts which, in particular at an overhead line to underground cable transition, may induce ground currents. These currents may partly find their way along the metallic screens of the measurement cables. Depending on the measurement system design and layout, this can disturb signal recordings.

In Figure 1 the configuration for a D/I system is compared to a capacitive divider configuration. The D/I sensor consisted of two 30×30 cm^2 plates, with the top plate sensing the phase voltages (C_{HV} = 0.1–1 pF) and the bottom plate connected to earth. For the D/I method, the capacitance between the plates is relatively low (here $C_E \approx 300$ pF). The measurement cable is characteristically terminated and its impedance R_m is far lower than $|Z_E| = 1/(\omega C_E)$. Up to the cut-off frequency, $1/(2\pi R_m C_E)$, the time derivative $U'(t)$ arises at the cable end. The signal waveform is restored by integration with an integrator time constant, τ_{int}, depending on the integrator design. The active/passive integrator employed for the measurements is described in [11]. The over-all system response $U(t)$ is given by:

$$U(t) = \tfrac{1}{\tau_{int}} \int U'(t)\mathrm{d}t \text{ where } U'(t) = \tau_{dif}\frac{\mathrm{d}U_{HV}(t)}{\mathrm{d}t}$$
$$\tau_{dif} = R_m C_{HV} \tag{2}$$

For a divider topology the situation is reversed and the impedance R_m is large compared to $|Z_E|$ in order not to load the low voltage branch of the divider. This difference has repercussions on EMC aspects from ground currents as demonstrated in Figure 2. A part of the ground current will find its way along the measurement cable screens through magnetic induction. This common mode current I_{CM} will cause a differential mode voltage U_{DM} which adds to the intended signal. Its value depends on the quality of the measurement cable, which can be expressed as its transfer impedance Z_t [12]:

$$U_{DM} = I_{CM} \cdot Z_t \tag{3}$$

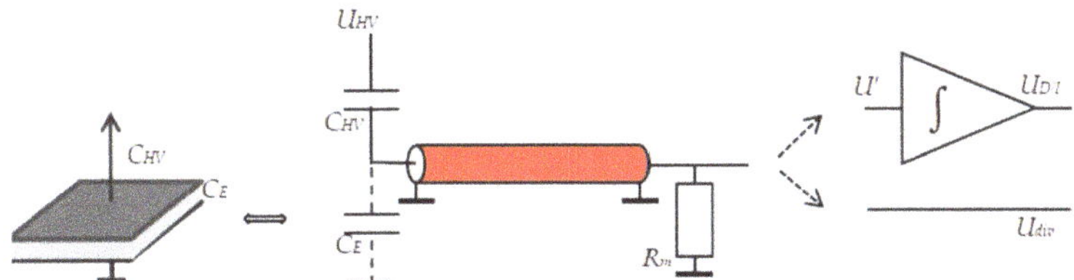

Figure 1. Measurement topologies for D/I and divider methods. Component values for D/I are such that the receiving cable end represents the time derivative of phase voltage U_{HV} and the waveform is restored by integration (the open-air sensor design is shown on the left); For standard dividers this signal directly represents the phase voltage U_{HV}.

The fraction of U_{DM} arising at the receiving end of the cable is a division over the impedances at both ends as shown in the right side of Figure 2. For the D/I concept this is rather a small fraction, whereas for a divider it is the major part. In addition, the integration step in the D/I system reduces high frequency interference, which may have entered the measurement cable. Results will be presented comparing the D/I approach with results from pre-installed dividers near underground power cables at their transition points to overhead lines.

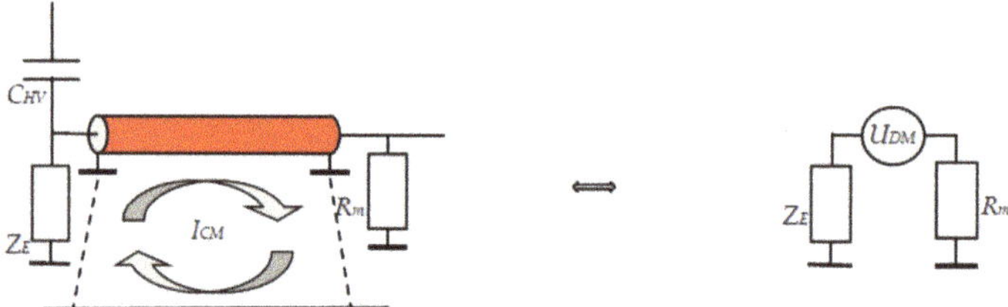

Figure 2. Effect of induced ground currents along the measurement cable (**left**); Through the cable transfer impedance the resulting common mode current I_{CM} causes a differential mode voltage U_{DM} which divides over the impedances at the cable ends (**right**).

3. Results

The challenge with open-air sensors is to entangle the cross-coupling. The approach advocated in this paper is to employ only the stationary power frequency component from symmetric phase voltages. This requires a minimal installation effort to set up the measurement system, which can be done in typically one to one and a half hour. The option to sequentially inject a signal in each phase and determine the responses as in [5] would provide sufficient information to derive the complete coupling matrix, but it takes time and requires extensive safety precautions. Using the first arriving travelling wave fronts as in [4] assumes that these remain sufficiently clear and recognizable after travelling along the lines and cables. The cross-bonding applied along the underground power cable system modifies the waveform further and therefore such a method is mainly feasible when measurements are conducted near the substation where the switching activities take place.

Different causes for uncertainty are analysed in [10]. The measurement accuracy was checked by determining the variation of the coupling matrix throughout a measurement session at a single location. As the sensors remain at fixed positions, variations are related to noise, stability of the D/I measurement system (in particular the active integrator part) and fitting technique. The overall variation remained within 2%. Deviation from assumed symmetric phase voltages will translate into similar magnitude variations in the recorded signals. However, it was shown that in terms of per unit values, i.e., the percentage of the overvoltage related to the amplitude of the power frequency component, such deviation is negligible. The uncertainty analysed in Section 3.1 concerns the assumptions made in order to reduce the number of independent matrix components in (1). Furthermore, imprecise sensor positioning or metal structures in the vicinity may contribute to deviations from model assumptions.

Section 3.1 details the methodology to establish the coupling matrix for two distinct measurement locations. Section 3.2 elaborates on comparison with system simulation and comparison between D/I and divider measurement.

3.1. Accuracy of Decoupling Procedure

The left side of each row equation in (1) is fitted with the steady state power frequency part of the recorded signal. A single frequency sinusoidal function is determined by two parameters, meaning each row equation in (1) is underdetermined. Since the phase voltages are symmetric, with their sum equal to zero, a constant Δ_i added to each component in row i ($i = 1, \ldots, 3$) will result in the same function and therefore will still satisfy the fitted relation on the left hand side: all components within a row are determined up to the same constant. Usually, the far end couplings (M_{13} and M_{31}) are relatively small with respect to the direct couplings (M_{ii}). Neglecting these or assigning an estimated (small) value will therefore only have a minor effect. For the middle row in (1), in case of a reflection symmetric configuration, the additive value is principally undetermined. However, deviations from symmetry should be accountable to provide for a margin in ϕ_0. Therefore, a fourth uncertainty, Δ_0, is introduced. The coupling matrix with four uncertainties related to the same number of lacking parameter information after fitting can be formulated as:

$$\mathbf{M} = \begin{pmatrix} M_{11} & M_{12} & M_{13} \\ M_{21} & M_{22} & M_{23} \\ M_{31} & M_{32} & M_{33} \end{pmatrix} \pm \begin{pmatrix} \Delta_1 & \Delta_1 & \Delta_1 \\ \Delta_2 - \tfrac{1}{2}\Delta_0 & \Delta_2 & \Delta_2 + \tfrac{1}{2}\Delta_0 \\ \Delta_3 & \Delta_3 & \Delta_3 \end{pmatrix} \tag{4}$$

For distinct locations, site specific assumptions will be made with confidence intervals and the consequence on the reconstructed transient waveforms is determined as discussed below.

The components in coupling matrix M_{ij} in (4) can be re-ordered such that they constitute a linear set of nine parameters indicated as $x_{k=3(i-1)+j} = M_{ij}$. For these parameters the error matrix $\mathbf{E_x}$ needs to be established, which contains all variances and covariances:

$$E_{x,kl} = \langle \Delta x_k \Delta x_l \rangle = \left\langle (x - \bar{x})_k (x - \bar{x})_l \right\rangle \tag{5}$$

To this end, normal distributions are assigned to the four uncertainties in (4) and 1000 simulations are made to evaluate (5). The reconstructed phase voltages can be found by inverting matrix $\mathbf{M}$ and applying it to recorded waveforms containing the transient events. The components of the inverted coupling matrix $\mathbf{M}^{-1}$ can be arranged in a linear set $y_{l=3(i-1)+j} = M_{ij}^{-1}$ as well, and its error matrix $\mathbf{E_y}$ can be evaluated according [13,14]:

$$\mathbf{E_y} = \mathbf{O_1 E_x O_1^T} \quad \text{with} \quad O_{1,lk} = \frac{\partial y_l}{\partial x_k} \tag{6}$$

The matrix $\mathbf{O_1}$ represents a linearization of how component y_l of the inverted matrix depends on a variation in the value x_k of the original coupling matrix. Its calculation can conveniently be implemented by slightly varying numerically each coupling matrix component M_{ij} separately. Next, the phase voltage waveforms are reconstructed by matrix multiplication for each sample in the measurement recordings. The propagation of the error is described by means of matrix $\mathbf{O_2}$ which provides information on how each of the three reconstructed phase voltages varies upon variation in each of the inverted matrix components. The reconstructed phase waveforms are evaluated with [13,14]:

$$\mathbf{E_U} = \mathbf{O_2 E_y O_2^T} \quad \text{with} \quad O_{2,jl} = \frac{\partial U_j}{\partial y_l} \Rightarrow$$

$$\mathbf{O_2} = \begin{pmatrix} u_1 & u_2 & u_3 & 0 & 0 & 0 & 0 & 0 & 0 \\ 0 & 0 & 0 & u_1 & u_2 & u_3 & 0 & 0 & 0 \\ 0 & 0 & 0 & 0 & 0 & 0 & u_1 & u_2 & u_3 \end{pmatrix} \tag{7}$$

This equation is applied for each measured sample point. The square roots of the diagonal components in $\mathbf{E_U}$ provide the standard deviations for each phase (per sample point). Adding and subtracting the deviations provide the confidence intervals of the reconstructed waveforms.

3.1.1. Substation

The configuration with a line ending at a GIS substation is depicted in Figure 3a. The chosen sensor positioning leaves a reflection symmetric configuration. This leads to equal coupling of the central sensor to the outer phase voltages. Similarly, the coupling of the outer sensors to the central phase conductors are equal. A fundamental problem, as mentioned above, relates to the fact that from calibrating by means of symmetric phase voltages it is impossible to distinguish the contribution of the coupling to the central sensor by the central phase from the combined coupling with the outer phases. An estimate is needed including sufficient margin to accommodate for its uncertainty [10]:

$$M = \begin{pmatrix} M_1 & M_4 & f_0 M_4 \\ M_5 - 1/2\Delta_0 & M_2 & M_5 + 1/2\Delta_0 \\ f_0 M_6 & M_6 & M_3 \end{pmatrix} \text{ with } M_5 = 1/2(M_4 + M_6) \tag{8}$$

The main couplings in Figure 3a, diagonal elements in (8), are taken independent. Although they are expected to be similar, the effect of imprecise sensor positioning on the main coupling can be accounted for. The couplings of the side sensors to the middle phase are taken independent as well. The coupling to the phase furthest away is roughly estimated to be about 30% of the neighbour couplings M_4 and M_6 with an uncertainty of 50% (i.e., $f_0 = 0.30 \pm 0.15$, so between a fraction of 0.15 to 0.45 of the value of M_4 and M_6). The estimate is based on the distance ratio sensor line, which was just over a factor two. As the sensor couples to a relatively short line, its sensed electric flux is expected to decay between $1/r$ (long line) and $1/r^2$ (point source). The chosen range with uncertainty covers both extremes. Electrostatic field simulation [7] provided a value of 40%. For M_5 an average value of M_4 and M_6 is taken as the distances are similar. With Δ_2 an uncertainty is assigned in this assumption of 50% of its value ($\Delta_2 = 1/2 M_5$). As precise positioning of the sensor in front of the central phase could relatively easy be judged, the value of Δ_0 was taken 20% of M_5 ($\Delta_0 = 1/5 M_5$).

Solving (1) with (8) as coupling matrix results in the following set of equations:

$$M_4 = \frac{1/3\sqrt{3}(a_1 \sin\varphi_0 - b_1 \cos\varphi_0) + (a_1 \cos\varphi_0 + b_1 \sin\varphi_0)}{1 - f_0} \quad M_1 = 2/3\sqrt{3}(a_1 \sin\varphi_0 - b_1 \cos\varphi_0) + f_0 M_4$$

$$M_6 = \frac{1/3\sqrt{3}(-a_3 \sin\varphi_0 + b_3 \cos\varphi_0) + (a_3 \cos\varphi_0 + b_3 \sin\varphi_0)}{1 - f_0} \quad M_3 = 2/3\sqrt{3}(-a_3 \sin\varphi_0 + b_3 \cos\varphi_0) + f_0 M_6 \tag{9}$$

$$M_2 = 1/2(M_4 + M_6) + (a_2 \cos\varphi_0 + b_2 \sin\varphi_0)$$

$$2/3\sqrt{3}(-a_2 \sin\varphi_0 + b_2 \cos\varphi_0) - \Delta_0 = 0 \tag{10}$$

The parameters a_i and b_i are obtained from fitting sinusoidal functions to the last five cycles in the recordings as shown in Figure 4a (left). Equation (10) is solved numerically in ϕ_0 and subsequently all matrix components M_i can be calculated. The solution depends on the stochastic variable Δ_0. The other stochastic variables do not affect the solution of (1) and their effect on the coupling matrix elements can be added afterwards.

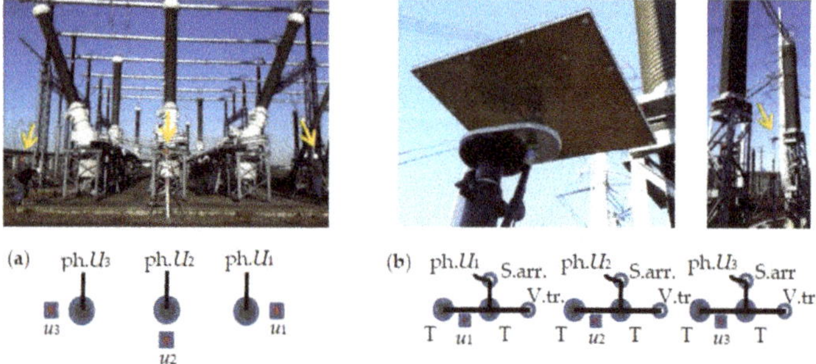

Figure 3. Topologies and sensor (u_i) positioning with respect to phase voltages U_j: (**a**) Overhead line ending at a substation, photo shows the installation of the sensors; (**b**) Overhead line to underground power cable transition (with termination T, surge arrester S.arr. and voltage transformer V.tr.), photos show the sensor (as seen from below) and its positioning in between two terminations belonging to the same phase. The sensor heights (indicated with arrows) are always chosen such, that they remain below the height reached by the supporting structures of the terminations.

Figure 4 shows the results upon line energization from the far end of the connection and from the substation where the measurements are taken. The black lines indicate one standard deviation confidence margin and the dash-dotted lines are extrapolated steady state phase voltages. The complete waveform in Figure 4a includes the recording of the steady state power frequency established after the switching transient. The last five cycles are employed for determining the coupling matrix. The zoomed figure shows that within the confidence bounds the overvoltage magnitude remains clearly below 2 pu. The switching event in Figure 4b contains steep transients as energization takes place near the measurement location. The zoomed waveform indicates that these relate to travelling waves, which reflect on the line to cable transition point at 6.8 km distance. Also here no serious overvoltages were observed.

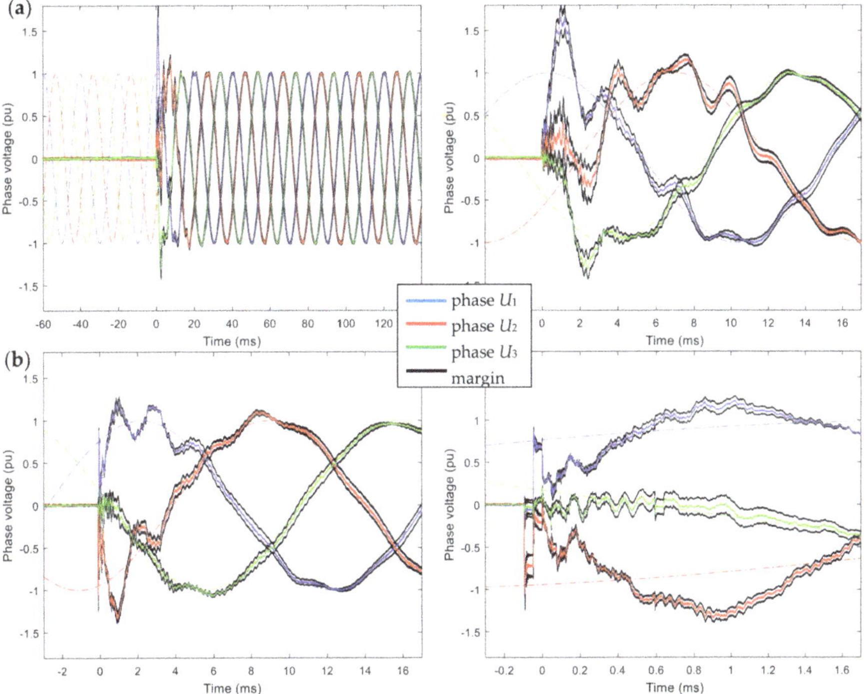

Figure 4. Reconstructed waveforms for substation measurements recorded with 5 Msample/s over a duration corresponding to ten power cycles. The curves in colour represent the three phases with confidence margins in black: (**a**) Full and 10 times zoomed waveform showing overvoltages during the first cycle from energization at the far end of the connection; (**b**) Waveform with overvoltages during the first power cycle from energization at the near end and 10 times zoomed waveform revealing initial travelling waves along the connection. The extrapolated power cycles (dash-dotted lines) indicate the 50 Hz phase angles at the moments of contact.

3.1.2. Transition Point

The configuration at a transition point is depicted in Figure 3b. Each overhead line is connected with two underground cables to match transmission capacity. The sensors are placed in between the cable terminations belonging to the same phase. The huge terminations provide shielding from coupling to the other phases. Therefore, the far end couplings (u_1 to U_3 and u_3 to U_1) are small and could in principle be neglected. The configuration also suggests that the couplings of sensor u_i to phase U_{i+1} have similar magnitudes as is the case for the couplings of sensor u_{i+1} to phase U_i ($i = 1, 2$):

$$\mathbf{M} = \begin{pmatrix} M_1 & M_4 & f_0 M_1 \\ M_5 - \tfrac{1}{2}\Delta_0 & M_2 & M_4 + \tfrac{1}{2}\Delta_0 \\ f_0 M_3 & M_5 & M_3 \end{pmatrix} \tag{11}$$

From Figure 3b the main couplings, diagonal elements in (11), are expected to be close, but in case of imprecise sensor positioning deviations can be accounted for by allowing independent values. Symmetry in the configuration allows to define only two distinct parameters for coupling of a sensor to a neighbouring phase. A further assumption relates to the far end couplings, which are minor contributions due to the shielding by the terminations. Electrostatic field analysis provided an estimate of 2% [7], which is implemented with an uncertainty margin of 50%: $\Delta_1 = \Delta_3 = \tfrac{1}{4}f_0(M_1+M_3)$ with $f_0{=}0.02$. In the second row the uncertainty is taken 50% of the average value of the neighbour couplings, $\Delta_2{=}\tfrac{1}{4}(M_4{+}M_5)$. The additional uncertainty Δ_0, as the measurement equipment was not synchronized with the phase voltages, is taken equal to Δ_2 and accounts for the uncertainty caused by ϕ_0. This reduces $\mathbf{M}$ to five independent components.

Solving (1) with (11) results in

$$M_1 = \frac{\tfrac{2}{3}\sqrt{3}(a_1 \sin\varphi_0 - b_1 \cos\varphi_0)}{1-f_0} \quad M_4 = \tfrac{1}{2}M_1(1+f_0) + (a_1 \cos\varphi_0 + b_1 \sin\varphi_0)$$
$$M_3 = \frac{\tfrac{2}{3}\sqrt{3}(-a_3 \sin\varphi_0 + b_3 \cos\varphi_0)}{1-f_0} \quad M_5 = \tfrac{1}{2}M_3(1+f_0) + (a_3 \cos\varphi_0 + b_3 \sin\varphi_0) \tag{12}$$
$$M_2 = \tfrac{1}{2}(M_4 + M_5) + (a_2 \cos\varphi_0 + b_2 \sin\varphi_0)$$

$$\tfrac{2}{3}\sqrt{3}(-a_2 \sin\varphi_0 + b_2 \cos\varphi_0) - M_4 + M_5 - \Delta_0 = 0 \tag{13}$$

Equations (12) and (13) can be solved numerically after fitting the power frequency parts of the recorded waveforms.

Figure 5 shows the results obtained at the transition points. In both cases the overhead line leaving from the measurement location was energized at its far end. The confidence bounds are significantly smaller compared to those obtained in the substation measurements. Apparently, shielding by the terminations at either side of the sensors reduce cross-coupling significantly. Uncertainties in the cross-coupling therefore have minor effect on the reconstructed waveforms. Also here it was confirmed that the maximum voltages remain within safe limits.

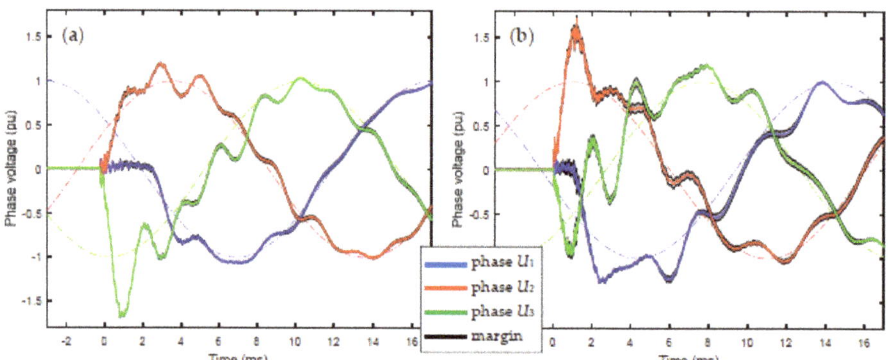

Figure 5. Part of the reconstructed waveforms (covering 10 cycles recorded with 5 Msample/s) around the switching moment for measurement at transition points. Coloured curves represent the phase voltages with one standard deviation confidence margins indicated in black: (a) Overvoltages from energization at the substation connected by 4.4 km overhead line to the transition point; (b) Overvoltages from energization at the substation connected via 6.8 km overhead line. Moments of contact can be observed from the extrapolated power frequency waveforms (dash-dotted lines).

3.2. Reliability of the Reconstructed Waveforms

The D/I measurement results are compared with power system simulations obtained using PSCAD. The studied connection basically consists of a double circuit with 10.8 km cable in between two overhead line sections (4.4 km and 6.8 km length). However, both the magnitude and the harmonic content of switching transient responses depend on a much more extensive part of the grid [9]. The complete Dutch EHV grid was modelled in PSCAD using frequency dependent transmission line models [7,15]. A second measurement campaign was arranged aiming for simultaneously recording using the D/I measurement system and the RC dividers present at the transition points. The comparison allows to judge the sensitivity of different techniques in terms of EMC.

3.2.1. Comparison with PSCAD Simulation

Measurements were conducted at both line to cable transition points and at one of the substations. At each location six switching actions were performed differing in the side from which the connection was energized, the operation state of the parallel circuit and the choice of the parallel circuits to be energized. A selection is presented in Figure 6, corresponding to the results in Figures 4 and 5, with the parallel circuit de-energized.

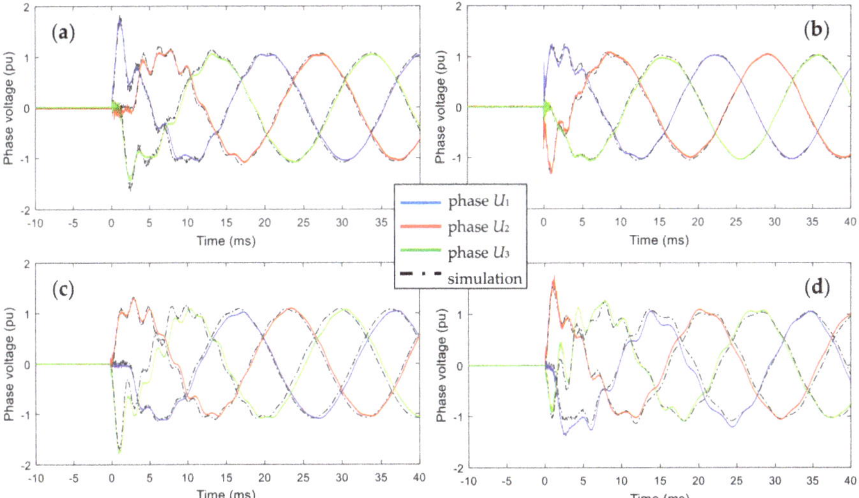

Figure 6. Comparison between D/I measurement (continuous lines in colour for the three phases) and PSCAD simulations (dash-dotted black lines). Generally, the maximum overvoltage value observed for each switching event and phase agree well with calculations for: (**a**) Recording at substation with energization from far end substation; (**b**) Recording and energization at the same substation; (**c**) Energization at substation 4.4 km from the observed transition point; (**d**) Energization from substation at 6.8 km from the observed transition point.

Figure 6a,b show the result obtained at the substation, where the circuit is excited from the far side and from the measurement side, respectively. Figure 6c,d provide the associated results for the transition points. The excitation of the connection is performed at the overhead line ending at the transition point where measurements are conducted. It can be concluded that measurement results confirm the simulations in large detail.

3.2.2. Comparison with RC Divider Measurement

Figure 7 shows the result from simultaneous measurement with the D/I method (top figures) and an RC divider (bottom figures). In between, the PSCAD simulations are presented. The measurements are taken at both transition points with the parallel circuit in service. It is observed that oscillations occur in the divider response, which are far more severe than found in both the simulation and the D/I response. It should be noted that, although simulation and D/I response are quite similar, also here the measured oscillation immediately after energizing is somewhat stronger than simulated.

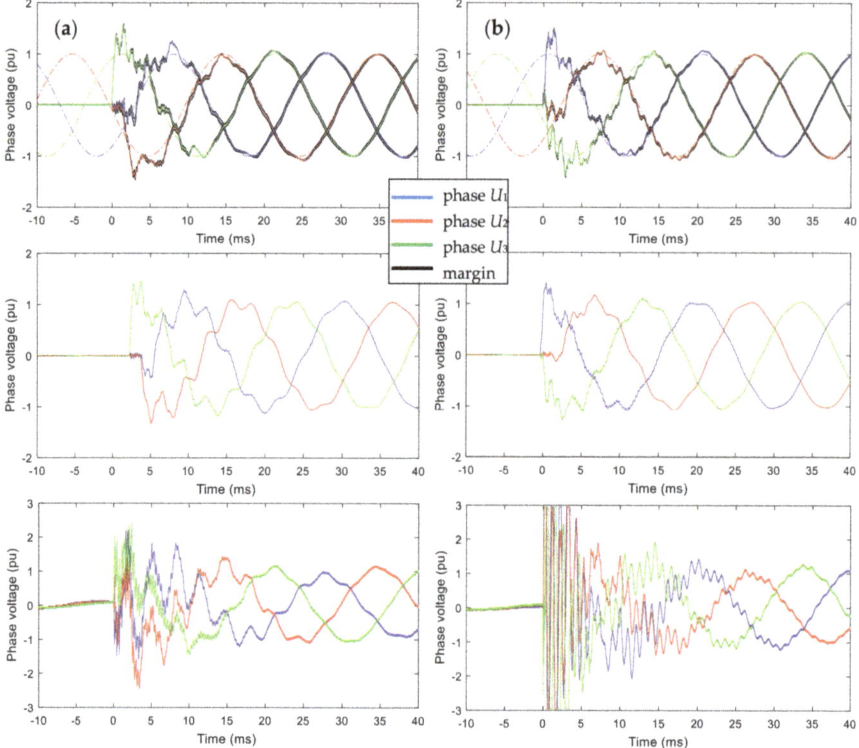

Figure 7. Comparison of results from D/I methodology (top figures, black lines indicate the uncertainty margins caused by uncertainty in decoupling), PSCAD simulation (middle figures) and RC-divider results (bottom figures): (**a**) Energization from substation connected by 4.4 km overhead line to the transition point; (**b**) Energization from substation connected via 6.8 km overhead line. For both switching events the high frequency oscillations are overrepresented in the divider measurement as compared to both D/I measurement and simulation.

4. Discussion

The confidence margins depend on the strength of the cross-couplings and on the accuracy by which they can be determined. For the transition points the contribution of the model uncertainties is about 2%, whereas for the substation it reaches 8% [10]. These uncertainties should be considered when comparing measured and simulated overvoltage magnitudes. Figure 8 shows the statistical distribution of relative differences between simulated and observed maximum overvoltages. The data include six switching actions at all measurement locations, each providing three overvoltage values for the three phases. The width of the fitted normal distributions found for the substation measurement

(6.8%, Figure 8a) is only slightly larger as compared to the width for the two transition points (6.0%, Figure 8b). On one hand, this shows that the modelling method is adequate and predicts observations well. On the other hand, it does not represent the difference in measurement accuracy expected from both locations. Apparently, the simulations also contribute to deviations. This is not surprising as, although the complete Dutch EHV grid was simulated in large detail, a number of simplifications were made. Connections to downstream networks (110, 150, 220 kV) were replaced by equivalent impedances based on active and reactive power consumption by the loads according to powerflow calculations [7]. Moreover, connections abroad were not modelled in detail.

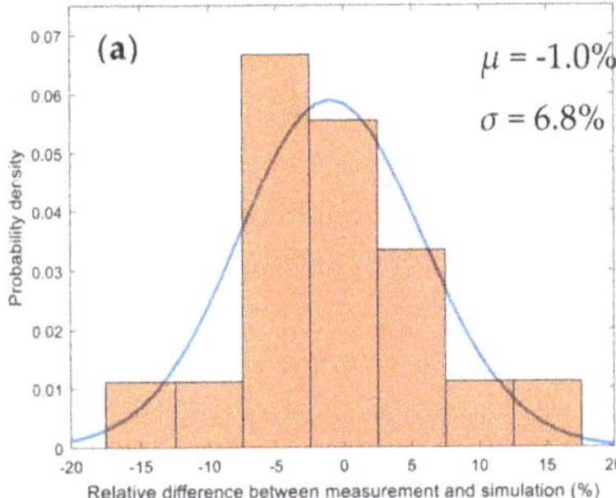
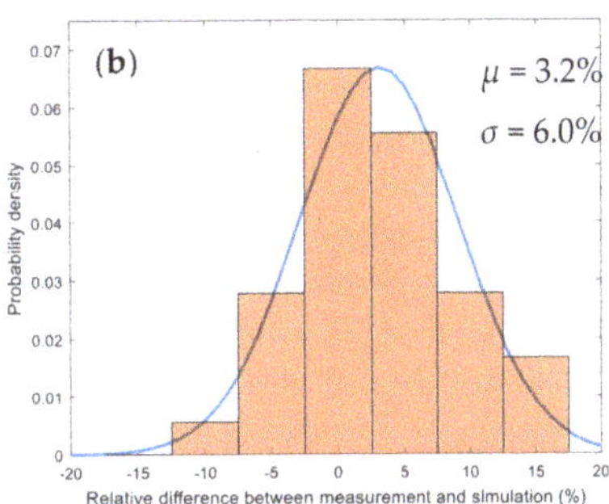

Figure 8. Comparison of measurement and simulation for transition overvoltages upon energizing from nearby substation: (**a**) Results for the substation; (**b**) Results for the two transition points.

The comparison with divider measurements revealed serious differences and apparently EMC is an issue. High frequency components are overrepresented in the divider waveforms, which points to interference related to the signal time derivative. Common mode currents along measurement cables, as discussed in Section 2, can be induced by the time varying magnetic flux caused by ground currents. Poor measurement cable quality and cable layout are suspected to translate the common mode currents into disturbance adding to the recorded waveforms. Also, slight disturbance seems to be picked up by the D/I system. For the measurements presented in Section 3.2 the ground plates of each sensor were connected to the metal support of the termination near the sensor. This causes ground loops formed by the earth screens of the measurement cables, which can pick up magnetic flux from a switching event. For the measurements presented in Section 3.1 all sensor ground plates were connected to a central point, avoiding the occurrence of such loops to a large extent. It shows, that even with the D/I methodology, with its intrinsic good EMC characteristics, every connection detail is important in order to perform disturbance free measurements.

5. Conclusions

The D/I methodology with open-air capacitive sensor is capable of recording switching transients in an EMC harsh environment. However, the decoupling inherent to open-air sensors is a tedious process and may limit its application to specific situations:

- Verification of installed measurement systems in case of doubt or when fast transient overvoltages are expected to occur outside the (R) C divider bandwidth. The D/I method can be designed to have a bandwidth in excess of 5 MHz [2,7].
- Commissioning of a new connection to avoid the need of permanent installation of a costly divider. After an operation time, e.g., to confirm that overvoltages remain within safe limits, there might be no further need for having such dedicated equipment installed.
- Upon a system fault the D/I method can be installed in a short time to verify whether voltage transients occur in the connection and may be involved.

The need of pre-assumptions is an obvious drawback when dealing with significant cross-coupling, which may especially occur at large substations. Future research is directed to reduce the number of assumptions to be made in establishing the coupling matrix. Complete information can be extracted by employing the responses from different sensors upon the front of a switching event (as in [4]) in combination with information provided by the power frequency responses as presented here. Having the topologies of Figure 3 in mind, upon an initial travelling wave on one of the lines, only the response ratios of neighbouring sensors are needed. There are four combinations which, together with information from power frequency fitting, would completely determine the coupling matrix.

Author Contributions: F.B. and P.W. prepared and performed the experiments, designed the analysis methods and wrote the paper. F.B. performed the power system modelling in PSCAD. F.S. provided technical feedback and reviewed the paper.

Funding: This research was funded by TenneT TSO B.V. within the framework of the Randstad380 cable research project.

Acknowledgments: TenneT is acknowledged for performing the switching actions in their 380 kV grid. Movares Nederland B.V. is acknowledged for providing their set of integrators with EMC shielding. Marcel Hoogerman, Hennie van der Zanden, Frank Beckers and Armand van Deursen from Eindhoven University of Technology are acknowledged for assisting with the measurement campaigns and their preparations.

Conflicts of Interest: The authors declare no conflict of interest.

References

1. Irwin, T.; Ryan, H.M. Insulation co-ordination for AC transmission and distribution systems. In *High-Voltage Engineering and Testing*, 3rd ed.; Ryan, H.M., Ed.; IET Power and Energy Series; IET: Stevenage, UK, 2013; Volume 66, pp. 61–63.
2. Barakou, F.; Wouters, P.A.A.F.; Mousavi Gargari, S.; Smit, J.; Steennis, E.F. Merits and challenges of a differentiating-integrating measurement methodology with air capacitors for high-frequency transients. In Proceedings of the CIGRE 2018, Paris, France, 26–31 August 2018; C4-203.
3. Van Heesch, E.J.M.; van Deursen, A.P.J.; van Houten, M.A.; Jacobs, G.A.P.; Kersten, W.F.J.; van der Laan, P.C.T. Field tests and response of the D/I H.V. measuring system. In Proceedings of the 6th International Symposium on High-Voltage Engineering, New Orleans, LA, USA, 28 August–1 September 1989.
4. Van Heesch, E.J.M.; Caspers, R.; Gulickx, P.F.M.; Jacobs, G.A.P.; Kersten, W.F.J.; van der Laan, P.C.T. Three phase voltage measurements with simple open air sensors. In Proceedings of the 7th International Symposium on High-Voltage Engineering, Dresden, Germany, 26–30 August 1991.
5. Smeets, R.P.P.; van der Linden, W.A.; Achterkamp, M.; Damstra, G.C.; de Meulemeester, E.M. Disconnector switching in GIS: Three-phase testing and phenomena. *IEEE Trans. Power Deliv.* **2000**, *15*, 122–127. [CrossRef]
6. Chen, K.-L.; Guo, Y.; Ma, X. Contactless voltage sensor for overhead transmission lines. *IET Gener. Transm. Distrib.* **2018**, *12*, 957–966. [CrossRef]
7. Barakou, F.; Wouters, P.A.A.F.; Mousavi Gargari, S.; de Jong, J.P.W.; Steennis, E.F. Online transient measurements of EHV cable system and model validation. *IEEE Trans. Power Deliv.* **2019**, *34*, 532–541. [CrossRef]
8. *EMTDC User's Guide*, 5th ed.; Manitoba HVDC Research Centre Inc.: Winnipeg, MB, Canada, 2010.
9. Barakou, F.; Haverkamp, A.R.A.; Wu, L.; Wouters, P.A.A.F.; Steennis, E.F. Investigation of necessary modeling detail of a large scale EHV transmission network for slow front transients. *Electr. Power Syst. Res.* **2017**, *147*, 192–200. [CrossRef]
10. Wouters, P.A.A.F.; Barakou, F.; Mousavi Gargari, S.; Smit, J.; Steennis, E.F. Accuracy of switching transients measurement with open-air capacitive sensors near overhead lines. In Proceedings of the IEEE International Conference on High Voltage Engineering and Application, Athens, Greece, 10–13 September 2018.
11. Van Deursen, A.P.J.; Smulders, H.W.M.; de Graaff, R.A.A. Differentiating/integrating measurement setup applied to railway environment. *IEEE Trans. Instrum. Meas.* **2006**, *55*, 316–326. [CrossRef]
12. Van der Laan, P.C.T.; van Deursen, A.P.J. Reliable protection of electronics against lightning: Some practical applications. *IEEE Trans. Electromagnet. Compat.* **1998**, *40*, 513–520. [CrossRef]
13. Kuperus, J.; Meten in de fysika IIb. Course material in Dutch 1976. Utrecht University, The Netherlands.

14. Brandt, S. *Data analysis—Statistical and Computational Methods for Scientists and Engineers*, 4th ed.; Springer International Publishing: Cham, Switzerland, 2014; pp. 15–40.
15. Paul, C.R. *Analysis of Multiconductor Transmission Lines*, 2nd ed.; John Wiley & Sons Inc.: Hoboken, NJ, USA, 2008.

Article

A New Approach to Include Complex Grounding System in Lightning Transient Studies and EMI Evaluations [†]

Vegard Steinsland *, Lasse Hugo Sivertsen, Emil Cimpan and Shujun Zhang

Department of Electrical Engineering, Western Norway University of Applied Sciences, 5063 Bergen, Norway
* Correspondence: v.steinsland@me.com; Tel.: +47-951-62-405
† This paper is an extended version of our paper published in and presented at the 6th IEEE International Conference on High Voltage Engineering and Application (IEEE ICHVE2018), Athens, Greece, 10–13 September 2018; pp. 1–4.

Received: 7 May 2019 ; Accepted: 12 August 2019; Published: 15 August 2019

Abstract: A new approach to lightning transient studies including complex grounding grids is presented in this paper. The grounding system is modeled in Matlab/Simulink based on the transmission line theory. Using a bottom-up approach and considering the properties of the fundamental elements, a detailed view of measurement values will be presented and analyzed. The Matlab/Simulink grounding system models are interfaced for co-simulation with EMTP-RV trough Functional Mock-up Interface (FMI) 2.0. This modeling approach allows the use of the full component library and network design by EMTP-RV to evaluate and analyze the effects of the grounding system and transmission network simultaneously in Matlab/Simulink. The results present a simplified transmission system where a surge is injected, Conseil International des Grands Réseaux Électriques (CIGRE) 1 kA 1.2/50, in far-end of a transmission line. When reaching a substation, the surge is injected into the grounding system through a surge arrester.

Keywords: grounding system; substation; lightning; transmission system; surge arrester performance

---

1. Introduction

A grounding system, which is essential for proper, reliable and safe operation of a power system, can be accomplished by providing a true reference to the electrical system that controls the discharge path of high energy faults. The performance of the grounding system during power frequency faults is a present key driver in general design of substation grounding system [1]. The transient behavior of the grounding system during a lightning discharge is characterized by a steep front that induces inductive effects in the grounding system. Thus, a large short-term voltage rises within the grounding system region, close to the injection point. The uneven voltage distribution can be explained in terms of current and voltage waves traveling along the grounding grid conductors that can be modeled by the telegrapher's equation [2]. Problematic lightning surge behavior has long been recognized by the industry and several models to describe these transient events. Related topics have been proposed in the relevant research literatures [3–7]. The reviewed literatures tend to treat high energy faults originating from direct lightning strikes discharged trough the grounding system. A method of implementation to evaluate the effect associated with lightning transient in the transmission system and the corresponding effect on the grounding system is not publicly available.

Lightning surges on overhead transmission lines may introduce travelling waves which have the potential to penetrate deep into a substation and present hazards ElectroMagnetic Interference (EMI) for sensitive equipment. To ensure reliable operation in the design of transmission systems, the grounding system should rather be considered as an integrated part. This paper presents an approach whereby the combined simulation of the grounding and transmission system is performed based on software implementation. Since the substation surge arrester is evaluated as the injection point to the grounding system, the surge arrester performance is considered. Simultaneously, the time and spatial distribution of the potential rise in the grounding itself is presented to consider EMI preventative action in the substation area.

The grounding system is implemented in Matlab/Simulink in combination with the specialized commercial software for power system transients studies, EMTP-RV. Detailed software specification is found in Appendix A.

2. Parameters of the Grounding System

Based on the classic work by Sunde, the grounding system in this work has been modelled as a lossy transmission line [2]. The soil medium with electrical resistivity and permittivity surrounds the grounding wires that are characterized by their electrical parameters thus forming a unified system. The electrical parameters in per-unit length are defined through Equation (1):

$$Z = j\omega L \tag{1a}$$

$$Y = G + j\omega C \tag{1b}$$

where the per-unit length G is defined as the grounding system conductance (S), C is the capacitance (F) and L is the inductance (H).

With relatively short conductor length and large cross section of grounding wires, the internal resistance (including the skin effect) and inductance are significantly smaller than the external self-inductance. With this consideration, a simplification that only includes the self-inductance is made [8]. The connection between the soil properties and the grounding system elements defined above is represented in Equations (2)–(4) [2] for a horizontal buried grounding wire.

$$G_i = G_j = \frac{\pi}{\rho_{soil}\left[ln\left(\frac{2l}{\sqrt{2ad}} - 1\right)\right]} \tag{2}$$

$$C_i = C_j = G_i\rho_{soil}\left(\epsilon_0 \times \epsilon r_{soil}\right) \tag{3}$$

$$L_i = L_j = \frac{\mu_0}{2\pi}\left(ln\frac{2l}{a} - 1\right) \tag{4}$$

where ρ_{soil} is the soil resistivity (Ωm), ϵr_{soil} is the soil relative permittivity (-), a is the conductor radius (m) and d is the buried depth (m).

A grounding wire conducting a lightning impulse current will exert a time-varying electric field, outwards through the wire and into the surrounding soil. The soil itself, depending on soil resistivity and properties, will conduct a current from the grounding wire, dissipated into the soil. Depending on the dissipated current density , the electric field and the current density are given in Equation (5a). The surface current density of a round wire is found in Equation (5b), which exerts the electric field on the soil [9] (p. 1586). The linear behaviour between current density and electric field are valid below the soil critical breakdown value, E_c.

$$E_{soil} = J_{soil}\rho_{soil} \tag{5a}$$

$$J_{soil} = \frac{I_{soil}}{2\pi al} \tag{5b}$$

3. Modeling the Grounding System

The element length in per-unit model of the grounding wire is implemented in Matlab as a T-section of the transmission line model. The grounding wire implementation is illustrated with fundamental electrical elements and measurement nodes in Figure 1 which forms a horizontal buried grounding wire. From the grounding wire t-section development, the grounding system is implemented by an appropriate number of T-sections and additional nodes using Simulink graphical block elements. The grounding system layout with formation strategy is illustrated Figure 2 which gives references to the simulation database is found in Appendix B. To assure a smooth representation of the interconnection between elements, and to possibly extend the model functionality to include mutual couplings, the T-section represent one meter of ground wire (in per-unit) [4]. Consequently, the model can be extended to arbitrary lengths and configurations.

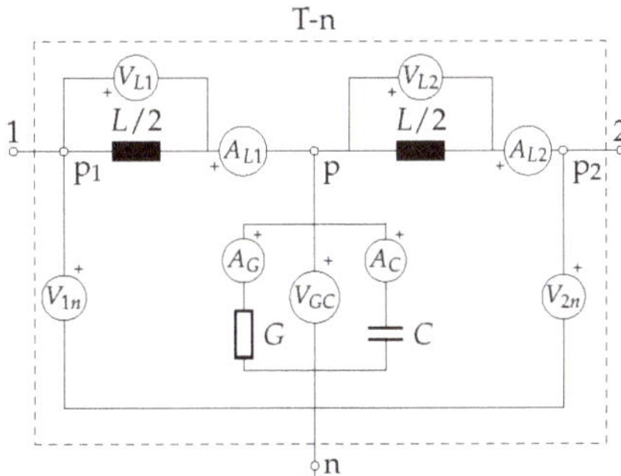

Figure 1. Element lenght in per-unit grounding wire illustration of implementation.

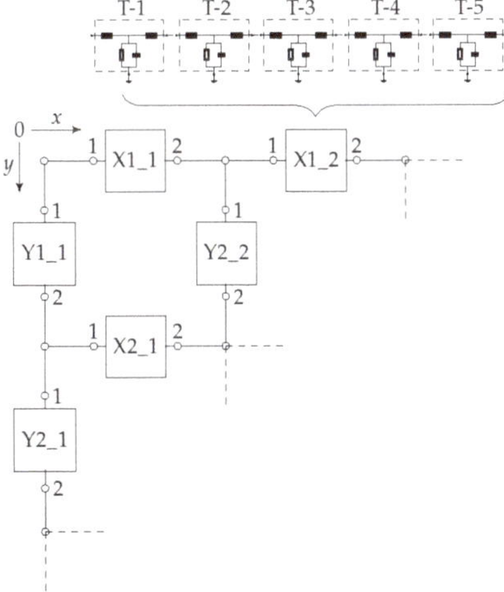

Figure 2. The grounding system layout with formation strategy.

This consideration gives the total number of required logged variables for the formed horizontal buried grounding grid square meshes as an indication for two different mesh sizes and total area in Table 1.

Table 1. Required number of logged variables for grounding grid of two different square mesh sizes and total area values.

Grounding Grid	Area	Variables
5 × 5 m mesh	1600 m^2	6480
	3600 m^2	14040
10 × 10 m mesh	1600 m^2	3600
	3600 m^2	7560

As described trough Section 2 the grounding system is modelled as a transmission line, connecting the soil and grounding conductor properties. The matrices of the grounding system may be expressed trough the coupled telegraphers equations in frequency domain trough Equation (6). Where $V(z)$ and $I(z)$ are the line phasor of voltage and current [10]:

$$\frac{d^2}{dz^2}\hat{V}(z) = \hat{Z}\hat{Y}\hat{V}(z) \tag{6a}$$

$$\frac{d^2}{dz^2}\hat{I}(z) = \hat{Y}\hat{Z}\hat{I}(z) \tag{6b}$$

In transient analysis, considering the spatial distribution of voltage potential and current flowing in the grounding system, it is mandatory to simulate each element in time domain. With the layout of grounding grids, as is illustrated Equation (2), several connection points exists. Consequently, the time domain model is required to account for the coupling between elements as feeding points. For J feeding points the impact of the grounding system, as an electrical network exited by an injected lightning current, is global. Each element in the grounding system matrices is numerically solved independently by the Matlab ODE23t solver, which implements the trapezoidal rule using a "free" interpolation for the time domain solution. The implementation account for the injected current characteristics and corresponding frequency response of the grounding system.

4. Grounding Model Verification

The grounding system model is validated by the work based on the ElectroMagnetic Field (EMF) theory first performed by Grcev [11] and later by Jardines et al. [7] who introduced a variant of the Multi-conductor Transmission Line (MTL) approach. In these cases of model verifications, the current source is implemented using the double exponential waveform, $i(t) = \hat{I}(e^{-\alpha t} - e^{-\beta t})$, and the source parameters where adjusted to fit the given stroke function. The referenced work is reproduced from manual reading.

4.1. Grounding Rod of 15 m Length

A grounding rod of 15 m was simulated and measured by Grcev [11] and later evaluated by Jardines et al. [7] as shown in Figure 3a. Using two independent modeling approaches and relative similarities in results may justify the present modeling accuracy. The grounding rod is horizontally buried in soil of $\rho_{soil} = 70$ Ωm and $\epsilon r_{soil} = 15$, with a wire radius of $a = 0.012$ m at depth $d = 0.6$ m. The current source was set with the amplitude of $\hat{I} = 36$ A and the stroke of 0.36/12 µs which leads to the new parameter values $\alpha = 32 \times 10^3$ and $\beta = 7.6 \times 10^6$. The results from the implemented model are shown in Figure 3b. As it can be observed from Figure 3a in the referenced work, the peak values at the injection point are approximately 560 V. While using the implemented model, Figure 3b, the peak value shows 569 V. The second point for comparison is at 7 m from the injection point where Grcev simulations show approximate 200 V while the results in this work gives 172 V after 0.7 µs. Besides the point value readings, the surge wave propagation along the grounding rod and voltage distribution are of similar character.

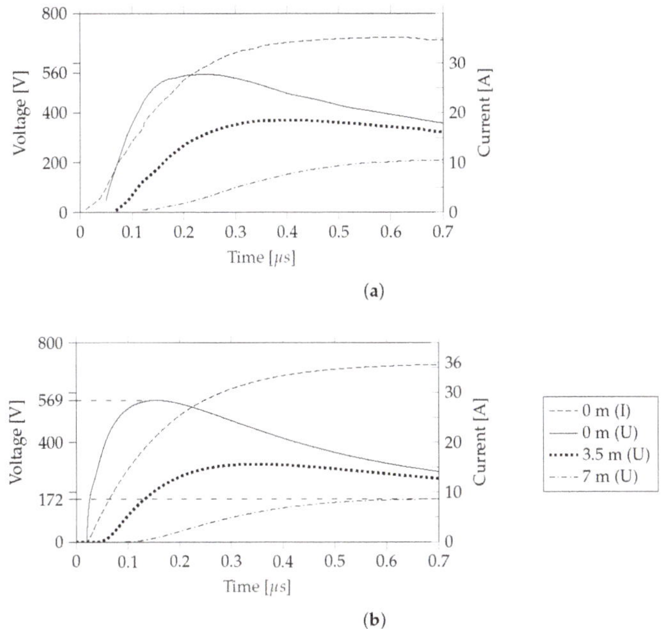

Figure 3. Voltage distribution along an horizontal copper wire in ground (length l = 15 m, conductor radius a = 0.012 m, at soil depth d = 0.6 m) in soil of ρ_{soil} = 70 Ωm and ϵr_{soil} = 15: (**a**) the reproduced results for comparison from the research given in [11] (p. 818); and (**b**) results from implemented model with excitation current $\hat{I}$ = 36 A of double exponential waveform (α = 32 $\times$ 10^3, β = 7.6 $\times$ 10^6).

4.2. Grounding Grid of 10 m Meshes with Total Size 3600 m²

A grounding system of 10 $\times$ 10 m meshes size and a total square area of 3600 m² was simulated by Jardines et al. [7] using a variant of the MTL approach (see Figure 4a). The grounding grid was buried in soil of ρ_{soil} = 100 Ωm and ϵr_{soil} = 36, with a wire radius of American Wire Gauge (AWG) 2/0 ($a \approx$ 0.004126 m) at depth d = 0.6 m. The current source was set with amplitude of $\hat{I}$ = 1 kA and a 1/20 μs which gave adjusted parameters to α = 38 $\times$ 10^3 and β = 2.54 $\times$ 10^6. The results from the implemented model are given in Figure 4b. Since the injected current was not given in the referenced work (see Figure 4a) it is worth noting that a small deviation in the injected current will have a large impact on the grounding system response, especially for the region contributing to limiting the peak voltage. As it can be observed, both the voltage distribution and the propagation characteristics correspond well. When comparing the simulation to Grcev [11] which was based on EMF against the simulation in the presented work, the peak value and the propagation characteristics are of comparable values. However, the distributed voltage at the given nodal points is more conservative in the transmission line approach, both in the implemented model and in Jardines et al. work [7]. This unveils a model accuracy difference compared to the EMF. As reviewed, Liu [4] developed the non-uniform transmission line approach to compensate for the inaccuracy of this method and are treated in details in her work.

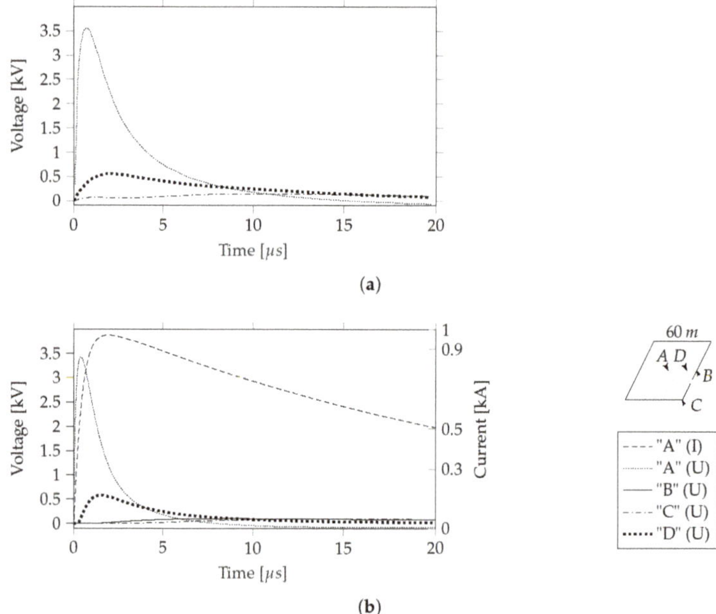

Figure 4. Voltage distribution in a grounding grid consisting of 6 × 6 meshes of 10 m size. The grounding grid consist of copper conductors of AWG 2/0, buried at d = 0.6 m in soil of ρ_{soil} = 100 Ωm and ϵr_{soil} = 36: (**a**) the reproduced results for comparison from research given in [7] (p. 31); and (**b**) results from implemented model with excitation current $\hat{I}$ = 1 kA of double exponential waveform (α = 38 × 10³, β = 2.54 × 10⁶).

5. Integration of Grounding and Transmission System

The Matlab/Simulink grounding grid are integrated with EMTP-RV through a newly developed FMI software, which was released by Powersys Solutions in early 2018 [12,13]. The FMI package gives possibilities for co-simulation with information exchange at a per simulation time-step interval (sequentially processed). The specific FMI interface used in this study is found in Appendix A.2. From the transmission system surge arrester, the injection point impedance describes the grounding system response trough Equation (7).

$$Z_{InjectPoint}(j\omega) = \frac{U_{InjectPoint}(j\omega)}{I_{arrester}(j\omega)} \tag{7}$$

When the transmission system surge arrester reaches the breakdown voltage a current surge is injected into the grounding system. The current injection value is exchanged from EMTP-RV to Matlab, which simulates the grounding system response. The grounding system initial impedance is set to the power frequency value corresponding to the grounding system resistance [14]. From the injection point current and the induced voltage potential rise, the impulse impedance is calculated and exchanged from Matlab to EMTP-RV, which connects the dynamic response of the grounding system to the transmission system. With a large number of logged variables required by the grounding system (see Table 1) and presented implementation strategy, the additional measured values of the transmission system were exchanged from EMTP-RV to Matlab to provide a common simulation log for pre-processing. The advanced functionality offered by the Matlab/Simulink modeling of the grounding system lies in the preprocessing of large data-sets. A schematic overview of the model integration is illustrated in Figure 5.

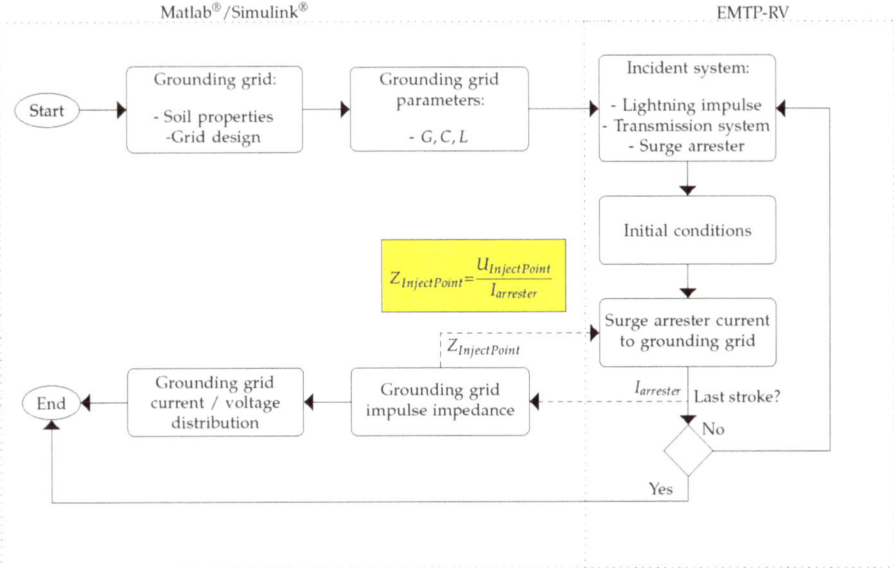

Figure 5. Schematic overview of the integration between grounding and transmission system.

6. An Example of Results: Case Study

A simplified transmission system network is shown in Figure 6 with the grounding system model interfaced in Matlab. In this model the transmission line, the cable, the transformer and the surge arrester are selected from the standard EMTP-RV software library. At 10 km distance from a substation, a lightning strikes the 300 kV overhead transmission line (Zc_l = 400 Ω). A shielding failure causes an injected current with a magnitude of 1 kA and 1.2/50 µs of CIGRE waveform stroke to flow towards the substation. In the substation, the cable (Zc_c) between the surge arrester (Appendix C) and transformer are 10 m. Figure 7 shows the simulation results of the transmission system when the grounding system is ignored, thus giving a peak voltage of 171 kV at the transformer.

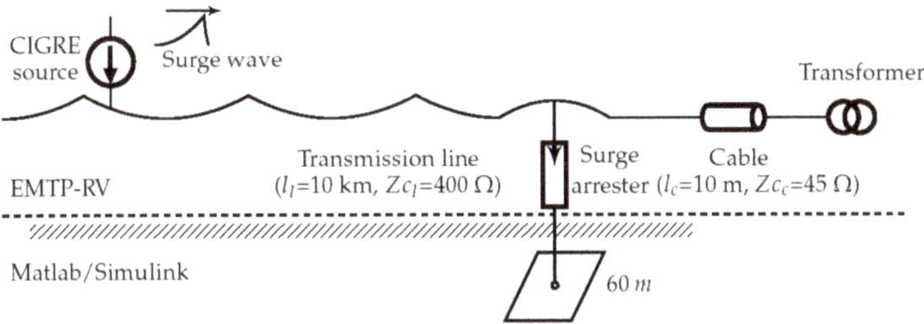

Figure 6. Simplified illustration of implemented transmission and grounding system.

A grounding system with 10 × 10 m mesh size is added with a total grid area of 3600 m^2 in soil of ρ_{soil} = 2000 Ωm, ϵr_{soil} = 16, with a wire radius of a = 0.04126 m at depth d = 0.6 m is obtained. The transmission system conditions are similar. The surge arrester is connected in center of the grid. The simulation results of the transmission system are given in Figure 8. As it can be observed, the surge arrester performance is reduced to give a peak transformer voltage of 190 kV.

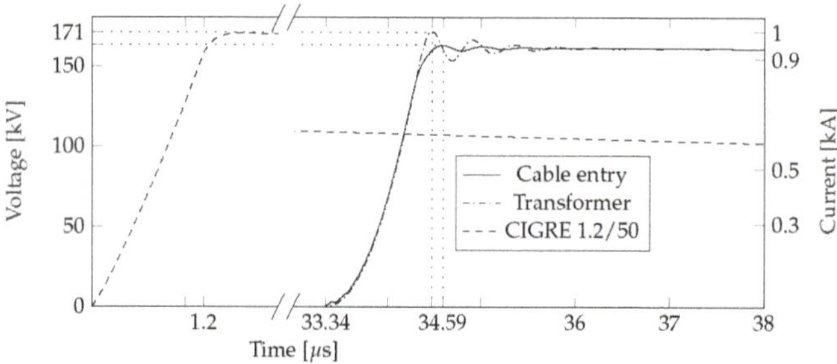

Figure 7. Ignoring grounding system: transmission system nodal voltages and CIGRE 1.2/50 μs injected current stroke in far-end (l_l = 10 km).

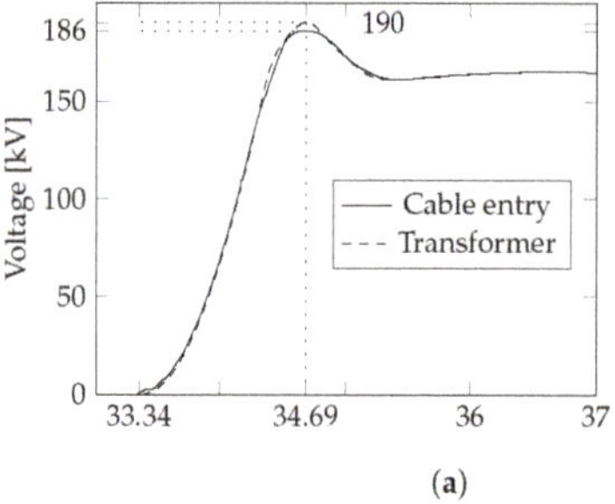

(a)

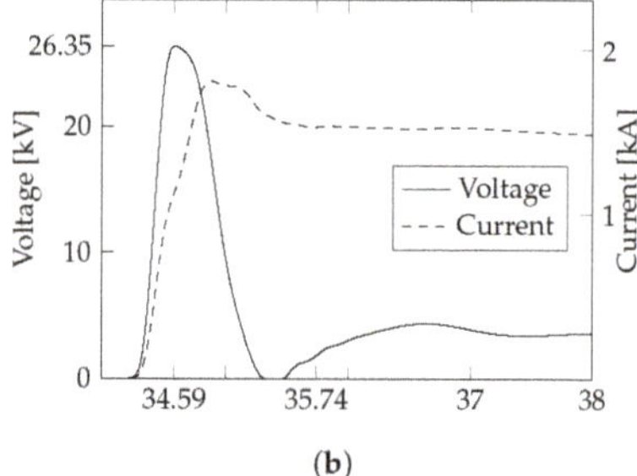

(b)

Figure 8. Application case when the grounding system is added: (**a**) the effect in the transmission system; and (**b**) the results at the surge arrester injection point.

Then, the EMI analysis of the substation could be performed based on the overall voltage distributions in the grounding system as shown in Figure 9. Figure 9a shows measurments of the grounding system, with nodal points selected diagonaly outwards from the center injection point. For the selected events an overall voltage distribution is presented. Figure 9b shows the distribution at center peak voltage and Figure 9c when the voltage potential in the grid corners are at peak.

With the comprehensive log dataset further analysis is exemplified in Figures 10 and 11. The impulse effective area of the grounding grid defined in [15] is shown in Figure 10. This is the total area of the grounding grid which limits the peak voltage in the grid. There exists several definitions and empirical formulas for estimating the effective length when optimizing the grounding grid design that was recently evaluated [5]. However, these approaches have not taken the transmission line network itself as an integrated element into consideration. In addition, the electric field exerted on the soil close to the grounding wires that was based on the current density leaked to the soil is shown in Figure 11.

Institute of Electrical and Electronics Engineers (IEEE) indicates (quoted in standard) a critical breakdown value of E_c = 1000 kV/m [16] (p. 1263). This definition gives reference to the above mentioned experimental relation between soil resistivity and E_c [17]. Further evaluation of IEEE standards gives a value of E_c = 400 kV/m, a level which are referred without reference in [18] (p. 38) (refereed standard is currently under review for update). If the ionization level is reached the electric field in the soil has pronounced influence on the impulse peak voltage. Moreover, ionization has a positive effect by lowering the peak voltage due to arcing or puncturing in the soil. Soil ionization can be included in dynamic simulations as it has been proposed in [9]. This would modify the apparent grounding conductor radius in Equations (2) and (3). For this application case when considering strokes in the transmission system, soil ionization phenomena has been evaluated to have minor deviation of the results due to relative small currents injected trough the surge arrester.

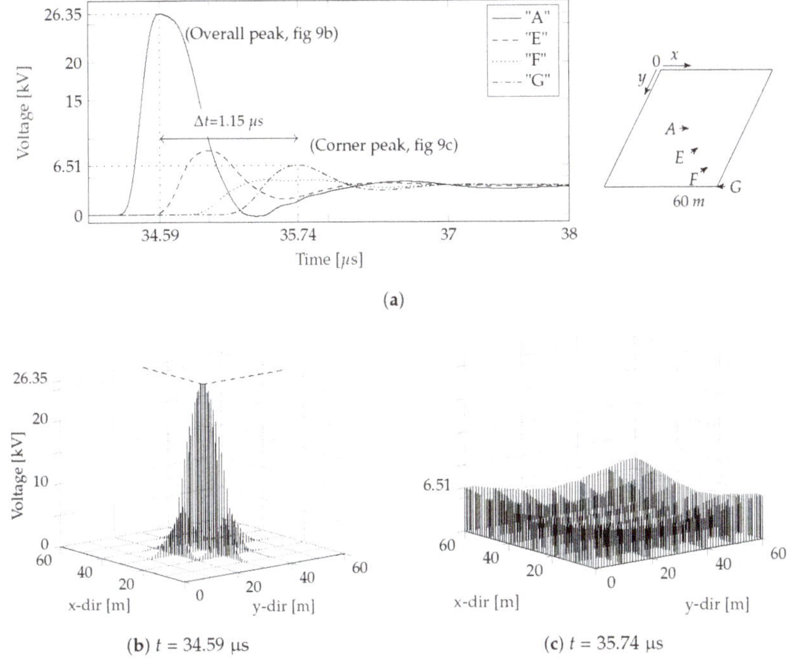

Figure 9. Application case when the grounding system is added: (**a**) nodal voltage measurement values for selected points in the grounding grid; (**b**) the overall voltage distribution in the grounding grid at peak; and (**c**) the overall voltage distribution in the grounding grid at corner peak.

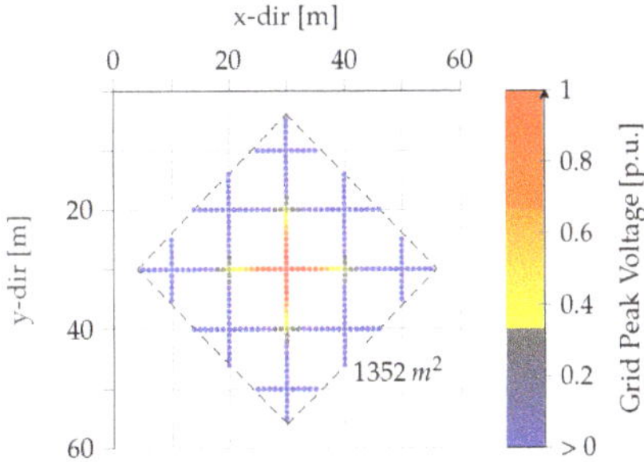

Figure 10. Application case: used area.

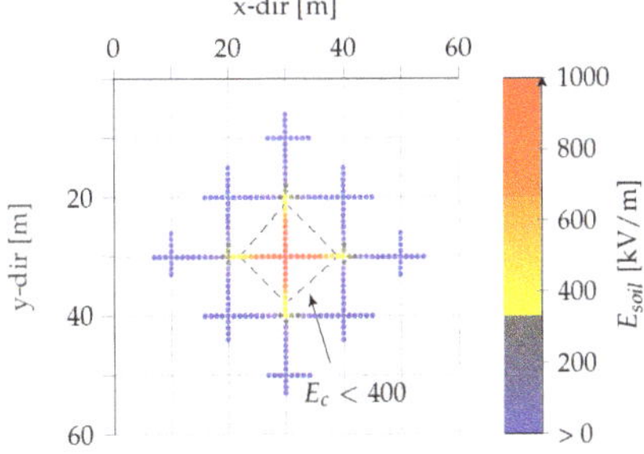

Figure 11. Application case: peak field.

7. Conclusions

The new modeling approach allows integration of the grounding system into the transmission system when analyzing the lightning surge performance of the transmission system. By using the more detailed Matlab/Simulink model presented in this work, large data-sets are processed to extract overall measured values in EMI analysis. Different functions and parameters may be processed by the simulation log of the grounding system and parameters such as the effective length and the electric field distribution. Moreover, by taking advantage of the newly developed FMI interface, the grounding grid model itself is integrated as an element in the transmission system modeling and analysis by EMTP-RV. Lastly, the surge arrester performance is better assessed when the grounding system is included and consequently, the corresponding effects of all parts in the transmission system could be more accurately analyzed.

The grounding system model presented in this work is simplified and neglect significant factors of a physical system. Moreover, with the implementation of the grounding system in Matlab/Simulink features flexibility and significant potential for further development so that the accuracy could be improved.

Author Contributions: V.S. developed the method, performed simulation/data analysis and prepared the manuscript as the first author. L.H.S., S.Z. and E.C. assisted the project with insights in the method development phase and review process. All authors discussed the results and approved the publication.

Funding: This research received no external funding

Acknowledgments: The authors thank for the support given by Western Norway University of Applied Sciences and the University of Bergen to carry out this work. A special thank to International Conference on High Voltage Engineering and Application (ICHVE) 2018 for the invitation to publish in this special issue entitled "Selected Papers from 2018 IEEE International Conference on High Voltage Engineering".

Conflicts of Interest: The authors declare no conflict of interest.

Glossaries

This document is incomplete. The external file associated with the glossary 'main' (which should be called `energies-511550-addedDoi.gls`) hasn't been created.

Check the contents of the file `energies-511550-addedDoi.glo`. If it's empty, that means you haven't indexed any of your entries in this glossary (using commands like `\gls` or `\glsadd`) so this list can't be generated. If the file isn't empty, the document build process hasn't been completed.

If you don't want this glossary, add `nomain` to your package option list when you load `glossaries-extra.sty`. For example:

```
\usepackage[nomain]{glossaries-extra}
```

Try one of the following:

- Add `automake` to your package option list when you load `glossaries-extra.sty`. For example:

  ```
  \usepackage[automake]{glossaries-extra}
  ```
- Run the external (Lua) application:

  ```
  makeglossaries-lite.lua "energies-511550-addedDoi"
  ```
- Run the external (Perl) application:

  ```
  makeglossaries "energies-511550-addedDoi"
  ```

  Then rerun LaTeX on this document.

  This message will be removed once the problem has been fixed.

Abbreviations

This document is incomplete. The external file associated with the glossary 'abbreviations' (which should be called `energies-511550-addedDoi.gls-abr`) hasn't been created.

Check the contents of the file `energies-511550-addedDoi.glo-abr`. If it's empty, that means you haven't indexed any of your entries in this glossary (using commands like `\gls` or `\glsadd`) so this list can't be generated. If the file isn't empty, the document build process hasn't been completed.

Try one of the following:

- Add `automake` to your package option list when you load `glossaries-extra.sty`. For example:

  `\usepackage[automake]{glossaries-extra}`

- Run the external (Lua) application:

  `makeglossaries-lite.lua "energies-511550-addedDoi"`

- Run the external (Perl) application:

  `makeglossaries "energies-511550-addedDoi"`

Then rerun LaTeX on this document.

This message will be removed once the problem has been fixed.

Appendix A. Software and Integration

Appendix A.1. Software Versions

All simulations were performed on a standard consumer laptop computer with Intel Core I7-2640M (dual-core, 2.70GHz, 4MB Cache) CPU and 8 GB (1333 MHz) RAM. The Matlab/Simulink grounding grid models was developed and tested on 64-bit Microsoft Windows 10 Pro version 1709. Windows version:

- MathWorks Matlab version 9.3, R2017b, 64-bit (14 September 2017)
- MathWorks Simulink version 9.0, R2017b, 64-bit (24 July 2017)
- MathWorks Simscape version 4.3 (18 November 2017)

Powersys EMTP-RV

- Powersys EMTP-RV 3.5 32-bit (17 January 2017)
- Powersys FMI Add-On for Matlab/Simulink (15 March 2018)

Appendix A.2. Matlab/Simulink and EMTP-RV FMI Interface

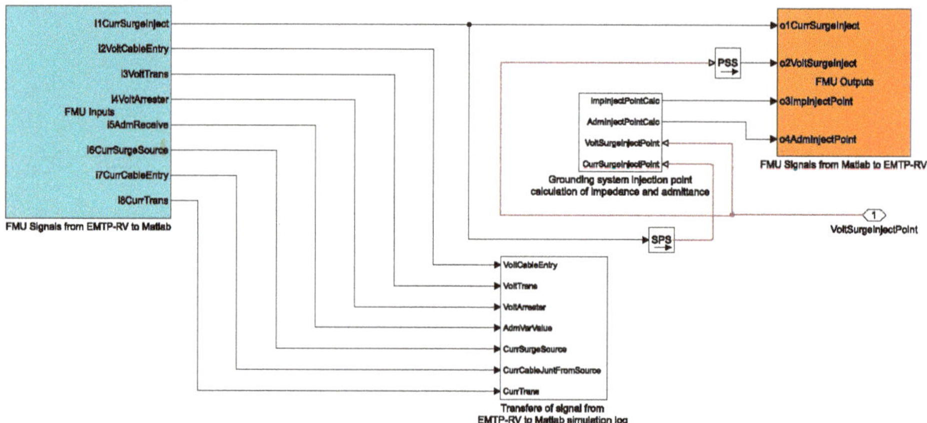

Figure A1. The Matlab/Simulink FMI interface to EMTP-RV with the signals transferred between the softwares and block representation of the integration.

Appendix B. Matlab/Simulink Simulation Log Definition

Values from the implemented measurement nodes, for each grounding wire t-section as illustrated by Section 3 and Figure 1, are stored in a simulation-log database. The simulation-log database is organized with value identifications from the t-section measurements definitions in addition to the grounding grid formation strategy, Figure 2, and forms nodal points connection to physcial properties.

Table A1. Simulation-log nodal measurement overview and grounding system variable definitions.

Node (Figure 1)	Log-File Name	Grid Segment (Figure 2)	Element	Series Selection
V_{l1}	simlog_scc_grounding_system.	X/Y'y-dir'_'x-dir'.	l1.v.	series.values/time
A_{l1}	simlog_scc_grounding_system.	X/Y'y-dir'_'x-dir'.	l1.i.	series.values/time
V_{l2}	simlog_scc_grounding_system.	X/Y'y-dir'_'x-dir'.	l2.v.	series.values/time
A_{l2}	simlog_scc_grounding_system.	X/Y'y-dir'_'x-dir'.	l2.i.	series.values/time
V_{1n}	simlog_scc_grounding_system.	X/Y'y-dir'_'x-dir'.	p1.v.	series.values/time
V_{2n}	simlog_scc_grounding_system.	X/Y'y-dir'_'x-dir'.	p2.v.	series.values/time
A_g	simlog_scc_grounding_system.	X/Y'y-dir'_'x-dir'.	g.i.	series.values/time
V_{gc}	simlog_scc_grounding_system.	X/Y'y-dir'_'x-dir'.	p.v.	series.values/time
A_c	simlog_scc_grounding_system.	X/Y'y-dir'_'x-dir'.	c.i.	series.values/time

Appendix C. EMTP-RV Surge Arrester Parameterization *V-I* characteristics

Surge arrester switching and damping interaction is according to standard by EMTP-RV model number: 865630. The surge arrester is parameterized according to the overview below. Where k_{seg} is a fitting constant (-), α_{arr} is a coefficient of non-linearity (-) and U_{arr} (pu) is a factor of the arrester system voltage (-).

Table A2. EMTP-RV ZnO surge arrester *V-I* parameters, adopted from [19] (p. 9).

Segment	k_{seg}	α_{arr}	U_{arr} (pu)
1	$4.23208099271728 \times 10^9$	$2.40279296219991 \times 10^1$	$2.98198270953446 \times 10^{-1}$
2	$2.81773645053899 \times 10^{10}$	$2.66219333383972 \times 10^1$	$4.81500000000002 \times 10^{-1}$
3	$4.15144087019289 \times 10^8$	$2.00870413085783 \times 10^1$	$5.24450368910346 \times 10^{-1}$
4	$2.63271405014350 \times 10^{12}$	$3.52906710089596 \times 10^1$	$5.62230840318764 \times 10^{-1}$
5	$3.21774149817822 \times 10^6$	$1.11310570543270 \times 10^1$	$5.69192036592734 \times 10^{-1}$
6	$1.93774621300766 \times 10^5$	5.36270125014300	$6.14408449804349 \times 10^{-1}$

References

1. IEEE. *80-2013—IEEE Guide for Safety in AC Substation Grounding*; IEEE Std 80-2013 (Revision of IEEE Std 80-2000/Incorporates IEEE Std 80-2013/Cor 1-2015); IEEE: New York, NY, USA, 2015; pp. 1–226. [CrossRef]
2. Sunde, E.D. *Earth Conduction Effects in Transmission Systems*, 1st ed.; The Bell Telephone Laboratories, Dover Publications: New York, NY, USA, 1949.
3. Grcev, L.; Dawalibi, F. An Electromagnetic Model for Transients in Grounding Systems. *IEEE Trans. Power Deliv.* **1990**, *5*, 1773–1781. [CrossRef]
4. Liu, Y. Transient Response of Grounding Systems Caused by Lightning: Modelling and Experiments. Ph.D. Thesis, Uppsala University, Uppsala, Sweden, 2004.
5. Grcev, L. Lightning Surge Efficiency of Grounding Grids. *IEEE Trans. Power Deliv.* **2011**, *26*, 1692–1699. [CrossRef]
6. Alipio, R.; Schroeder, M.A.O.; Afonso, M.M. Voltage Distribution Along Earth Grounding Grids Subjected to Lightning Currents. *IEEE Trans. Ind. Appl.* **2015**, *51*, 4912–4916. [CrossRef]
7. Jardines, A.; Guardado, J.L.; Torres, J.; Chávez, J.J.; Hernández, M. A Multiconductor Transmission Line Model for Grounding Grids. *Int. J. Electr. Power Energy Syst.* **2014**, *60*, 24–33. [CrossRef]
8. Liu, Y.; Theethayi, N.; Thottappillil, R. An Engineering Model for Transient Analysis of Grounding System under Lightning Strikes: Nonuniform Transmission-Line Approach. *IEEE Trans. Power Deliv.* **2005**, *20*, 722–730. [CrossRef]
9. He, J.; Gao, Y.; Zeng, R.; Zou, J.; Liang, X.; Zhang, B.; Lee, J.; Chang, S. Effective Length of Counterpoise Wire under Lightning Current. *IEEE Trans. Power Deliv.* **2005**, *20*, 1585–1591. [CrossRef]
10. Alipio, R.; Costa, R.M.; Dias, R.N.; Conti, A.D.; Visacro, S. Grounding Modeling Using Transmission Line Theory: Extension to Arrangements Composed of Multiple Electrodes. In Proceedings of the 2016 33rd International Conference on Lightning Protection (ICLP), Estoril, Portugal, 25–30 September 2016; pp. 1–5. [CrossRef]
11. Grcev, L.D. Computer Analysis of Transient Voltages in Large Grounding Systems. *IEEE Trans. Power Deliv.* **1996**, *11*, 815–823. [CrossRef]
12. Modelica Association c/o PELAB. Functional Mock-Up Interface. 2018. Available online: http://fmi-standard.org/ (accessed on 23 March 2018).
13. Cornau, J. *Guide EMTP-RV Matlab/Simulink FMI Export*; Powersys: Paris, France, 2018.
14. IEEE. *IEEE Guide for Measuring Earth Resistivity, Ground Impedance, and Earth Surface Potentials of a Grounding System*; IEEE Std 81-2012 (Revision of IEEE Std 81-1983); IEEE: New York, NY, USA, 2012; pp. 1–86. [CrossRef]
15. Gupta, B.R.; Thapar, B. Impulse Impedance of Grounding Grids. *IEEE Trans. Power Appar. Syst.* **1980**, *PAS-99*, 2357–2362. [CrossRef]
16. Estimating Lightning Performance of Transmission Lines. II. Updates to Analytical Models. *IEEE Trans. Power Deliv.* **1993**, *8*, 1254–1267. [CrossRef]
17. Oettle, E.E. A New General Estimation Curve for Predicting the Impulse Impedance of Concentrated Earth Electrodes. *IEEE Trans. Power Deliv.* **1988**, *3*, 2020–2029. [CrossRef]

18. IEEE. *IEEE Guide for the Application of Insulation Coordination*; IEEE Std 1313.2-1999; IEEE: New York, NY, USA, 1999. [CrossRef]
19. Hagen, S.T. *Lightning Strike Analysis*; Western Norway University of Applied Sciences: Porsgrunn, Norway, 2016.

MDPI

St. Alban-Anlage 66

4052 Basel

Switzerland

Tel. +41 61 683 77 34

Fax +41 61 302 89 18

www.mdpi.com

Energies Editorial Office

E-mail: energies@mdpi.com

www.mdpi.com/journal/energies